Lecture Notes in Statistics

138

Edited by P. Bickel, P. Diggle, S. Fienberg, K. Krickeberg,
I. Olkin, N. Wermuth, S. Zeger

Springer
New York
Berlin
Heidelberg
Barcelona
Budapest
Hong Kong
London
Milan
Paris
Singapore
Tokyo

Peter Hellekalek
Gerhard Larcher (Editors)

Random and Quasi-Random Point Sets

With contributions by

József Beck
Peter Hellekalek
Fred J. Hickernell
Gerhard Larcher
Pierre L'Ecuyer
Harald Niederreiter
Shu Tezuka
Chaoping Xing

 Springer

Peter Hellekalek
Institut für Mathematik
Universität Salzburg
Hellbrunner Str. 34
A-5020 Salzburg
Austria

Gerhard Larcher
Institut für Mathematik
Universität Salzburg
Hellbrunner 34
A-5020 Salzburg
Austria

CIP data available.
Printed on acid-free paper.

Camera ready copy provided by the editors.
Printed and bound by Braun-Brumfield, Ann Arbor, MI.
Printed in the United States of America.

9 8 7 6 5 4 3 2 1

ISBN 0-387-98554-9 Springer-Verlag New York Berlin Heidelberg   SPIN 10681028

Preface

This volume is a collection of survey papers on recent developments in the fields of quasi-Monte Carlo methods and uniform random number generation. We will cover a broad spectrum of questions, from advanced metric number theory to pricing financial derivatives.

The Monte Carlo method is one of the most important tools of system modeling. Deterministic algorithms, so-called uniform random number generators, are used to produce the input for the model systems on computers. Such generators are assessed by theoretical ("a priori") and by empirical tests. In the a priori analysis, we study figures of merit that measure the uniformity of certain high-dimensional "random" point sets. The degree of uniformity is strongly related to the degree of correlations within the random numbers.

The quasi-Monte Carlo approach aims at improving the rate of convergence in the Monte Carlo method by number-theoretic techniques. It yields deterministic bounds for the approximation error. The main mathematical tool here are so-called low-discrepancy sequences. These "quasi-random" points are produced by deterministic algorithms and should be as "super"-uniformly distributed as possible.

Hence, both in uniform random number generation and in quasi-Monte Carlo methods, we study the uniformity of deterministically generated point sets in high dimensions. By a (common) abuse of language, one speaks of random and quasi-random point sets.

The central questions treated in this book are (i) how to generate, (ii) how to analyze, and (iii) how to apply such high-dimensional point sets.

The contribution of József Beck addresses readers interested in metric number theory. It continues, in a certain sense, the well-known monograph of the author and Chen on irregularities of distribution. The object of this theory is to measure the uniformity of point sequences relative to natural classes of geometric objects like rectangles, balls, and convex sets. Questions of probabilistic diophantine approximation lead to the study of distribution problems like estimating the number of lattice points in tilted rectangles. The classical diophantine series $\sum_{k=1}^{n}(\{k\alpha\} - 1/2)$, where α is a fixed irrational and $\{\cdot\}$ denotes the fractional part, is the starting point for a thorough treatment of relations between quadratic fields and continued fractions.

Peter Hellekalek discusses the theory and practice of figures of merit for uniform random number generators and low-discrepancy point sets. The first part of this contribution addresses practitioners. It deals with the importance and the reliability of the a priori assessment of random number generators and its relation to the quasi-Monte Carlo approach. This topic is illustrated by examples from numerical practice. The second part of this contribution is devoted to the presentation of a new concept of measuring uniformity of sequences and point sets, the general spectral test. The author proves the fundamental properties of this numerical quantity and establishes the relation to well-known measures like diaphony and discrepancy.

Fred J. Hickernell deals with the assessment of a particular class of quasi-random point sets, the family of integration lattices. Popular quadrature error bounds involve certain figures of merit that are only defined for lattices. This fact strongly hinders the comparison of lattice rules to other types of Monte Carlo node sets. For this reason, the author extends these numerical quantities to general figures of merit and discusses various forms of discrepancy from a common viewpoint. Additional information is given in the form of discrepancy bounds.

Gerhard Larcher presents a survey on recent developments concerning the analysis and the application of digital (t, m, s)-nets and digital (t, s)- and $(\mathbf{T}, s)$-sequences. The concept of digital point sets was introduced by Harald Niederreiter. It was studied in a series of papers by Niederreiter and by the author, and it has proved to be the most powerful general concept for the generation of low-discrepancy point sets at the present stage of research. In this contribution, new (metrical) discrepancy estimates for digital nets are given and the so-called digital lattice rule, which is a method for multivariate numerical integration based on digital point sets, is surveyed.

Pierre L'Ecuyer and Peter Hellekalek treat the question how to select good uniform random number generators for practical purposes. The authors explain the design principles behind such algorithms, the techniques for theoretical support, and the basic ideas of empirical statistical testing. Serial tests of equidistribution and tests based on close points in space serve as an illustration for the general discussion and as tools in an empirical study of the minimal sample size at which the tests decisively reject generators. Numerical evidence for various phenomena is provided by the systematic testing of "small" generators belonging to the families of linear, explicit inversive, and compound cubic congruential generators.

With his well-known monograph on random number generation and quasi-Monte Carlo methods, which surveys the state of the art in this field up to the year 1992, Harald Niederreiter has provided the basis for intensive research on low-discrepancy point sets. Concerning the generation of such point sets, the most important progress since then is provided by the method of Niederreiter and Chaoping Xing for the construction of digital (t, s)-sequences based on tools from algebraic geometry. In this volume, Harald Niederreiter and Chaoping Xing give a survey of their method and present new results concerning generation strategies.

Shu Tezuka gives an introduction to the interesting topic of pricing financial derivatives. It is one of the most fascinating applications of Monte

Carlo and quasi-Monte Carlo methods nowadays. The author surveys recently developed speed-up techniques that allow to accelerate the computation of financial parameters considerably when compared to simple Monte Carlo simulation. He further discusses important future issues relevant for financial applications of Monte Carlo and quasi-Monte Carlo methods.

We would like to express our gratitude to several collegues. First of all, we would like to thank our contributors, whose competence has allowed to cover a broad range of subjects. John Kimmel, with his precision and patience, made the cooperation with Springer-Verlag a remarkable experience. Special thanks go to Karl Entacher and Wolfgang Ch. Schmid for the assistance in processing some of the contributions. The first editor would also like to thank his (former) students Hannes Leeb, Otmar Lendl, Karin Schaber, and Stefan Wegenkittl for numerous interesting discussions.

Peter Hellekalek
Gerhard Larcher

Contents

Lattice Rules: How Well Do They Measure Up?   109
Fred J. Hickernell

Digital Point Sets: Analysis and Application   167
Gerhard Larcher

Random Number Generators: Selection Criteria and Testing   223
Pierre L'Ecuyer and Peter Hellekalek

Nets, (t, s)-Sequences, and Algebraic Geometry   267
Harald Niederreiter and Chaoping Xing

Financial Applications of Monte Carlo and Quasi-Monte Carlo Methods   303
Shu Tezuka

From Probabilistic Diophantine Approximation to Quadratic Fields

József Beck

1 PART I: Super Irregularity

The theory of Irregularities of Distribution (the counterpart of Uniform Distribution) was initiated by the classical papers of H. Weyl (1916), Van der Corput (1935), van Aardenne-Ehrenfest (1945), and was developed to a coherent theory by K.F. Roth and W.M. Schmidt (1954-75). The object of this theory is to measure the uniformity and non-uniformity of point sequences and point distributions (or integer sequences and sets of integers) relative to natural geometric classes like aligned and tilted rectangles, squares, boxes, circles, balls, convex sets, etc. (or residue classes, etc.). It has important applications in number theory, combinatorics and computational geometry.

Ramsey theory is one of the most extensively developed theories in discrete mathematics, originated from the fundamental works of Van der Waerden (1926) and Ramsey (1930), and was greatly extended by Paul Erdős and his "school". It became a deep theory – it is enough to refer to such contributions as Szemerédi's theorem (1974), Furstenberg's ergodic method (1976-90), and Shelah's new upper bound on Van der Waerden's function (1988). Ramsey theory has many important applications beyond combinatorics, mainly in number theory and geometry.

This Part I is a first step to extend the theory of Irregularities of Distribution in the direction of Ramsey theory.

Roth's theorem on long arithmetic progressions. *For any partition of the integers* $1, 2, \ldots, N$ *into two sets* S_1 *and* S_2, *there exists an arithmetic progression* $P = \{a, a + d, a + 2d, \ldots, a + (k-1)d\}$ *in* $1, 2, \ldots, N$ *such that*

$$\left| |P \cap S_1| - |P \cap S_2| \right| > \frac{1}{20} N^{1/4}.$$

Compare Roth's theorem to the following fundamental result of Ramsey theory.

Van der Waerden's theorem on short arithmetic progressions. *For any integer* k *there exists an* N *such that for any partition of the integers* $1, 2, \ldots, N$ *into two sets* S_1 *and* S_2, *there exists an arithmetic progression* $P = \{a, a + d, a + 2d, \ldots, a + (k-1)d\}$ *of length* k *which is* *entirely contained in either S_1 or in S_2. In other words,*

$$\left| |P \cap S_1| - |P \cap S_2| \right| = |P| = k.$$

Though these two theorems have pretty much the same structure, there is a fundamental difference: in Van der Waerden's theorem the discrepancy is as large as possible. On the other hand, the length k of the "monochromatic" arithmetic progression in terms of the length N of the underlying interval $[1, N]$ is extremely short: it is not known whether $k > \log^* N$ where $\log^*$ is the inverse of the "exponential tower function" (i.e. the number of iterations needed to achieve $\log \left(\log \left(\cdots (\log N) \cdots \right) \right) < 1$).

Next consider the following basic result from geometric discrepancy theory.

Schmidt's theorems on rectangles.
Let P be an arbitrary set of N points in the unit square.
(i) There exists an axis-parallel rectangle A in the unit square such that

$$\left| |P \cap A| - N \ area(A) \right| > \frac{1}{100} \log N.$$

(ii) There exists a tilted rectangle B in the unit square such that

$$\left| |P \cap B| - N \ area(B) \right| > N^{1/4 - \varepsilon}.$$

Both statements (i) and (ii) are best possible. We mention that circles have roughly the same discrepancy as tilted rectangles, and even for the class of *all* possible convex sets in the unit square the discrepancy is less than $N^{1/3 + \varepsilon}$. Is there any natural geometric shape which has "extra large" discrepancy? Well, the answer is yes if we change the normalization, and instead of taking N-element point sets in the unit square, we consider infinite point sets of density one in the whole plane.

"EXTRA LARGE DISCREPANCY" PROBLEM: *Let λ be an arbitrarily large real number. Does there exist a set S of area λ such that for any point set $\mathcal{P}$ of density one on the plane there is a congruent copy S' of S such that*

$$|S' \cap \mathcal{P}| > (1 + c)\lambda, \tag{1}$$

and there is another congruent copy S'' of S such that

$$|S'' \cap \mathcal{P}| < (1 - c)\lambda? \tag{2}$$

Here $c > 0$ is an absolute constant independent of $\lambda, S, \mathcal{P}$.

We gave a positive answer to this question. We proved that hyperbola-segments like

$$S_\lambda = \{(x, y) : \ 1 \le x \le e^{\lambda/2}, -1/x < y < 1/x\} \tag{3}$$

satisfy requirements (1) and (2). The proof uses a version of the Roth-Halász method from the theory of Irregularities of Distribution. What is so special about hyperbolas? Well, the shape of the hyperbola "resembles" to a lacunary Fourier series with Hadamard gap condition. For these gap series a well-known theorem of S. Sidon states that the maximum is "almost as big as possible", more precisely, the maximum is bigger than an absolute constant multiple of the sum of the absolute values of the Fourier coefficients (which is of course the absolute limit). Lacunary Fourier series with Hadamard gap condition behave like "random series", and Sidon theorem corresponds to "extra large deviation" in the following sense: tossing a coin n times it is extremely unlikely to have n heads, or even more than 51% heads, but the probability is *positive*, so it can happen on a very-very long run. The whole Part II is devoted to the probabilistic aspects of diophantine approximation.

OPEN PROBLEM 1. *Does there exist a set S in the plane such that its* **translates alone** *satisfy requirements (1) and (2), respectively?*

We can answer the analogous question for 2-colorings of the lattice points $\mathbb{Z}^2$: Given any 2-coloring $f : \mathbb{Z}^2 \to \{-1, +1\}$ of the lattice points and given an arbitrarily large number λ, there is a plane set S of area λ such that

$$\sup_{v \in \mathbb{R}^2} \left| \sum_{n \in \mathbb{Z}^2 \cap (S+v)} f(\mathbf{n}) \right| > c \cdot \lambda. \qquad (4)$$

Here S is a tilted copy of the hyperbola-segment S_λ with slope $\sqrt{2}$ (the slope can be any other quadratic irrational), $S + \mathbf{v}$ is the translated by vector $\mathbf{v}$ copy of S, and $c > 0$ is a universal constant.

(4) is an "almost Van der Waerden" theorem for *translated* copies. This is a new phenomenon, because in Ramsey theory magnification plays an absolutely crucial role in the proofs.

OPEN PROBLEM 2. *Does there exist a* **convex** *set S such that its congruent copies S' and S'' satisfy requirements (1) and (2), respectively?*

It is not hard to see that a positive answer to Open Problem 2 implies the positive solution of two famous unsolved problems in geometry.

DANZER'S CONJECTURE (early 1950's): *Let $\mathcal{P}$ be a point set on the plane such that any convex set of area one contains a point of $\mathcal{P}$. Then the density of $\mathcal{P}$ is ∞.*

"DUAL DANZER": *Let $\mathcal{P}$ be a point set of density one on the plane. Then for arbitrarily large n, there is a convex set of area one which contains (at least) n points of $\mathcal{P}$.*

We cannot help mentioning here an old "extra large fluctuation" type problems of Erdős and Turán in combinatorial number theory.

ERDŐS-TURÁN CONJECTURE (1930's): *Let $A = \{a_1, a_2, a_3, \ldots\} \subset$ $\mathbf{N}$ be an infinite sequence of integers. Let $f(n)$ denote the number of solutions $n = a_i + a_j$. If $f(n) \geq 1$ for all $n \in \mathbf{N}$, then $f(n)$ cannot be bounded,* *what is more, $f(n) > c \cdot \log n$ for infinitely many n.*

This conjecture was motivated by the highly irregular behaviour of the well-known arithmetical functions

$$r(n) = \sum_{n=x^2+y^2} 1 \quad \text{and} \quad d(n) = \sum_{n=x^2-y^2} 1.$$

Answering a 20-year-old problem of S. Sidon, in 1954 Erdős proved the *existence* of a sequence $A = \{a_1, a_2, a_3, \ldots\} \subset \mathbb{N}$ such that

$$c_1 \cdot \log n < f(n) < c_2 \cdot \log n, \quad \text{for all } n.$$

The proof was one of the first applications of the Probabilistic Method. Erdős realized that his method cannot guarantee the stronger requirement

$$\frac{f(n)}{\log n} \to c > 0.$$

This led him to the following

ERDŐS CONJECTURE (1950's). *Let A and $f(n)$ be as in the Erdős-Turán conjecture above. Then*

$$\frac{f(n)}{\log n} \to c > 0$$

is impossible.

Erdős offered \$ 500 for a first solution to any of the last two conjectures.

2 PART II: Probabilistic Diophantine Approximation

A classical problem of the theory of diophantine approximation is to understand the *local* and *global* properties of the irrational rotation $n\alpha$ mod 1 $(n = 1, 2, 3, \ldots)$.

A *local* question is to decide whether an inequality

$$\|n\alpha\| < \frac{1}{n\phi(n)} \iff \left\|\alpha - \frac{m}{n}\right\| < \frac{1}{n^2 \cdot \phi(n)}$$

or in general,

$$\|n\alpha - \beta\| < \frac{1}{n\phi(n)} \tag{1}$$

has infinitely many integral solutions, and if this is the case, to determine the solutions, or at least determine the asymptotic number of integral solutions. Here $\|x\|$ denotes, as usual, the distance of a real x from the nearest integer, and $\phi(n)$ is a positive increasing function of n.

A *global* question is to give a quantitative description of the equidistribution property of the irrational rotation $n\alpha \mod 1$.

We are going to emphasize the similarities between the local and global aspects by formulating *parallel* theorems.

Consider first the subclass of *quadratic irrationals*. They play a very special role in diophantine approximation: numbers like $\alpha = \sqrt{2}, \sqrt{3}, \cdots$ have the worst rational approximation property in the sense that

$$\left| \alpha - \frac{p}{q} \right| > \frac{\text{const}(\alpha)}{q^2} \quad \text{for all rationals } p/q$$

(called "badly approximable" numbers), and at the same time, they are the most uniformly distributed $n\alpha \mod 1$ sequences. From algebraic point of view the great advantage is the extremely simple cyclic group structure of the units in the corresponding quadratic field.

2.1 LOCAL CASE: inhomogeneous Pell inequalities-hyperbolas

Consider first an inhomogeneous Pell inequality like

$$-c \le (x + \omega)^2 - 2y^2 \le c \tag{2}$$

where $c > 0$ and $\omega \in [0, 1)$ are fixed constants. (Note that the restriction to a symmetric interval in (2) is just a matter of taste - everything works for the general case, too. Similarly, factor 2 can be replaced by any other square-free integer.) In view of the factorization

$$(x + \omega)^2 - 2y^2 = (x + \omega - y\sqrt{2})(x + \omega + y\sqrt{2}), \tag{3}$$

the asymptotic number of integral solutions of (2) heavily depends on the local behavior of $n\sqrt{2} \mod 1$. In fact, (2) and (1) with $\alpha = \sqrt{2}, \beta = \omega$ and $\phi = c/2\sqrt{2}$ are essentially equivalent.

Let $f_\omega(c; N)$ be the number of integral solutions $(x, y) \in \mathbb{Z}^2$ of (2) satisfying $1 \le x \le N$ and $1 \le y$.

In the special case $c = 1$ and $\omega = 0$, (2) becomes the classical Pell equation. The integral solutions form a cyclic group generated by the smallest solution, so trivially

$$f_0(1; N) = \frac{\log N}{\log(1 + \sqrt{2})} + O(1).$$

In general, by a simple theorem of Lang [La], for every *homogeneous* Pell inequality the asymptotic number of integral solutions behaves like $\text{const} \cdot \log N$ with uniformly bounded fluctuations.

In contrast to this "deterministic behaviour" in the homogeneous case $\omega = 0$, the *inhomogeneous* case turns out to be entirely different: for a "randomly chosen" ω the asymptotic number of solutions $f_\omega(c; N)$ as $N \to \infty$ possesses "perfect randomness".

The two basic parameters of a random variable are the *mean value* and the *variance*. The *mean value* of $f_\omega(c; N)$ as $0 \le \omega < 1$ is

$$\int_0^1 f_\omega(c; N) \, d\omega = \frac{c}{\sqrt{2}} \log N + O(1). \tag{4}$$

Formula (4) expresses the simple fact that the average number of lattice points contained in all the translated copies of a hyperbola domain is precisely the area of the domain. Since for any $1 < M < N$,

$$f_\omega(c; N) - f_\omega(c; M) \ll \log N - \log M = \log \frac{N}{M},$$

it is more natural to use the exponential scaling $f_\omega(c; e^N)$ instead of the linear one.

The *variance* comes from the following limit formula: for any $c > 0$ there is a positive effective constant $\sigma = \sigma(c) > 0$ such that

$$\lim_{N \to \infty} \frac{1}{N} \int_0^1 \left(f_\omega(c; e^N) - \frac{c}{\sqrt{2}} N \right)^2 \, d\omega = \sigma^2(c). \tag{5}$$

Formula (5) follows from Fourier analysis by using the arithmetic of $\mathbf{Q}(\sqrt{2})$. The first result is a central limit theorem, see [Be5].

THEOREM 1.1.   *The renormalized counting function*

$$\frac{f_\omega(c; e^N) - \frac{c}{\sqrt{2}} N}{\sigma(c)\sqrt{N}}, \quad 0 \le \omega < 1$$

has a standard normal limit distribution as $N \to \infty$.

What is our intuition behind Theorem 1.1? Write

$$g_j(\omega) = f_\omega(c; e^j) - f_\omega(c; e^{j-1}), \quad j = 1, 2, ..., N.$$

That is, $g_j(\omega)$ is the number of integral solutions $n \in \mathbf{N}$ of (2) satisfying $e^{j-1} < n \le e^j$. Our key observation is that function $g_j(\omega)$ resembles to the j-th Rademacher function, so the sum

$$f_\omega(c; e^N) - \frac{c}{\sqrt{2}} N = \sum_{j=1}^N \left(g_j(\omega) - \frac{c}{\sqrt{2}} \right),$$

as a function of $\omega \in [0, 1)$, behaves like a sum of $\approx N$ *independent* Bernoulli variables

$$f_\omega(c; e^N) - \frac{c}{\sqrt{2}} N \approx \pm 1 \pm 1 \pm \cdots \pm 1 \quad (\approx N \text{ terms}). \tag{6}$$

Let us go back to Theorem 1.1: what can we say about the *sequence* of those n's which satisfy inequality (7) below (note that $-\infty < \lambda_1 < \lambda_2 < \infty$ are fixed constants)

$$\frac{c}{\sqrt{2}}n + \lambda_1 \sigma \sqrt{n} < f_\omega(c; e^n) < \frac{c}{\sqrt{2}}n + \lambda_2 \sigma \sqrt{n} \qquad (7)$$

for a "randomly chosen" ω?

We proved that the "logarithmic density" of these n's does exist for every fixed $-\infty < \lambda_1 < \lambda_2 < \infty$, and equals to $(2\pi)^{-1/2} \int_{\lambda_1}^{\lambda_2} e^{-u^2/2}\, du$.

THEOREM 1.2. [Be5] *For almost every* $\omega \in [0, 1)$,

$$\lim_{N \to \infty} \frac{1}{\log N} \sum_{\substack{1 \le n \le N: \\ n \text{ satisfies } (2)}} \frac{1}{n} = \frac{1}{\sqrt{2\pi}} \int_{\lambda_1}^{\lambda_2} e^{-u^2/2}\, du \qquad (8)$$

holds for all $-\infty < \lambda_1 < \lambda_2 < \infty$. *Moreover, the "logarithmic density" in (8) is necessary: (8) fails for the ordinary density and an "Arc sine law" holds instead.*

Note that the classical results for independent random variables are due to Paul Lévy.

As an analogue of Khintchine's Law of the iterated logarithm, we can show that the number of solutions $f_\omega(c; e^n)$ of (2) oscillates between the sharp bounds

$$\frac{c}{\sqrt{2}}n - \sigma\sqrt{n}\sqrt{(2+\epsilon)\log\log n} < f_\omega(c; e^n) < \frac{c}{\sqrt{2}}n + \sigma\sqrt{n}\sqrt{(2+\epsilon)\log\log n}$$

as $n \to \infty$ for almost every ω. What we actually formulate is the "ultimate" Kolmogorov-Erdős form which contains the Khintchine's form as a corollary.

THEOREM 1.3. [Be5] *Let $\phi(n)$ be an arbitrary positive increasing function of n. For almost every ω,*

$$f_\omega(c; e^n) > \frac{c}{\sqrt{2}}n + \phi(n)\sigma\sqrt{n} \qquad (9)$$

holds for infinitely many n's if and only if the series

$$\sum_{n=1}^{\infty} \frac{\phi(n)}{n} e^{-\phi^2(n)/2} \quad \text{diverges.}$$

Exactly the same holds for the other inequality

$$f_\omega(c; e^n) < \frac{c}{\sqrt{2}}n - \phi(n)\sigma\sqrt{n}. \qquad (9')$$

It is well-known that both the Law of the iterated logarithm and Lévy's law (8) are distant consequences of the central limit theorem. In particular, the Law of the iterated logarithm corresponds to the case $\lambda = \lambda(N) \approx \sqrt{2 \log \log N}$ in the central limit theorem. Note that for even moderately large values of λ like $\lambda \approx \log N$, the central limit theorem "becomes" the following limit problem:

$$\frac{\text{measure}\left\{\omega \in [0,1): \frac{c}{\sqrt{2}}N + \lambda\sigma\sqrt{N} < f_\omega(c; e^N)\right\}}{\frac{1}{\sqrt{2\pi}} \int_\lambda^\infty e^{-u^2/2} \, du} \to 1 \qquad (10)$$

where both λ and N tend to infinity. Results like (10) are called "Large deviation" theorems in probability theory.

THEOREM 1.4. [Be5] *If $\lambda = \lambda(N)$ varies with N in such a way that $0 < \lambda < N^{\text{const}}$ (where const means a "small" positive absolute constant), then (10) is true. Changing λ into $-\lambda$ we obtain exactly the same result for the "left tail".*

However, if λ is around $\sqrt{N}$, (10) fails even in the "ideal case" of independent random variables. So the following exiting question arises: Let $c > 0$ be fixed. Is it possible that the "Law of large numbers"

$$\frac{f_\omega(c; e^n)}{n} \to \frac{c}{\sqrt{2}} \quad \text{as} \quad n \to \infty \qquad (11)$$

holds for *every single* $\omega \in [0,1)$? Of course, Theorem 1.3 implies (11) for *almost* every ω. The answer to this question is negative.

THEOREM 1.5. [Be5] *For every $c > 0$ there are continuum many "divergence points" $\omega^* = \omega^*(c) \in [0,1)$ such that*

$$\limsup_{n \to \infty} \frac{f_{\omega^*}(c; e^n)}{n} > \frac{c}{\sqrt{2}} \qquad (12)$$

and

$$\liminf_{n \to \infty} \frac{f_{\omega^*}(c; e^n)}{n} < \frac{c}{\sqrt{2}}. \qquad (12')$$

It is easy to see that the fluctuation in (12) is as large as possible, so (12) can be interpreted as an "extra large deviation" result (see Part I).

2.2 *Beyond quadratic irrationals*

In view of factorization (3), (2) and the inhomogeneous inequality

$$\|n\alpha - \beta\| < \frac{\gamma}{2n} \tag{13}$$

are essentially equivalent. So Theorems 1.1-1.5 describe the randomness of the asymptotic number of integral solutions of (13) when $\alpha = \sqrt{2}$, or any other quadratic irrational, and β is the *variable*. These results are corollaries of (6), which is an "almost independence" property in the inhomogeneous local case. Of course, the analogues of Theorems 1.1-1.5 (and of (6)) must be true for a much bigger class of α's. To obtain a "simple" norming factor (see (5)), we need some kind of "regularity" for the sequence of partial quotients $a_1, a_2, a_3, ...$ of α. For example,

$$e = [2; 1, 2, 1, 1, 4, 1, 1, 6, 1, 1, 8, 1, ...] = [2; ..., 1, 2i, 1, ...] \tag{14}$$

$$\sqrt{e} = [1; 1, 1, 1, 5, 1, 1, 9, 1, 1, 13, 1, 1, ...] = [1; ..., 1, 4i + 1, 1, ...] \tag{15}$$

$$e^2 = [7; 2, 1, 1, 3, 18, 5, 1, 1, 6, 30, ...] = [7; ..., 3i - 1, 1, 1, 3i, 12i + 6, ...]$$

and so on. For these special numbers we can prove the perfect analogoues of the probabilistic theorems above.

What about the class of "badly approximable" numbers (i.e., $a_i = O(1)$)? The only technical, or aesthetic, problem with badly approximable numbers is as follows: though the central limit theorem and the other results hold true, the norming factor oscillates between two constant multiples of $\sqrt{N}$, and the limit in (5) not necessarily exists. In other words, for this class the central limit theorem (and the others) are not so "elegant".

2.3 *GLOBAL CASE: lattice points in tilted rectangles*

Theorems 1.1–1.5 describe the "randomness" of the local behavior of $n\sqrt{2}$ mod 1. We have surprisingly similar results on the global behavior of $n\sqrt{2}$ mod 1, too.

The global version of a local inequality $\|n\alpha - \beta\| < \frac{\gamma}{2n}$ is

$$\|n\alpha - \beta\| < \frac{\gamma}{2} \tag{16}$$

where $0 < \gamma < 1$ is a constant. Let $G(\alpha, \beta, \gamma; N)$ denote the number of integral solutions $n \in \mathbf{N}$ of (16) satisfying $1 \leq n \leq N$. The well-known equidistribution theorem states that for every irrational α and for every interval $I = (\beta - \frac{\gamma}{2}, \beta + \frac{\gamma}{2})$ of length $|I| = \gamma < 1$,

$$\frac{G(\alpha, \beta, \gamma; N)}{N} \to |I| = \gamma \quad \text{as} \quad N \to \infty.$$

It is, therefore, natural to study the discrepancy

$$D(\alpha, I; N) = D(\alpha, \beta, \gamma; N) = G(\alpha, \beta, \gamma; N) - \gamma N \qquad (17)$$

where $I = (\beta - \frac{\gamma}{2}, \beta + \frac{\gamma}{2})$ runs over all the subintervals of $[0, 1)$.

The classical works of Hardy-Littlewood and Ostrowski (see [Ha-Li1 and 2],[Os]) give an upper bound on $D(\alpha, I; N)$ in terms of the partial quotients a_i of α. Let p_i/q_i be the i-th convergent of α, let s be defined by $q_s \leq N < q_{s+1}$, and write $a_{s+1}(N) = N/q_s$. Now the upper bound is

$$\max_I |D(\alpha, I; N)| < 3(a_1 + a_2 + \cdots + a_s + a_{s+1}(N)). \qquad (18)$$

Our intuition about the *global* behavior of $n\alpha$ mod 1 is as follows: For a *fixed* irrational α, the discrepancy function $D(\alpha, I; N)$ behaves like a sum of *independent* random variables

$$D(\alpha, I; N) \approx \pm a_1 \pm a_2 \pm \cdots \pm a_s \pm a_{s+1}(N) \qquad (19)$$

as both n and I vary in $1 \leq n \leq N$ and $I \subset [0, 1)$, respectively. (That is, we have 3 variables.) Heuristic (19) is a sort of global analogue of the local heuristic (6).

Let us emphasize that the "almost independence" in (6) and (19) are completely different from the well-known almost independence property of the distribution of the partial quotients $a_i = a_i(\alpha)$ as α *varies* in an interval (see the classical works of Kusmin, Paul Lévy and Khintchine). In our case α is fixed (like $\alpha = \sqrt{2}$).

It is an important program to justify intuition (19) for every single α. We worked out the details in the special case $\alpha = \sqrt{2}$ (see [Be3]), and as a corollary obtained that the renormalised discrepancy function

$$\frac{D(\sqrt{2}, \beta, \gamma; N)}{\sqrt{\frac{\log N}{12\sqrt{2}\log(1+\sqrt{2})}}} \qquad (20)$$

has a standard normal limit distribution as $0 \leq \beta < 1, 0 \leq \gamma < 1, 1 \leq n \leq N$. In other words, we have a 3-parameter central limit theorem. Observe that (20) is a sort of global version of Theorem 1.1. Of course the proof of the special case $\alpha = \sqrt{2}$ easily extends to the whole class of quadratic irrationals.

Can we get a similar 3-parameter central limit theorem for a "randomly chosen" α? The answer is rather surprising: in contrast to the local case, in the global case we cannot expect a normal limit distribution. The reason is that a necessary condition for normal limit distribution is the relation

$$\frac{\max_{1 \leq i \leq s} a_i^2}{a_1^2 + a_2^2 + \cdots + a_s^2} \to 0 \quad \text{as} \quad s \to \infty. \qquad (21)$$

This means that we have a sum of "individually negligible components", but by the Gauss-Kusmin theorem, (21) fails for almost every α.

As we already said, (20) is a global version of Theorem 1.1. The global version of the "large deviation" Theorem 1.4 is as follows. The problem is to prove the limit of the ratio

$$\frac{\text{volume}\left\{(\beta,\gamma,\delta) \in [0,1)^3 : D(\sqrt{2},\beta,\gamma;\delta N) > \lambda\sqrt{\frac{\log N}{12\sqrt{2}\log(1+\sqrt{2})}}\right\}}{\frac{1}{\sqrt{2\pi}}\int_\lambda^\infty e^{-u^2/2}\,du} \to 1$$

(22)

where both λ and N tend to infinity.

THEOREM 2.4. [Be6] *If $\lambda = \lambda(N)$ varies with N in such a way that $0 < \lambda < (\log N)^{\text{const}}$ (where* const *means a "small" positive absolute constant), then (22) is true. Changing λ into $-\lambda$ we obtain exactly the same result for the "left tail".*

Next we formulate the analogues of Theorems 1.2-1.3. What can we say about the *set* of those positive *real numbers* $x \in \mathbf{R}$ which satisfy inequality (23) below

$$\lambda_1 < \frac{D(\sqrt{2},I;e^x)}{\sqrt{\frac{x}{12\sqrt{2}\log(1+\sqrt{2})}}} < \lambda_2$$

(23)

for a "randomly chosen" subinterval I of $[0,1)$?

THEOREM 2.2. [Be6] *For almost every subinterval I of $[0,1)$,*

$$\lim_{N\to\infty} \frac{1}{\log N} \int_{\substack{0\le x\le N: \\ x \text{ satisfies } (23)}} \frac{dx}{x} = \frac{1}{\sqrt{2\pi}}\int_{\lambda_1}^{\lambda_2} e^{-u^2/2}\,du$$

(24)

holds for all $-\infty < \lambda_1 < \lambda_2 < \infty$. Moreover, the "logarithmic density" in (24) is necessary: (24) fails for the ordinary density and an "Arc sine law" holds instead.

Why did we switch to real numbers (and to integral) in Theorem 2.2? The reason is a basic difference between the local and global counting functions. We recall: if $1 \le M < N$ then

$$f_\omega(c;N) - f_\omega(c;M) \ll \log N - \log M = \log\frac{N}{M},$$

and so $f_\omega(c;n)$ does not change more than $O(1)$ as (say) $N/e < n \le N$. It is, therefore, enough to study the asymptotic behavior of $f_\omega(c;N)$ on an exponentially rare subsequence like $N = e^n$, $n \in \mathbf{N}$. On the other hand, for the global function $G(\sqrt{2},\beta,\gamma;N)$: if $0 < M < N$ then

$$G(\sqrt{2},\beta,\gamma;N) - G(\sqrt{2},\beta,\gamma;M) \ll \log(N - M).$$

Hence the global discrepancy function $D(\sqrt{2}, \beta, \gamma; n)$ might change as much as its *maximum* $\log N$ on a short interval like $N - \sqrt{N} < n < N$. So in the global case we cannot restrict ourselves to an exponentially rare subsequence of integers.

The "global law of the iterated logarithm" goes as follows.

THEOREM 2.3. [Be6] *Let $\phi(n)$ be an arbitrary positive increasing function of n. Consider the set*

$$\{x \in [0, \infty) : x \text{ satisfies } (25)\}$$

where

$$D(\sqrt{2}, I; e^x) > \phi(x) \sqrt{\frac{x}{12\sqrt{2} \log(1 + \sqrt{2})}}. \tag{25}$$

For almost every subinterval I of $[0, 1)$,

$$\text{measure}\{x \in [0, \infty) : x \text{ satisfies } (25)\} = \infty$$

if and only if the integral

$$\int_1^\infty \frac{\phi(x)}{x} e^{-\phi^2(x)/2} \, dx \quad \text{diverges.}$$

Exactly the same holds for the other inequality

$$D(\sqrt{2}, I; e^x) < -\phi(x) \sqrt{\frac{x}{12\sqrt{2} \log(1 + \sqrt{2})}}. \tag{25'}$$

Remark. Switching back to the *linear* scaling we have:

$$\text{measure}\{x \in [0, \infty) : x \text{ satisfies } (25) \text{ with } e^x = y\} = \infty \Leftrightarrow \int_1^\infty \frac{dy}{y} = \infty.$$

Now if we drop the rather strong logarithmic requirement $\sum_1^\infty 1/n = \infty$, and just want infinitely many integers n for which $D(\sqrt{2}, I; n)$ is "large", then the fluctuation becomes much bigger than $\approx \sqrt{\log n \cdot \log \log \log n}$ (i.e., what the law of the iterated logarithm gives), namely as large as $\log n$ (i.e., as large as possible). Indeed T. Sós [So] proved that for *every* α (not just for quadratic irrationals) and for almost every subinterval I of $[0, 1)$,

$$\lim_{n \to \infty} \sup \frac{D(\alpha, I; n)}{\log n} > 0. \tag{26}$$

Note that Halász [Ha] and independently Tijdeman-Wagner [T-W] proved far-reaching generalizations of (26) for *arbitrary* sequences (not just for $n\alpha$ mod 1 sequences). Sós' theorem can be interpreted as a *global* "extra large deviation" result. In other words, it is a sort of global version of Theorem 1.5.

2.4 *SIMULTANEOUS CASE*

The main difficulty is that the theory of continued fractions does not seem to extend to higher dimensions. Some of the most natural problems, like the famous Littlewood's conjecture, are still completely hopeless.

Here we just mention one result which extends a 70-year-old theorem of Khintchine [Kh1] to higher dimensions (see [Be4]). Khintchine proved, by using continued fractions, that for almost every α, the usual interval discrepancy of the $n\alpha$ mod 1 sequence is between $\log n \cdot (\log\log n)^{1\pm\epsilon}$. We proved, by using completely different approach, that for almost every $(\alpha_1,...,\alpha_k) \in \mathbf{R}^k$, the usual box discrepancy of the $(n\alpha_1,...,n\alpha_k)$ mod 1 sequence is between $(\log n)^k \cdot (\log\log n)^{1\pm\epsilon}$. Similarly to Khintchine, we in fact proved a precise "convergence-divergence criterion" (i.e., a Borel-Cantelli type theorem).

Note that one can easily formulate the analogues of the above-mentioned "one-dimensional" probabilistic local and global theorems in the simultaneous case. We are pessimistic: we do not believe in any normal limit distribution in higher dimensions. The reason is that the same difficulty appears here as in Littlewood's notoriously intractable $n\|n\alpha\|\|n\beta\|$ problem.

Note that there were some earlier attempts to give a systematic study of probabilistic methods in Diophantine approximations (see e.g. Kac [Ka], Kesten [Kes2] and Kemperman [Kem]), but they were discussing $n\alpha$ (mod 1) for *almost every* α only, and didn't say anything about *concrete* sequences like $n\sqrt{2}$ (mod 1).

3 PART III: Quadratic Fields and Continued Fractions

In Part II we studied the local and global behaviour of the sequence $n\alpha$ (mod 1) for concrete values of α. So far it didn't make any real difference that $\alpha = \sqrt{2}$ or $\sqrt{3}$. But there is a substantially difference between the real quadratic fields $\mathbf{Q}(\sqrt{2})$ and $\mathbf{Q}(\sqrt{3})$: the corresponding fundamental units have norm 1 and -1, respectively (or in plain language, in contrast to $x^2 - 2y^2 = -1$, the equation $x^2 - 3y^2 = -1$ doesn't have integer solution). This fact will play an important role in what follows.

Consider now the classical diophantine series (α is a fixed irrational and $\{...\}$ stands for the fractional part)

$$S_\alpha(n) = \sum_{k=1}^{n}(\{k\alpha\} - 1/2). \tag{1}$$

This series has been thoroughly discussed by Hardy-Littlewood, Hecke, Ostrowski, Behnke, and more recently by Vera T. Sós and others. They concentrated on the maximum fluctuations as $n \to \infty$. In the spirit of probabilistic diophantine approximation we focused on the typical fluctuations, and managed to prove an elegant central limit theorem for individual α's, including the class of quadratic irrationals. In particular for $\alpha = \sqrt{2}$ this central limit theorem goes as follows, see [Be5].

THEOREM 3.1.   *There is a positive absolute constant c such that*

$$\frac{1}{N} \left| \left\{ 1 \leq n \leq N : \frac{S_{\sqrt{2}}(n)}{c\sqrt{\log n}} \leq \lambda \right\} \right| \longrightarrow \frac{1}{\sqrt{2\pi}} \int_{-\infty}^{\lambda} e^{-u^2/2} \, du$$

for every fixed value of λ as $N \to \infty$.

The reason behind a central limit theorem is usually some kind of "independence". Where does the "independence" in Theorem 3.1 come from? To understand it, consider the jth convergent denominator q_j of $\sqrt{2} = 1 + \frac{1}{2+\frac{1}{2+\cdots}} = [1; 2, 2, 2, \cdots]$:

$$q_j = \frac{1}{2\sqrt{2}} \left((1 + \sqrt{2})^j - (1 - \sqrt{2})^j \right). \tag{2}$$

Note that replacing $\sqrt{2}$ with the golden ratio $\frac{\sqrt{5}-1}{2}$, q_j becomes the jth Fibonacci number. It is well-known that every positive integer can be *uniquely* represented as a sum of distinct Fibonacci numbers if we don't allow to take two consecutive ones.

Similarly, every positive integer n can be uniquely written in the form

$$n = \sum_{j=0}^{l} b_j q_j, \tag{3}$$

where $b_j = b_j(n) \in \{0, 1, 2\}$, if we make the extra restriction $b_{i+1} = 2 \Rightarrow b_i = 0$.

Let

$$t_1 = t_1(n) = \sum_{k=1}^{b_l q_l} \left(\{k\sqrt{2}\} - 1/2 \right),$$

$$t_2 = t_2(n) = \sum_{k=b_l q_l + 1}^{b_l q_l + b_{l-1} q_{l-1}} \left(\{k\sqrt{2}\} - 1/2 \right),$$

$$t_3 = t_3(n) = \sum_{k=b_l q_l + b_{l-1} q_{l-1} + 1}^{b_l q_l + b_{l-1} q_{l-1} + b_{l-2} q_{l-2}} \left(\{k\sqrt{2}\} - 1/2 \right),$$

and so on. The last term is

$$t_{l+1} = t_{l+1}(n) = \sum_{k=n-b_0 q_0 + 1}^{n} \left(\{k\sqrt{2}\} - 1/2 \right).$$

Thus by (3) we get the following decomposition of sum (1):

$$S_{\sqrt{2}}(n) = t_1(n) + t_2(n) + t_3(n) + \cdots + t_{l+1}(n). \tag{4}$$

Note that here $l \approx \log n$, and each term $t_i = t_i(n)$ is between -2 and 2. But the crucial observation is that the terms $t_i = t_i(n)$ are "almost independent" in the precise sense that the correlation between $t_i = t_i(n)$ and $t_{i+d} = t_{i+d}(n)$ is going down exponentially fast (as a function of d) as n runs through the integers $1, 2, \cdots, N$.

In this sense decomposition (4) resembles to the partial sums of a lacunary Fourier series like

$$\sum_{i=1}^{\infty} \sin(\lambda_i x), \qquad \frac{\lambda_{i+1}}{\lambda_i} \geq 1 + \sqrt{2},$$

for which a central limit theorem was proved in the 1940-50s. An adaptation of the proof of this classical result gives Theorem 3.1.

Now what happens if in Theorem 3.1 we replace $\sqrt{2}$ with $\sqrt{3}$? Well, the big difference is that the *mean value* of

$$S_{\sqrt{3}}(n) = \sum_{k=1}^{n}(\{k\sqrt{3}\} - 1/2)$$

is *not* zero:

$$\frac{1}{N+1}\sum_{n=0}^{N} S_{\sqrt{3}}(n) = \frac{\log N}{12\log(2 + \sqrt{3})} + O(1). \tag{5}$$

The left hand side of (5) is called the *Cesaro mean* of the series $\sum(\{n\sqrt{3}\} - 1/2)$.

In general, let us define the Cesaro mean of (1):

$$\Xi_\alpha(N) = \frac{1}{N+1}\sum_{n=0}^{N} S_\alpha(n) = \sum_{n=1}^{N}\left(1 - \frac{n}{N+1}\right)(\{n\alpha\} - 1/2). \tag{6}$$

What is the reason that the Cesaro mean $\Xi_\alpha(N)$ is $O(1)$ for $\alpha = \sqrt{2}$, but it is constant times $\log N$ for $\alpha = \sqrt{3}$? What can we say about the Cesaro mean $\Xi_\alpha(N)$ for an arbitrary real α?

Well, this question has a surprisingly simple and elegant answer.

THEOREM 3.2. [Be7]   *If $N = q_l - 1$ then*

$$\left|\Xi_\alpha(N) - \frac{-a_1 + a_2 - a_3 \pm \cdots + (-1)^l a_l}{12}\right| < 20$$

where α has the continued fraction expansion

$$\alpha = a_0 + \cfrac{1}{a_1 + \cfrac{1}{a_2 + \cdots}} = [a_0; a_1, a_2, a_3, \cdots],$$

and q_l is the lth convergent denominator of α.

For a proof, see Part V.

From Theorem 3.2 easily follows

COROLLARY 3.3. [Be7]   *For arbitrary N*

$$\Xi_\alpha(N) = \frac{-a_1 + a_2 - a_3 \pm \cdots + (-1)^l a_l}{12} + O(\max_{1 \le i \le l} a_i),$$

where l is the first index for which $q_l \ge N$.

Now we have a perfect understanding of the difference between $\alpha = \sqrt{2}$ and $\sqrt{3}$.

For $\sqrt{3} = [1; 1, 2, 1, 2, 1, 2, \cdots]$ the period is $1, 2$. Since

$$p_{2i-1} \pm \sqrt{3} q_{2i-1} = (2 \pm \sqrt{3})^i,$$

we have

$$q_{2i-1} = \frac{1}{2\sqrt{3}}\left((2 + \sqrt{3})^i - (2 - \sqrt{3})^i\right).$$

Therefore,

$$q_{2i-1} \approx \frac{1}{2\sqrt{3}}(2 + \sqrt{3})^i \approx N$$

implies that $i = \log N / \log(2 + \sqrt{3}) + O(1)$. By Corollary 3.3 the Cesaro mean equals

$$\Xi_{\sqrt{3}}(N) = \frac{-1 + 2 - 1 + 2 \mp \cdots - 1 + 2}{12} = \frac{-1 + 2}{12} \frac{\log N}{\log(2 + \sqrt{3})} + O(1),$$

proving (5).

On the other hand, for $\sqrt{2} = [1; 2, 2, 2, \cdots]$ the alternating sum $-2 + 2 - 2 + 2 \mp \cdots - 2 + 2 \cdots$ in Theorem 3.2 *cancels out*. This is the reason why for any quadratic irrational α for which the length of the continued fraction period is *odd*, the Cesaro mean $\Xi_\alpha(N) = O(1)$.

Is this observation useful? Yes, for the subclass $\sqrt{p}$, p prime, we know exactly the parity of the length of the period: *odd* if $p \equiv 1 \pmod 4$, *even* if $p \equiv 3 \pmod 4$. Unfortunately, we don't have a good characterization like this for the whole class of quadratic irrationals.

How about special numbers like

$$e = [2; 1, 2, 1, 1, 4, 1, 1, 6, 1, 1, 8, 1, \cdots, 1, 2i, 1, \cdots]?$$

Well, the alternating sum $(-1 + 2 - 1) + (1 - 4 + 1) + (-1 + 6 - 1) + \cdots + (-1)^i(1 - 2i + 1)$ equals $i - 1$ if i is odd, and $-i$ if i is even. Thus by Corollary 3.3 we have

$$\Xi_e(N) = O(\log N / \log \log N), \tag{7}$$

which is the true order of magnitude.

3.1   CESARO MEAN OF $\sum \{n\alpha^{\frac{1}{2}}\}$ AND QUADRATIC FIELDS

In view of Theorem 3.2 and Corollary 3.3, the Cesaro mean $\Xi_\alpha(N)$ has a surprisingly simple formula. The proof is not easy, but it was equally difficult to *find* the right conjecture. What was our motivation to guess the right formula?

To explain it, we discuss an alternative approach to figure out the Cesaro mean $\Xi_\alpha(N)$. We start with the well-known Fourier series expansion of the fractional part function (warning: it is not absolutely convergent)

$$\{x\} = \frac{1}{2} - \sum_{n=1}^{\infty} \frac{\sin(2\pi n x)}{\pi n}. \tag{8}$$

Substituting it back to (6), after some straightforward manipulations we obtain

$$\Xi_\alpha(N) = -\frac{1}{2\pi} \sum_{n=1}^{N} \frac{1}{n \tan(\pi n \alpha)} + O(\max_{1 \le i \le l} a_i), \tag{9}$$

where l is the least index for which $q_l \ge N$.

Let $\alpha = \sqrt{d}$, d square-free positive integer. We clearly have

$$\frac{1}{\pi} \tan(\pi n \sqrt{d}) \approx \pm ||n\sqrt{d}|| = n\sqrt{d} - m \approx \frac{-(m^2 - dn^2)}{2n\sqrt{d}}. \tag{10}$$

In view of (9) and (10), the following formula is not too surprising:

$$\Xi_{\sqrt{d}}(N) = \frac{\sqrt{d}}{\pi^2} \sum_{x=1}^{N} \sum_{y=1}^{N} \frac{1}{x^2 - dy^2} + O(1). \tag{11}$$

If $d \equiv 3 \pmod 4$ then $x^2 - dy^2$ is the norm of the algebraic integer $x + \sqrt{d}y$ in $\mathbf{Q}(\sqrt{d})$.

Moreover, if we make the extra hypothesis that for the class number $h(d)$ of $\mathbf{Q}(\sqrt{d})$ we have $h(d) = 1$, then the right-hand side of (11) becomes

$$\frac{\sqrt{d}}{\pi^2} L(1, \chi^*) \frac{\log N}{\log \eta_d} + O(1), \tag{12}$$

where $L(1, \chi^*)$ is a Dirichlet's L-function, χ^* is the "norm-sign" character, and η_d is the fundamental unit of $\mathbf{Q}(\sqrt{d})$.

It is a well-known fact (a corollary of the Euler product and the quadratic reciprocity theorem) that

$$L(1, \chi^*) = L(1, \chi) L(1, \chi\chi') \text{ where } \chi(n) = \left(\frac{-4}{n}\right), \ \chi'(n) = \left(\frac{d}{n}\right). \tag{13}$$

By Dirichlet's class number formula,

$$L(1, \chi) = \frac{\pi}{4} \quad and \quad L(1, \chi\chi') = \frac{\pi h(-d)}{\sqrt{d}}. \tag{14}$$

Now this is where the famous Hirzebruch-Meyer-Zagier formula (HMZ-formula, in short) enters the story: $h(-p)$ can be expressed in terms of an alternating sum of the digits in the period of $\sqrt{p}$.

But before formulating the HMZ-formula, let us speak first about continued fractions and quadratic fields in general. The basic fact is that quadratic irrationals are perfectly characterized by the *periodicity* of their continued fraction expansions. It is well-known how to read out the least solution of the Pell equation $x^2 - dy^2 = 1$ from the period of $\sqrt{d}$. Moreover, the *parity* of the length of the period describes the sign of the norm of the fundamental unit: odd length means $+1$, even length means -1.

Combining Dirichlet's class number formulas with the *ineffective* Siegel theorem (or Deuring-Heilbronn-Siegel theorem), we obtain the deep asymptotic formulas

$$h(d) \log \eta_d = d^{1/2 \pm \epsilon}, \tag{15'}$$

$$h(-d) = d^{1/2 \pm \epsilon}, \tag{15"}$$

where $h(d)$ and $h(-d)$ are the class numbers of the real and complex quadratic fields $\mathbf{Q}(\sqrt{d})$ and $\mathbf{Q}(\sqrt{-d})$, respectively, and η_d is the fundamental unit of $\mathbf{Q}(\sqrt{d})$. Note that the order of magnitude of $\log \eta_d$ is roughly around the *length* of the period of the continued fraction of $\sqrt{d}$.

The beautiful Hirzebruch-Meyer-Zagier formula (HMZ-formula) was discovered in the 1970's:

$$h(-p) = \frac{-a_1 + a_2 - a_3 \pm \cdots + a_{2s}}{3},$$

where $p \equiv 3 \pmod 4$ prime, $h(p) = 1$, and $a_1, a_2, \cdots, a_{2s}$ forms the period of $\sqrt{p}$ (since $p \equiv 3 \pmod 4$ prime, the length $2i$ of the period has to be even).

Combining the HMZ-formula with (11)-(14), we conclude

$$\Xi_{\sqrt{p}}(N) = \frac{-a_1 + a_2 \mp \cdots + a_{2s}}{12} \frac{\log N}{\log \eta_d} + O(1)$$

$$= \frac{-a_1 + a_2 - a_3 \pm \cdots + (-1)^l a_l}{12} + O(1), \qquad (16)$$

where l is the least index for which $q_l \geq N$.

Now from (16) it was very natural to guess Corollary 3.3 (and Theorem 3.2) for arbitrary α, and this is exactly how we figured out the right conjecture.

Because our proof of Theorem 3.2 is completely elementary, reversing the previous argument one obtains an elementary, alternative proof of the HMZ-formula.

3.2 HARDY-LITTLEWOOD LEMMA 14

A few days after completing the proof of Theorem 3.2 we noticed the following somewhat technical lemma in Hardy-Littlewood [Ha-Li2] published in 1922.

"Lemma 14."   If $\alpha = [a_0; a_1, a_2, \cdots]$ then

$$\Xi_\alpha(N) = \frac{1}{12} \sum_{i=1}^{l} (-1)^k \left(\alpha_i + \frac{1}{\alpha_i} \right) + O\left((\max_{1 \leq i \leq l} a_i)^2 \right),$$

where l is the least index such that $q_l \geq N$, and

$$\alpha_i = a_i + \cfrac{1}{a_{i+1} + \cfrac{1}{a_{i+2} + \cfrac{1}{a_{i+3} + \cdots}}} = [a_i; a_{i+1}, a_{i+2}, a_{i+3}, \cdots].$$

By using the trivial identity $\alpha_i = a_i + \frac{1}{\alpha_{i+1}}$, the alternating sum in "Lemma 14" becomes

$$-\left(\alpha_1 + \frac{1}{\alpha_1} \right) + \left(\alpha_2 + \frac{1}{\alpha_2} \right) - \left(\alpha_3 + \frac{1}{\alpha_3} \right) \pm \cdots$$

$$= -a_1 + a_2 - a_3 \pm \cdots + (-1)^i a_i \pm \cdots.$$

The surprising conclusion is that from "Lemma 14" one can obtain a somewhat weaker version of Corollary 3.3 *in one line*. This is weaker because the error term $O\left((\max_{1 \leq i \leq l} a_i)^2 \right)$ is the square of the error term $O(\max_{1 \leq i \leq l} a_i)$ in Corollary 3.3.

I made this funny observation in November'95, but I was informed in the March'96 Oberwolfach Number Theory Meeting that some other mathematicians made the same observation much earlier, and the trivial cancellation in (16) was "folklore" among a group of continued fraction experts. So all what is left for me is the credit for the best possible error term.

Note that the weaker error term is still more than enough to derive the HMZ-formula. Since the proof of "Lemma 14" is elementary, in fact it is substantially shorter than our proof of Theorem 3.2, this gives the shortest and most elementary way to prove the HMZ-formula. (The proof of "Lemma 14" is based on a beautiful "reciprocity identity" instead of Ostrowski's Lemma 1 - see Part V)

A sure sign that Hardy and Littlewood overlooked the trivial cancellation in (16) and were not aware of a simple answer like Corollary 3.3 is that much later, around 1930, they returned to the following very similar question. In [Ha-Li3] they studied the "diophantine" series

$$\sum_{n=1}^{\infty} \frac{1}{n \sin(\pi n \alpha)} \ . \tag{17}$$

Though the terms of series (17) don't tend to zero for any α, they made the interesting discovery that for the special value $\alpha = \sqrt{2}$ the partial sums of (17) are $O(1)$, that is, they remain uniformly bounded from above. In general, if $\alpha = \sqrt{a^2 + 1}$, a is *odd*, then the partial sums are $O(1)$.

On the other hand, for numbers like $\alpha = \sqrt{6}/2 - 1$ the Nth partial sum is $c \log N + O(1)$ with $c \neq 0$.

What is going on here? The proof of the "$O(1)$-theorem" for $\alpha = \sqrt{a^2 + 1}$, a is *odd*, was so complicated, mysterious and ad hoc that in his Introduction to the Collected Papers of G.H. Hardy, Vol. 1, Davenport listed the "real understanding" of this paper as a major research problem in diophantine approximation.

But this "$O(1)$-theorem" becomes almost trivial by using Corollary 3.3 (even with the weaker error term, it is in fact irrelevant in this particular application). Indeed, all what we need is the following simple identity (which is not so surprising, in view of (9))

$$\sum_{n=1}^{\infty} \frac{1}{n \sin(\pi n \alpha)} = 4\pi \Xi_{\alpha/2}(N) - 2\pi \Xi_{\alpha}(N) + O(\max_{1 \leq i \leq l} a_i), \tag{18}$$

where l is the least index such that $q_l \geq N$.

By formula (18), to evaluate series (17) we have to apply Corollary 3.3 to both α *and* $\alpha/2$. This means we have to understand the procedure of how to get the expansions of $\alpha/2$ from $\alpha = [a_0; a_1, a_2, \cdots]$.

Consider first a very special case. The quadratic irrationals $\alpha = \sqrt{a^2 + 1}$, a integer, have particularly simple expansions: $\alpha = [a; 2a, 2a, 2a, \cdots] = [a; \overline{2a}]$.

Case 1: if $a = 2b + 1$ is odd, then $\alpha/2 = [b; \overline{1, 1, 2b}]$;
Case 2: if $a = 2b$ is even, then $\alpha/2 = [b; \overline{8b, 2b}]$.

In Case 1 both α and $\alpha/2$ have periods of *odd* length, so by Corollary 3.3 the partial sums of series (17) are $O(1)$.

On the other hand, in Case 2, $\alpha/2$ has a period of *even* length, so the partial sums of series (17) have the form $c_\alpha \log N + O(1)$ where c_α is *never* zero. Now we understand why in the "$O(1)$-theorem" of Hardy and Littlewood the condition "a is odd" was necessary.

How about for general α's? Well, if $\alpha = [a_0; a_1, a_2, a_3, \cdots]$ then

$$\alpha/2 = [a_0/2; 2a_1, a_2/2, 2a_3, a_4/2, \cdots, a_{2i}/2, 2a_{2i+1}, \cdots]$$

if this *does make sense*, i.e. if a_{2i} is even for every $i \geq 0$. Under this "simplicity condition", by using (18) and Corollary 3.3, it is very easy to characterize those quadratic irrationals for which the partial sums of series (17) are $O(1)$.

Indeed, if the length s of the period $a_{j+1}, a_{j+2}, \cdots, a_{j+s}$ of α is *odd*, then the necessary and sufficient condition for an "$O(1)$-theorem" is

$$\sum_{i=j+1}^{j+s} (-1)^i a_i = 0.$$

On the other hand, if the length of the period is *even*, then there is no "$O(1)$-theorem" whatsoever.

The general case when the "simplicity condition" is violated is technically more complicated and somewhat unpleasant, so we don't discuss it here.

The next application of Corollary 3.3 is a far-reaching generalization of Theorem 3.1, see [Be7]. For what α's can we prove an analogous central limit theorem?

Well, the *mean value* (or first moment) is

$$\frac{1}{N+1} \sum_{n=0}^{N} S_\alpha(n) = \Xi_\alpha(N) = \frac{1}{12} \sum_{i:q_i \leq N} (-1)^i a_i + O(\max_{i:q_i \leq N} a_i).$$

For the *variance* (or second moment) we don't have a similar elegant formula, but we still have the analogue of (9):

$$
\begin{aligned}
V_\alpha(N) &= \frac{1}{N+1} \sum_{n=0}^{N} \left(S_\alpha(n) - \Xi_\alpha(N) \right)^2 \\
&= \frac{1}{4\pi^2} \sum_{n=1}^{N} \frac{1}{(n \tan(\pi n \alpha))^2} + O\left(\left(\max_{1 \leq i \leq l} a_i \right)^2 \right).
\end{aligned}
\tag{19}
$$

Note that the right-hand side of (19) is between two absolute constant multiples of $\sum_{i:q_i \leq N} a_i^2$.

THEOREM 4.1.(General CLT)   *Let* $\alpha = [a_0; a_1, a_2, \cdots]$, *and assume*

$$\frac{a_l^2}{\sum_{i=1}^{l} a_i^2} \longrightarrow 0 \quad as \quad l \to \infty. \tag{20}$$

Then

$$\frac{1}{N} \left| \left\{ 1 \le n \le N : \frac{S_\alpha(n) - \Xi_\alpha(N)}{\sqrt{V_\alpha(N)}} \le \lambda \right\} \right| \longrightarrow \frac{1}{\sqrt{2\pi}} \int_{-\infty}^{\lambda} e^{-u^2/2} \, du$$

for every fixed value of λ *as* $N \to \infty$.

Observe that (20) expresses the apparently necessary condition that the components are "individually negligible" (Lindeberg condition). This is why Theorem 4.1 is the most general result what we can hope for.

Note that in Theorem 4.1 we heavily used the fact that, in view of (20), the error term $O(\max_{1 \le i \le l} a_i)$ in Corollary 3.3 is negligible compared to $\sqrt{\sum_{i:q_i \le N} a_i^2}$. The weaker error term $O\left((\max_{1 \le i \le N} a_i)^2\right)$ which comes out of Hardy-Littlewood Lemma 14 is not enough to prove Theorem 4.1 in general.

Similarly, this weaker error term is not enough to prove the Cesaro mean formula (7) for e.

4   PART IV: Class Number One Problems

It is a well-known experimental observation that up to the first hundred million primes about 76% of the real quadratic fields with prime discriminant have class number one, i.e. $h(p) = 1$. But even the modest conjecture that there are infinitely many primes p with $h(p) = 1$ is beyond hope right now.

In sharp contrast with the real case, for *imaginary* quadratic fields there are only a finite number of discriminants with class number one. The difference comes from the effect of "big" fundamental units, see (15') and (15") in Part III. The imaginary class number one problem was completely solved in the 1960's, see e.g. Chapter 5 in Baker's book [Ba]. In view of the famous Baker-Heegner-Stark theorem,

$$h(-d) = 1 \implies d \le 163. \tag{1}$$

When the fundamental unit is "small" in a real quadratic field, i.e. when the continued fraction period of $\sqrt{d}$ is "short", then in view of (15') and (15") in Part III, the situation is a perfect analogue of the imaginary case. Therefore, one can come up with analogous conjectures.

The well-known examples for short periods are (Richaud-Degert type, see e.g. Mollin [Mo]):
$\sqrt{a^2 + 1} = [a; \overline{2a}]$ (a is even),   Chowla's conjecture:

$$h(a^2 + 1) = 1 \implies a \leq 26. \tag{2}$$

$\sqrt{a^2 + 4} = [a; \overline{(a-1)/2, 1, 1, (a-1)/2, 2a}]$ (a is odd),   Yokoi's conjecture:

$$h(a^2 + 4) = 1 \implies a \leq 17. \tag{3}$$

$\sqrt{a^2 - 4} = [a - 1; \overline{1, (a-3)/2, 2, (a-3)/2, 1, 2a-2}]$ ($a \geq 5$ odd), Mollin's conjecture:

$$h(a^2 - 4) = 1 \implies a \leq 21. \tag{4}$$

Of course, the fundamental problem in (1)-(4) is that Siegel's theorem is *ineffective*. This means there is at most one exceptional value only, and the difficulty is how to exclude it. (In 1934 Heilbronn and Linfoot proved that besides the nine values $d = -1, -2, -3, -7, -11, -19, -43, -67, -163$ there is at most one more possible negative discriminant with class number one. It took more than thirty years to exclude this "one more possible value", which clearly illustrates the non-trivial nature of the problem.)

Note that in (2)-(4) $d = a^2 + 1$ (or $a^2 \pm 4$) $\equiv 1 \pmod 4$, and then in $\mathbf{Q}(\sqrt{d})$ the algebraic integers have the form $x + y(\sqrt{d} + 1)/2$ (and also the discriminant is d). So what really matters is the expansion of $(\sqrt{d} + 1)/2$. Since $(\sqrt{a^2 + 4} + 1)/2 = [(a+1)/2; \overline{a}]$ has the simplest form, Yokoi's conjecture (3) turns out to be the simplest problem to attack.

Gauss' genus theory reduces (3) to the special case $d = p^2 + 4$ where *both* p and d are *primes*.

Mollin and Williams proved that the Riemann hypothesis for the zeta-function $\zeta_K(s)$ of the corresponding quadratic field implies relations (2)-(4). Moreover, it is known that a hypothetical counter-example to the Yokoi's conjecture has to be bigger than 10^{300000}.

We are going to prove that Yokoi's conjecture is true for several *explicit* residue classes of p (mod 71) (in fact 44 from the 70, which is about 63 %).

Note that it is trivial to exclude 2 residue classes like $p \not\equiv \pm 1 \pmod 5$ (indeed, then $d = p^2 + 4 \equiv 0 \pmod 5$) which contradicts to the fact that p is a prime). But anything more than that seems to be fairly non-trivial.

THEOREM 5.1.
If prime p belongs to the following 44 residue classes (mod 71):

$$p \equiv \pm 6, \pm 9, \pm 10, \pm 14, \pm 15, \pm 16, \pm 18, \pm 19, \pm 20, \pm 22, \pm 24, \pm 25,$$

$$\pm 26, \pm 27, \pm 28, \pm 29, \pm 30, \pm 31, \pm 32, \pm 33, \pm 34, \pm 35$$

then $h(d) \geq 2$ with $d = p^2 + 4$.

Here we outline the proof. For a complete account of the details, see [Be8]. The starting point is exactly the same kind of product formula for Dirichlet's L-functions as formula (13) in Part III.

Product Formula. Let d and m_0 be relatively prime and positive, $d \equiv 1 \pmod 4$, $m_0 \equiv 3 \pmod 4$, $\chi(n) = \left(\frac{-m_0}{n}\right)$, $\chi'(n) = \left(\frac{d}{n}\right)$. Then for $s > 1$

$$L(s,\chi)L(s,\chi\chi') = \sum_F \sum_{n=1}^{\infty} \frac{\chi(n)REP_F(n)}{n^s} \tag{5}$$

where $F = F(x,y)$ runs over a representative set of quadratic forms of discriminant d, and $REP_F(n)$ is the number of primary representations of n by $F(x,y)$.

This product formula can be proved the same way as (13) in Part III, or as the well-known Euler product of the zeta-function of the quadratic field $K = \mathbf{Q}(\sqrt{d})$: $\zeta_K(s) = \zeta(s)L(s,\chi')$.

Taking the limit $s \to 1+0$ in (5), by Dirichlet's class number formula

$$L(s,\chi) \to L(1,\chi) = \frac{\pi h(-m_0)}{\sqrt{m_0}}, \tag{6'}$$

$$L(s,\chi\chi') \to L(1,\chi\chi') = \frac{\pi h(-m_0 d)}{\sqrt{m_0 d}}. \tag{6''}$$

Let $\alpha = (1 + \sqrt{d})/2 = [(p+1)/2; \overline{p}]$, and recall that $\|x\|$ stands for the distance of x from the nearest integer. If $h(d) = 1$ then the only quadratic form is

$$
\begin{aligned}
Norm(x + y(\sqrt{d}-1)/2) &= \frac{(2x-y)^2 - dy^2}{4} \\
&= \left(x + \frac{\sqrt{d}-1}{2}y\right)\left(x - \frac{\sqrt{d}+1}{2}y\right) \\
&= (x + (\alpha-1)y)(x - \alpha y) \\
&\approx (2x - y)(x - \alpha y) \approx \sqrt{d}y(\pm\|\alpha y\|).
\end{aligned}
$$

We obtain, therefore, the following crucial formula *if $h(d) = 1$*:

$$h(-m_0)h(-m_0 d) = \lim_{N\to\infty} \frac{\log \eta_d}{\log N} \frac{m_0\sqrt{d}}{\pi^2} \sum_{x=1}^{N}\sum_{y=1}^{N} \frac{\chi(((2x-y)^2 - dy^2)/4)}{((2x-y)^2 - dy^2)/4}$$

$$= \lim_{N\to\infty} \frac{\log \eta_d}{\log N} \frac{m_0\sqrt{d}}{\pi^2} \sum_{n=1}^{N} \frac{\chi\left(nearint\left(n\|n\alpha\|\sqrt{d}\right)\right)}{n\|n\alpha\|\sqrt{d}}, \tag{7}$$

where *nearint(x)* stands for the *nearest integer to x*.

Note that in (7) $m_0 = 71$ and 79 are good choices (among many others). Why? Well, the key property is that $h(-m_0)$ has to divide $(m_0 - 1)(m_0 + 1)/6$ (which is true for both $m_0 = 71$ and 79 since $h(-71) = 7$ and $h(-79) = 5$). By choosing $m_0 = 71$ we get that the right-hand side of (7) has to be divisible by 7. This requirement will exclude the given 44 residue classes (mod 71), and it goes as follows.

It is easy to see that the main contribution of the right-hand side of (7) comes from those n's for which $\|n\alpha\|$ is "roughly around $1/n$". It turns out that these n's actually belong to not more than three simple classes. This is where we employ the fact that the period of α is so short, that is, this part is pure diophantine approximation. We recall that q_i is the ith convergent denominator of $\alpha = (1 + \sqrt{p^2 + 4})/2 = [(p+1)/2; \overline{p}]$. Now, with $\rho := (p+1)/2$, the three classes are

1. $n = jq_i$, $1 \le j \le \rho$;
2. $n = jq_i \pm kq_{i-1}$, $1 \le j \le \rho, 1 \le k \le \rho$;
3. $n = jq_i \pm kq_{i-1} \pm lq_{i-2}$, $1 \le j \le \rho, 1 \le k \le \rho, 1 \le l \le \rho$.

These classes give the following contributions. Let

$$\Sigma_1(p) = \frac{1}{\pi^2} \sum_{1 \le j \le \rho:\ j \not\equiv 0\ (mod\ m_0)} \frac{1}{j^2};$$

$$\Sigma_2(p) = \frac{1}{\pi^2} \sum_{j=1}^{\rho} \sum_{1 \le \pm k \le \rho} \frac{\chi(pjk + k^2 - j^2)}{pjk + k^2 - j^2};$$

$$\Sigma_3(p) = \frac{1}{\pi^2} \sum_{j=1}^{\rho} \sum_{1 \le \pm k \le \rho} \sum_{1 \le \pm l \le \rho} \frac{\chi((p^2 + 2)jl + pk(l - j) + j^2 - k^2 + l^2)}{(p^2 + 2)jl + pk(l - j) + j^2 - k^2 + l^2}.$$

In these classes index i runs up to $\log N / \log \eta_d$, and since the factor $\log N / \log \eta_d$ actually cancels out in (7), we can rewrite (7) in the following short form:

$$h(-m_0)h(-m_0(p^2 + 4)) = m_0 p\,(\Sigma_1(p) + \Sigma_2(p) + \Sigma_3(p)) + \delta(p), \qquad (8)$$

where $\delta(p) \to 0$ as $p \to \infty$.

It is trivial to evaluate the first class:

$$\Sigma_1(p) = \frac{1}{\pi^2} \sum_{\substack{1 \le j \le \rho \\ j \not\equiv 0\ (mod\ m_0)}} \frac{1}{j^2} \approx \left(1 - \frac{1}{m_0^2}\right) \frac{1}{6}$$

if p is "large".

The contribution of the other two classes is $O(1/p)$, so even multiplied up by p in (8) it is not more than $O(1)$, in fact there is a limit constant here.

More precisely, let $p \equiv b_0$ (mod m_0). In contrast to $\Sigma_1(p)$, which is independent of b_0, both $\Sigma_2(p)$ and $\Sigma_3(p)$ heavily depend on b_0. In fact there is a limit constant $c_1(b_0, m_0)$ such that, with $p = m_0\widetilde{p} + b_0$,

$$h(-m_0)h(-m_0(p^2 + 4)) = m_0 p \left(1 - \frac{1}{m_0^2}\right)\frac{1}{6} + c_2(p; b_0, m_0)$$

where $c_2(p; b_0, m_0) \to c_1(b_0, m_0)$ as $p \to \infty$. Therefore,

$$h(-m_0)h(-m_0(p^2 + 4)) = \frac{(m_0 - 1)(m_0 + 1)}{6}\widetilde{p} + c_3(p; b_0, m_0) \qquad (9)$$

where

$$c_3(p; b_0, m_0) = c_2(p; b_0, m_0) + \frac{(m_0 - 1)(m_0 + 1)b_0}{6m_0}$$

tends to the limit constant

$$c_4(b_0, m_0) = c_1(b_0, m_0) + \frac{(m_0 - 1)(m_0 + 1)b_0}{6m_0}$$

as $p \to \infty$.

Now let $m_0 = 71$, then $h(-m_0) = 7$ divides the first term $(m_0 - 1)(m_0 + 1)/6 = 140$ on the right-hand side of (9). It follows that 7 has to divide the limit constant $c_4(b_0, 71) = c_4(b_0)$.

By estimating the speed of convergence and making extensive computations with computer (about 10^6 additions are required), we can verify that the limit constants $c_4(b_0)$ are *not* divisible by 7 for the given 44 values of b_0. This proves Theorem 5.1 for "large" values of p.

Finally, we don't have to worry about the "small" values since it is known that a counter-example to Yokoi's conjecture has to be bigger than 10^{300000} (again by making use of computer).

Remarks. It turns out that the right-hand side of (7) is *precisely*

$$\frac{(m_0 - 1)(m_0 + 1)}{6}\widetilde{p} + c_4(b_0, m_0) \qquad (10)$$

where $d = p^2 + 4$, and $p = m_0\widetilde{p} + b_0$ is an *integral variable* for all possible (not necessarily prime) values $p = 1, 2, 3, 4, \cdots$. In other words, in the right-hand side of (7) the *asymptotic* value is the *actual* value for every (not necessarily prime) value of variable p.

Moreover, (10) *always* turns out to be an integer for all p and m_0.

We have the optimistic feeling that a *simultaneous* application of the argument of the proof of Theorem 5.1 might be enough to settle the whole Yokoi's conjecture (and the other two ones as well). At the end of this section we describe such an attempt. The success of this attempt depends on the validity of some "randomness hypothesis" which can be checked by computer.

4.1 AN ATTEMPT TO REDUCE THE YOKOI'S CONJECTURE TO A FINITE AMOUNT OF COMPUTATION

The proof of Theorem 5.1 is basically a modulo 7 "sieve" argument by using formula (5) at $s = 1$. We restate it by using the following unusual notation: the $[s = 1, \bmod 7]$-*sieve* excludes 63% of the residue classes (mod 71) for the values of parameter p in $d = p^2 + 4$.

The new idea is to *simultaneously* use several $[s = 2r, \bmod p_0]$-*sieves* where $r = 1, 2, 3, \cdots$ and p_0 is a fixed small prime, in order to exclude all the "unavoidable" residue classes (mod m_0), where $m_0 = 3 \cdot 5 \cdot 7 \cdot 13 \cdot 17$, for the values of parameter p in $d = p^2 + 4$.

What does it mean? Well, in the proof of Theorem 5.1 we used product formula (5) at $s = 1$ because the class numbers $h(-m_0)$ and $h(-m_0 d)$ are *integers* (the factor π^2 cancels out in a natural way).

It has been known for a long time (Hecke, Siegel, etc.) that the values of the zeta-function of a real quadratic field at negative odd integers are *rationals*, in fact this is true for any totally real number field. By using the *functional equation* of the zeta-function (of real quadratic fields) relating the values at s and $1 - s$, we conclude that one can attempt to copy the argument of the proof of Theorem 5.1 by using the positive even integer values of s in product formula (5) instead if $s = 1$. The big advantage is that these values provide infinitely many "sieves". If these "sieves" behave more-or-less "independently" of each other then we have a very good chance to "sieve out" all possible residue classes (mod m_0) for p. This is our strategy to reduce the "global" Yokoi's conjecture to a "local problem", i.e. to a *finite* amount of computation.

Well, the first technical difficulty is that the residue classes $p \equiv \pm\{1, 3, 5, 7, 13, 17\}$ (mod m_0) are "unavoidable" in the sense that it is impossible to sieve them out. The explanation is that $h(p^2 + 4) = 1$ for precisely $p = 1$, 3, 5, 7, 13, 17. But this is exactly the reason why we choose the composite modulus $m_0 = 3 \cdot 5 \cdot 7 \cdot 13 \cdot 17$ (we also need $m_0 \equiv 1$ (mod 4)). Indeed, if $p \equiv \pm\{3, 5, 7, 13, 17\}$ (mod $3 \cdot 5 \cdot 7 \cdot 13 \cdot 17$) then p cannot be a prime, so these "unavoidable" residue classes become irrelevant. Similarly, if $p \equiv \pm 1$ then $d = p^2 + 4 \equiv 5$ (mod $3 \cdot 5 \cdot 7 \cdot 13 \cdot 17$), so this time d turns out to be a composite number. (We recall that by Gauss' genus theory, $h(d) = 1$ implies that both $d = p^2 + 4$ and p have to be primes). Therefore, with this composite modulus we have a chance to "sieve out" all possible relatively prime residue classes (mod m_0) for p.

Next question, what is the value of parameter p_0. Well, this is a fixed small prime like 11 (or 19). (The factors 3, 5, 7, $\cdots$ of m_0 don't seem to work as p_0.)

The starting point is the following product formula ($s > 1$)

$$L(s,\chi)L(s,\chi\chi') = \sum_F \left(\sum_{n=1}^{\infty} \frac{\chi(n)REP_F(n)}{n^s}\right) \qquad (12)$$

where d and m_0 are relatively prime, $d = p^2 + 4 \equiv 1 \pmod 4$, $m_0 = 3 \cdot 5 \cdot 7 \cdot 13 \cdot 17 \equiv 1 \pmod 4$, $\chi(n) = \left(\frac{m_0}{n}\right)$, $\chi'(n) = \left(\frac{d}{n}\right)$, $F = F(x,y)$ runs over a representative set of quadratic forms of discriminant d, and $REP_F(n)$ is the number of primary representations of n by $F(x,y)$.

By the *ineffective* Siegel's theorem $h(d) = d^{\frac{1}{2}\pm\varepsilon}$, that is, $\sum_F$ supposed to be a sum of $d^{\frac{1}{2}\pm\varepsilon}$ terms. We try to derive a contradiction from the assumption $h(d) = 1$. This assumption means that $((2x - y)^2 - dy^2)/4$ is the only quadratic form with discriminant d. Thus we have, with $s \geq 2$ even integer,

$$L(s,\chi)L(s,\chi\chi') = \lim_{N\to\infty} \frac{\log \eta_d}{\log N} \sum_{x=1}^{N}\sum_{y=1}^{N} \frac{\chi\left(((2x-y)^2 - dy^2)/4\right)}{(((2x-y)^2 - dy^2)/4)^s}$$

$$= \lim_{N\to\infty} \frac{\log \eta_d}{\log N} \sum_{n=1}^{N} \frac{\chi\left(nearint\left(n||n\alpha||\sqrt{d}\right)\right)}{\left(n||n\alpha||\sqrt{d}\right)^s}, \qquad (13)$$

where $\alpha = (1+\sqrt{d})/2 = [(p+1)/2; \overline{p}]$, and $nearint(x)$ stands for the *nearest integer* to x.

Our expectation is that the two sides of (13) are "independent" of each other (mod 11). In order to have integers, we involve the zeta-functions of real quadratic fields at negative odd integers (which are known to be rationals).

Let $K = \mathbf{Q}(\sqrt{D})$ where $D \equiv 1 \pmod 4$. The zeta-function of K has the factorization $\zeta_K(s) = \zeta(s)L(s,\chi_0)$ where χ_0 is the character of K. Furthermore, from the *functional equation* relating $\zeta_K(s)$ and $\zeta_K(1 - s)$, with $s = 2r \geq 2$ even integer,

$$\zeta_K(1 - 2r) = \frac{(2r - 1)!(2r - 1)!}{2^{4r-2}\pi^{4r}}D^{2r-1/2}\zeta_K(2r).$$

Finally,

$$\zeta(2r) = (-1)^{r-1}\frac{2^{2r}\pi^{2r}}{(2r - 1)!}\frac{B_{2r}}{r},$$

where B_i is the ith Bernoulli number ($B_0 = 1, B_1 = -1/2, B_2 = 1/6, B_3 = 0, B_4 = -1/30, \cdots$).

Combining these formulas, we can express the *rational* $\zeta_K(1 - 2r)$ in terms of $L(2r, \chi_0)$ as follows:

$$\zeta_K(1 - 2r) = \frac{(-1)^{r-1}(2r - 1)!B_{2r}}{2^{2r-2}\pi^{2r}r}D^{2r-1/2}L(2r, \chi_0).$$

Thus we have, with $K = K_1 = \mathbf{Q}(\sqrt{m_0})$ and $K = K_2 = \mathbf{Q}(\sqrt{m_0 d})$, $\chi_0 = \chi$ and $\chi_0 = \chi\chi'$, respectively,

$$\zeta_{K_1}(1 - 2r)\zeta_{K_2}(1 - 2r) = \frac{(2r - 1)!(2r - 1)!B_{2r}B_{2r}m_0^{4r-1}d^{2r-1/2}}{2^{4r-4}\pi^{4r}r^2} \cdot * \quad (14)$$

where

$$* = \lim_{N\to\infty} \frac{\log\eta_d}{\log N} \sum_{x=1}^{N}\sum_{y=1}^{N} \frac{\chi\left(((2x - y)^2 - dy^2)/4\right)}{\left(((2x - y)^2 - dy^2)/4\right)^{2r}}$$

$$= \lim_{N\to\infty} \frac{\log\eta_d}{\log N} \sum_{n=1}^{N} \frac{\chi\left(nearint\left(n||n\alpha||\sqrt{d}\right)\right)}{\left(n||n\alpha||\sqrt{d}\right)^{2r}}.$$

The left-hand side of (14) is a rational number. The following result describes the denominator – see the appendix of Zagier [Za4].

STATEMENT. $w_r(K)\zeta_K(1-2r) \in \mathbf{Z}$ where the integer $w_r(K)$ is defined as $j(r) = g.c.d. \left\{n^{r+2}(n^{2r} - 1) : n \in \mathbf{Z}\right\}$ (independent of K) unless K is one of the finitely many fields $\mathbf{Q}(\sqrt{q})$ with q a prime such that $(q - 1)$ divides $4r$ but $(q - 1)$ doesn't divide $2r$, in which case $w_r(K) = q^{\nu+1}j(r)$, where q^ν is the largest power of q dividing r.

Let us return to (14). Both $j(r)\zeta_{K_1}(1 - 2r)$ and $j(r)\zeta_{K_2}(1 - 2r)$ are integers.

Therefore,

$$j(r)\zeta_{K_1}(1 - 2r)j(r)\zeta_{K_2}(1 - 2r) \quad (15)$$

equals

$$\frac{j(r)u(r)(2r - 1)!(2r - 1)!B_{2r}B_{2r}m_0^{4r-1}d^{2r-1/2}}{2^{4r-4}r^2} \cdot * \quad (16)$$

where

$$* = \lim_{N\to\infty} \frac{\log\eta_d}{\log N}\pi^{-4r} \sum_{x=1}^{N}\sum_{y=1}^{N} \frac{\chi\left(((2x - y)^2 - dy^2)/4\right)}{\left(((2x - y)^2 - dy^2)/4\right)^{2r}} \quad (16a)$$

$$= \lim_{N\to\infty} \frac{\log\eta_d}{\log N}\pi^{-4r} \sum_{n=1}^{N} \frac{\chi\left(nearint\left(n||n\alpha||\sqrt{d}\right)\right)}{\left(n||n\alpha||\sqrt{d}\right)^{2r}}, \quad (16b)$$

$\chi(n) = \left(\frac{m_0}{n}\right)$, and $m_0 = 3 \cdot 5 \cdot 7 \cdot 13 \cdot 17$.

The crucial fact is that (16) is an *integral-valued* function of *parameter p*. We recall: $(1 + \sqrt{d})/2 = [(p + 1)/2; \bar{p}]$ and $d = p^2 + 4$. Moreover, just like in the proof of Theorem 5.1, we consider p as an *integral variable* (not necessarily prime). Therefore, if $p = m_0 x + b_0$ $(0 \le b < m_0)$, then (16) is an *integral-valued* polynomial $POL_{b_0, r}(x)$ of degree $4r - 1$. By a well-known general result about integral-valued polynomials, it can be written in the standard form

$$(16) = POL_{b_0,r}(x) = \sum_{j=0}^{4r-1} c_j(b_0)\binom{x}{j}, \qquad (17)$$

where the coefficients $c_j(b_0)$ are *efficiently computable* integers. Indeed, again we can restrict ourselves to the same three classes as in the proof of Theorem 5.1 (again let $\rho := (p+1)/2$):

1. $n = jq_i, \ 1 \le j \le \rho$;
2. $n = jq_i \pm kq_{i-1}, \ 1 \le j \le \rho, 1 \le k \le \rho$;
3. $n = jq_i \pm kq_{i-1} \pm lq_{i-2}, \ 1 \le j \le \rho, 1 \le k \le \rho, 1 \le l \le \rho$.

These classes give the following contributions. Let

$$\Sigma_{1,r}(p) = \frac{1}{\pi^{4r}} \sum_{\substack{1 \le j \le \rho: \\ g.c.d.(j,m_0)>1}} \frac{1}{j^{4r}};$$

$$\Sigma_{2,r}(p) = \frac{1}{\pi^{4r}} \sum_{j=1}^{\rho} \sum_{1 \le \pm k \le \rho} \frac{\chi(pjk + k^2 - j^2)}{(pjk + k^2 - j^2)^{2r}};$$

$$\Sigma_{3,r}(p) = \frac{1}{\pi^{4r}} \sum_{j=1}^{\rho} \sum_{1 \le \pm k \le \rho} \sum_{1 \le \pm l \le \rho} \frac{\chi((p^2+2)jl + pk(l-j) + j^2 - k^2 + l^2)}{((p^2+2)jl + pk(l-j) + j^2 - k^2 + l^2)^{2r}}.$$

In these classes index i runs up to $\log N / \log \eta_d$, and since the factor $\log N / \log \eta_d$ actually cancels out in (16a-b), we can rewrite (16) in the following short form:

$$\frac{j(r)u(r)\left((2r-1)!\right)^2 B_{2r}^2 m_0^{4r-1}}{2^{4r-4}r^2} \cdot d^{2r-1/2} \sum_{i=1}^{3} \Sigma_{i,r}(p) + \varepsilon_r(p), \qquad (18)$$

where $\varepsilon_r(p) \to 0$ as $p \to \infty$.

As an illustration we evaluate the first class. We have

$$\Sigma_{1,r}(p) = \frac{1}{\pi^{4r}} \sum_{\substack{1 \le j < \infty: \\ g.c.d.(j,m_0)>1}} \frac{1}{j^{4r}} - \frac{1}{\pi^{4r}} \sum_{\substack{\rho < j < \infty: \\ g.c.d.(j,m_0)>1}} \frac{1}{j^{4r}}. \qquad (19)$$

Clearly

$$\frac{1}{\pi^{4r}} \sum_{\substack{1 \le j < \infty: \\ g.c.d.(j,m_0)>1}} \frac{1}{j^{4r}}$$

equals

$$-\frac{B_{4r}2^{4r}}{(4r-1)!2r}\left(1 - \frac{1}{3^{4r}}\right)\left(1 - \frac{1}{5^{4r}}\right)\left(1 - \frac{1}{7^{4r}}\right)\left(1 - \frac{1}{13^{4r}}\right)\left(1 - \frac{1}{17^{4r}}\right)$$

$$= -\frac{B_{4r}2^{4r}}{(4r-1)!2r}\frac{(2^{4r}-1)(2^{4r}-1)(2^{4r}-1)(13^{4r}-1)(17^{4r}-1)}{m_0^{4r}}. \quad (20)$$

On the other hand, since

$$\sum_{n<j<\infty}\frac{1}{j^{4r}} = \int_n^\infty \frac{dx}{x^{4r}} + O(n^{-4r}) = \frac{1}{(4r-1)n^{4r-1}} + O(n^{-4r}),$$

the "tail"

$$\frac{1}{\pi^{4r}}\sum_{p<j<\infty:\ g.c.d.(j,m_0)>1}\frac{1}{j^{4r}}$$

equals

$$\frac{1}{(4r-1)\pi^{4r}}\left(\frac{2}{p}\right)^{4r-1}\left(1-\frac{1}{3}\right)\cdots\left(1-\frac{1}{17}\right) + O(p^{-4r}). \quad (21)$$

The last thing what we need is to express the factor $d^{2r-1/2}$ in terms of p. This is plain binomial series

$$d^{2r-1/2} = (p^2+4)^{2r-1/2} = p^{4r-1}\left(1-4p^{-2}\right)^{2r-1/2}$$

$$= p^{4r-1}\left(1 + \binom{2r-1/2}{1}4p^{-2} + \binom{2r-1/2}{2}16p^{-4} + \cdots\right). \quad (22)$$

Similar but more involved argument is required to evaluate the contribution of the second class. Note that the contribution of the third class goes to the constant term $j=0$.

Now we go back to (15). Since K_1 is fixed, $j(r)\zeta_{K_1}(1-2r)$ is an explicitly computable integer for "middle size" values of r. Among the first (say) 330 values $r = 1, 2, \cdots, 330$ consider those r's for which $j(r)\zeta_{K_1}(1-2r)$ is divisible by $p_0 = 11$. Let R denote the set of these r's. R is probably a "random set" of cardinality around $330/11 = 30$.

Next we use the following almost trivial

LEMMA 1: *If* $0 \le j < 11^3$ *then* $\binom{x}{j} \equiv \binom{x+11^3}{j}$ *(mod 11).*

Let $r' \in R$, then in view of this lemma, the divisibility by 11

$$POL_{b_0,r'}(x) = \sum_{j=0}^{4r'-1} c_j(b_0)\binom{x}{j} \equiv 0 \pmod{11}$$

means a condition on the residue class of x modulo 11^3, or equivalently, a condition on the residue class of p $(= m_0x + b_0)$ modulo $11^3 m_0$.

For a fixed power $r' \in R$, the "modulo 11 test" *probably* excludes more than half of the $11^3 m_0$ residue classes for p. The *probability* that a residue class "survives" all the $|R| \approx 30$ "modulo 11 tests" is less than

$$\left(\frac{1}{2}\right)^{30} 11^3 m_0 \approx 0,$$

which makes the success of this approach very likely. In these heuristic computations we heavily used the following principle.

"*Randomness hypothesis*": If we don't know something, we assume that it is uniformly distributed among the available options.

In particular we needed

"*Independence of sieves*": For $r \in R$, the $[s = 2r, \text{mod } 11]$-*sieves* are "mutually independent", or at least they are close to it.

Finally, a big computational saving comes from the fact that it is enough to deal with the following four small moduli

$$m = 5 \cdot 13, \ 13 \cdot 17, \ 3 \cdot 5 \cdot 7, \ 3 \cdot 7 \cdot 13,$$

instead of the single big modulus $m_0 = 3 \cdot 5 \cdot 7 \cdot 13 \cdot 17$.

We can make this substantial reduction because of the following technical LEMMA 2: *Assume that* $\pm 1, \pm 3, \pm 5, \pm 7, \pm 13, \pm 17$ *are the only "possible" reduced residue classes for* p *modulo* m *where* $m = 5 \cdot 13, 13 \cdot 17, 3 \cdot 5 \cdot 7, 3 \cdot 7 \cdot 13$. *Then the only "possible" reduced residue classes for* p *modulo* $m_0 = 3 \cdot 5 \cdot 7 \cdot 13 \cdot 17$ *are* ± 1 *(which implies the Yokoi's conjecture)*.

5   PART V: Cesaro Mean of $\sum_n \left(\{ n\alpha \} - \frac{1}{2} \right)$

This last section is devoted to a proof of Theorem 3.2.

First we recall the well-known recursive formulas on the denominators q_i of the convergents p_i/q_i of $\alpha = [a_0; a_1, a_2, \cdots]$:

$$q_0 = 1, q_1 = a_1, \text{ and for all } i \geq 0, \ q_{i+2} = a_{i+2} q_{i+1} + q_i.$$

In view of this, there is a unique way to express an arbitrary positive integer n as a linear combination of the q_i's as follows:

$$n = \sum_i^* b_i q_i, \qquad 0 \leq b_i = b_i(n) \leq a_{i+1},$$

where the * indicates the restriction that if $b_{i+1} = a_{i+2}$ then $b_i = 0$.

The starting point of the proof is the following somewhat complicated but very useful lemma of Ostrowski (see [Os]) expressing sum $S_\alpha(n)$ in terms of the basic parameters of the continued fraction expansion of α. The only new parameter is $\varepsilon_i = q_i \alpha - p_i$. Let $sign(\varepsilon_i) = \pm 1$ be the usual sign. It is well-known that this is an *alternating* sequence (as i is running) for every α.

LEMMA 1 (Ostrowski).   *Let* $q_l \leq n < q_{l+1}$, *and write* $n = \sum_{0 \leq i \leq l}^* b_i q_i$.
Then

$$S_\alpha(n) = \sum_{i=0}^{l} sign(\varepsilon_i) b_i \left(-\frac{1}{2} + \frac{b_i q_i |\varepsilon_i|}{2} + \left(\sum_{0 \leq j < i}^* b_j q_j \right) |\varepsilon_i| + \frac{|\varepsilon_i|}{2} \right).$$

For the sake of completeness, we include a proof.

Proof. Write $n = b_l q_l + n'$ where $n' = \sum_{0 \leq i < l}^{*} b_i q_i$. We have

$$S_\alpha(n) = S_1 + S_2, \qquad \text{where}$$

$$S_1 = \sum_{k=1}^{b_l q_l} \left(\left\{ k\alpha - \frac{1}{2} \right\} \right) \qquad \text{and}$$

$$S_2 = \sum_{m=b_l q_l + 1}^{n} \left(\left\{ m\alpha - \frac{1}{2} \right\} \right).$$

We recall $\alpha - p_i/q_i = \varepsilon_i/q_i$ where $|\varepsilon_i| < 1/q_{i+1}$. We can rewrite S_1 as follows:

$$S_1 = S_\alpha(b_l q_l) = -\frac{b_l q_l}{2} + \sum_{k=1}^{b_l q_l} \left\{ \frac{k p_l}{q_l} + \frac{k \varepsilon_l}{q_l} \right\}.$$

We distinguish two cases.

   CASE 1: $\varepsilon_l > 0$.   In this case

$$\sum_{k=1}^{b_l q_l} \left\{ \frac{k p_l}{q_l} + \frac{k \varepsilon_l}{q_l} \right\} = \sum_{k=1}^{b_l q_l} \left\{ \frac{k p_l}{q_l} \right\} + \sum_{k=1}^{b_l q_l} \frac{k \varepsilon_l}{q_l}$$

$$= b_l \frac{q_l(q_l - 1)}{2 q_l} + \frac{b_l q_l (b_l q_l + 1) \varepsilon_l}{2 q_l} = \frac{b_l q_l}{2} - \frac{b_l}{2} + \frac{b_l}{2}(b_l q_l + 1)\varepsilon_l.$$

In the last step we used the facts that $\frac{k \varepsilon_l}{q_l} < \frac{1}{q_l}$ for $k \leq b_l q_l (< q_{l+1})$, and also that the residue of $k p_l$ modulo q_l, as k runs, is a permutation of $0, 1, 2, \cdots, q_l - 1$. Therefore,

$$S_1 = -\frac{b_l}{2} + \frac{b_l}{2}(b_l q_l + 1)\varepsilon_l.$$

   CASE 2: $\varepsilon_l < 0$.   In this case, if $k \not\equiv 0 \pmod{q_l}$,

$$\left\{ \frac{k p_l}{q_l} + \frac{k \varepsilon_l}{q_l} \right\} = \left\{ \frac{k p_l}{q_l} \right\} + \frac{k \varepsilon_l}{q_l}$$

since $|k \varepsilon_l / q_l| < 1/q_l$ for $1 \leq k \leq b_l q_l$, and if $k \equiv 0 \pmod{q_l}$ and $1 \leq k \leq b_l q_l$, then

$$\left\{ \frac{k p_l}{q_l} + \frac{k \varepsilon_l}{q_l} \right\} = 1 + \frac{k \varepsilon_l}{q_l}.$$

Repeating the summations in Case 1, again we have

$$S_1 = -\frac{b_l}{2} + \frac{b_l}{2}(b_l q_l + 1)\varepsilon_l.$$

Next we rewrite S_2 :

$$S_2 = \sum_{m=1}^{n'} \{b_l q_l \alpha + m\alpha\} - \frac{n'}{2} \qquad \text{where} \quad n' = \sum_{0 \leq i < l}^{*} b_i q_i.$$

To evaluate S_2, we again distinguish two cases.

   CASE 1: $\varepsilon_l > 0$.   In this case,

$$\{b_l q_l \alpha + m\alpha\} = \{b_l q_l \alpha\} + \{m\alpha\}$$

since the sum of the fractional parts on the right-hand side is less than 1. Indeed, $\|b_l q_l \alpha\| \leq a_{l+1}\|q_l \alpha\|$, $\|m\alpha\| \geq \|q_{l-1}\alpha\|$ for all $m < q_l$, and $(a_{l+1}q_l + q_{l-1})\alpha = q_{l+1}\alpha$ has the property that

$$sign(\varepsilon_{l+1}) = sign(q_{l+1}\alpha - p_{l+1}) = sign(\varepsilon_{l-1}) = sign(q_{l-1}\alpha - p_{l-1}).$$

So we have

$$\{b_l q_l \alpha + m\alpha\} = \{b_l q_l (p_l/q_l + \varepsilon_l/q_l)\} + \{m\alpha\} = b_l \varepsilon_l + \{m\alpha\}\,,$$

and

$$S_2 = \sum_{m=1}^{n'} \{b_l q_l \alpha + m\alpha\} - \frac{n'}{2} = n' b_l \varepsilon_l + S_\alpha(n').$$

   CASE 2: $\varepsilon_l < 0$.   Then

$$\{b_l q_l \alpha + m\alpha\} = \{b_l q_l \alpha\} + \{m\alpha\} - 1.$$

Indeed,

$$\{b_l q_l \alpha\} = \{b_l q_l (p_l/q_l + \varepsilon_l/q_l)\} = b_l \varepsilon_l + 1,$$

$$\{m\alpha\} = \{m(p_l/q_l + \varepsilon_l/q_l)\} = \{mp_l/q_l\} + \frac{m}{q_l}\varepsilon_l + 1,$$

and

$$\{mp_l/q_l\} + \frac{(b_l q_l + m)}{q_l}\varepsilon_l + 1 > 1$$

since $b_l q_l + n' < q_{l+1}$ and $|\varepsilon_l| < 1/q_{l+1}$. Thus we have $\{b_l q_l \alpha + m\alpha\} = \{b_l q_l \alpha\} + \{m\alpha\} - 1$ which equals $(1 + b_l \varepsilon_l) + \{m\alpha\} - 1 = b_l \varepsilon_l + \{m\alpha\}$, and so again we have

$$S_2 = n' b_l \varepsilon_l + S_\alpha(n').$$

Summarizing,

$$S_\alpha(n) = S_\alpha(n') - \frac{b_l}{2}(1 - b_l q_l - 2n'|\varepsilon_l| - |\varepsilon_l|)sign(\varepsilon_l),$$

and Ostrowski's lemma, i.e. Lemma 1, follows by induction.

Now assume that $N = q_{s+1} - 1$. We are going to use a probabilistic interpretation, and according to this we introduce the "expected value"

$$\mathbf{E}b_j = \frac{1}{N+1} \sum_{n=0}^{N} b_j(n).$$

To evaluate the first, simplest term on the right-hand side of Lemma 1, we introduce the notation

$$\Sigma_1 = \sum_{0 \le j \le s} (-1)^j \mathbf{E}b_j.$$

We are going to prove the following elegant formula.

LEMMA 2.   If $N = q_{s+1} - 1$, then

$$\Sigma_1 = \sum_{0 \le j \le s} (-1)^j \mathbf{E}b_j = \frac{1}{2} \sum_{j=0}^{s} (-1)^j a_{j+1} - \frac{1 + (-1)^j}{4}.$$

Proof. To describe the distribution of $b_j = b_j(n)$ as n runs in $0 \le n \le N = q_{s+1} - 1$, we study the recursive formula. We have

$$
\begin{aligned}
q_{s+1} &= a_{s+1}q_s + q_{s-1} = a_{s+1}(a_s q_{s-1} + q_{s-2}) + q_{s-1} \\
&= (a_{s+1}a_s + 1)q_{s-1} + a_{s+1}q_{s-2} \\
&= (a_{s+1}a_s + 1)(a_{s-1}q_{s-2} + q_{s-3}) + a_{s+1}q_{s-2} \\
&= (a_{s+1}a_s a_{s-1} + a_{s+1} + a_{s-1})q_{s-2} + (a_{s+1}a_s + 1)q_{s-3} \\
&= \cdots \\
&= A_{s-j}q_{j+1} + A_{s-j-1}q_j = A_{s-j}(a_{j+1}q_j + q_{j-1}) + A_{s-j-1}q_j \\
&= \cdots
\end{aligned}
$$

where $A_0 = A_0(s) = 1$, $A_1 = A_1(s) = a_{s+1}$, $A_2 = A_2(s) = a_s A_1 + A_0$, and in general, $A_{s-j+1} = a_{j+1}A_{s-j} + A_{s-j-1}$, or equivalently, $A_{s-j} = a_{j+2}A_{s-j-1} + A_{s-j-2}$.

This means the following decomposition law: the interval $\{0, 1, 2, \cdots, q_{s+1} - 1\}$ of length q_{s+1} falls apart into subintervals, namely A_{s-j} subintervals of length q_{j+1} and A_{s-j-1} subintervals of length q_j. So the distribution of $b_j = b_j(n)$ as n runs in $0 \le n \le N = q_{s+1} - 1$ is as follows:

$$Prob(b_j = 0) = \frac{A_{s-j}q_{j+1}}{q_{s+1}} \frac{q_j}{q_{j+1}} + \frac{A_{s-j-1}q_j}{q_{s+1}};$$

next for $r = 1, 2, \cdots, a_{j+1} - 1$,

$$Prob(b_j = r) = \frac{A_{s-j}q_{j+1}}{q_{s+1}} \frac{q_j}{q_{j+1}};$$

and finally,

$$Prob(b_j = a_{j+1}) = \frac{A_{s-j}q_{j+1}}{q_{s+1}} \frac{q_{j-1}}{q_{j+1}}.$$

Therefore,

$$\sum_1 = \sum_{0 \le j \le s} (-1)^j \mathbf{E} b_j$$

$$= \sum_{j=0}^{s} (-1)^j \left(\frac{A_{s-j}q_{j+1}}{q_{s+1}} \frac{q_j}{q_{j+1}} (1+2+\cdots+(a_{j+1}-1)) + \frac{A_{s-j}q_{j+1}}{q_{s+1}} \frac{q_{j-1}}{q_{j+1}} a_{j+1} \right)$$

$$= \sum_{j=0}^{s} (-1)^j \left(\frac{A_{s-j}q_j}{q_{s+1}} \frac{a_{j+1}(a_{j+1}-1)}{2} + \frac{A_{s-j}q_{j-1}}{q_{s+1}} a_{j+1} \right)$$

$$= \sum_{j=0}^{s} (-1)^j \frac{a_{j+1}}{2q_{s+1}} \left(A_{s-j}q_j a_{j+1} + 2A_{s-j}q_{j-1} - A_{s-j}q_j \right).$$

Since $a_{j+1}q_j = q_{j+1} - q_{j-1}$ and $A_{s-j}q_{j+1} = q_{s+1} - A_{s-j-1}q_j$, we have

$$\sum_1 = \sum_{j=0}^{s} (-1)^j \frac{a_{j+1}}{2q_{s+1}} \left(A_{s-j}q_{j+1} - A_{s-j}q_{j-1} 2A_{s-j}q_{j-1} - A_{s-j}q_j \right)$$

$$= \sum_{j=0}^{s} (-1)^j \frac{a_{j+1}}{2q_{s+1}} \left(q_{s+1} - A_{s-j-1}q_j A_{s-j}q_{j-1} - A_{s-j}q_j \right)$$

$$= \sum_{j=0}^{s} (-1)^j \frac{a_{j+1}}{2} + \sum_{j=0}^{s} \frac{(-1)^j}{2q_{s+1}} \left(A_{s-j}q_{j-1}a_{j+1} - A_{s-j-1}q_j a_{j+1} - A_{s-j}q_j a_{j+1} \right).$$

By using $A_{s-j}a_{j+1} = A_{s-j+1} - A_{s-j-1}$ and $a_{j+1}q_j = q_{j+1} - q_{j-1}$, we obtain

$$\sum_1 - \sum_{j=0}^{s} (-1)^j \frac{a_{j+1}}{2}$$

$$= \frac{1}{2q_{s+1}} \sum_{j=0}^{s} (-1)^j \left(A_{s-j+1}q_{j-1} - A_{s-j-1}q_{j-1} + A_{s-j-1}q_{j-1} \right.$$

$$\left. - A_{s-j-1}q_{j+1} + A_{s-j}q_{j-1} - A_{s-j}q_{j+1} \right)$$

$$= \frac{1}{2q_{s+1}} \left(\sum_{j=0}^{s} (-1)^j A_{s-j+1}q_{j-1} - \sum_{j=0}^{s} (-1)^j A_{s-j-1}q_{j+1} \right.$$

$$\left. + \sum_{j=0}^{s} (-1)^j A_{s-j}q_{j-1} - \sum_{j=0}^{s} (-1)^j A_{s-j}q_{j+1} \right).$$

A big cancellation comes from the obvious fact that

$$\sum_{j=0}^{s}(-1)^j A_{s-j+1}q_{j-1} = \sum_{j=0}^{s}(-1)^j A_{s-j-1}q_{j+1},$$

which leads to

$$\sum_{1} - \sum_{j=0}^{s}(-1)^j \frac{a_{j+1}}{2} = \frac{1}{2q_{s+1}}\left(\sum_{j=0}^{s}(-1)^j A_{s-j}q_{j-1} - \sum_{j=0}^{s}(-1)^j A_{s-j}q_{j+1}\right).$$

Again, $A_{s-j}q_{j+1} = q_{s+1} - A_{s-j-1}q_j$ gives $\sum_{1} - \sum_{j=0}^{s}(-1)^j \frac{a_{j+1}}{2} =$

$$= \frac{1}{2q_{s+1}}\left(-\sum_{j=0}^{s}(-1)^j q_{s+1} + \sum_{j=0}^{s}(-1)^j\left(A_{s-j-1}q_j - A_{s-j}q_{j-1}\right)\right)$$

$$= -\frac{1+(-1)^j}{4} + \sum_{j=0}^{s}(-1)^j\left(A_{s-j-1}q_j + A_{s-j}q_{j-1}\right).$$

Since clearly $\sum_{j=0}^{s}(-1)^j\left(A_{s-j-1}q_j + A_{s-j}q_{j-1}\right) = 0$, Lemma 2 follows.

Very similar but technically more complicated long arguments lead to the following lemma.

LEMMA 3. *Let*

$$\sum_{2} = \sum_{j=0}^{s}(-1)^j \mathbf{E}\left(b_j|\varepsilon_j|\left(\sum_{0\leq i<j}^{*} b_i q_i\right)\right),$$

and

$$\sum_{3} = \sum_{j=0}^{s}(-1)^j \mathbf{E}\left(\frac{b_j^2|\varepsilon_j|q_j}{2}\right).$$

Then

$$\sum_{2} + \sum_{3} = \frac{1}{6}\sum_{j=0}^{s}(-1)^j a_{j+1} + O(1).$$

Proof. In order to estimate $\sum_{2}$ we recall first the well-known formula

$$\varepsilon_j = q_j\alpha - p_j = (-1)^j \frac{\alpha_{j+2}}{\alpha_{j+2}q_{j+1} + q_j}$$

where $\alpha_{j+2} = [a_{j+2}; a_{j+3}, a_{j+4}, \cdots]$. On the other hand, we have

$$q_{s+1} = A_{s-j}q_{j+1} + A_{s-j-1}q_j,$$

and so

$$q_{s+1} = \left(\frac{A_{s-j}}{A_{s-j-1}} q_{j+1} + q_j \right) A_{s-j-1}.$$

By using the identity $A_{s-j} = a_{j+2}A_{s-j-1} + A_{s-j-2}$, we conclude by induction

$$\frac{A_{s-j}}{A_{s-j-1}} = a_{j+2} + \frac{1}{\frac{A_{s-j-1}}{A_{s-j-2}}} = a_{j+2} + \frac{1}{a_{j+3} + \frac{1}{\frac{A_{s-j-2}}{A_{s-j-3}}}} = [a_{j+2}; a_{j+3}, a_{j+4}, \cdots].$$

It follows that if $(s - j)$ is "large" then

$$\frac{\alpha_{j+2}}{\alpha_{j+2}q_{j+1} + q_j} \approx \frac{A_{s-j}}{A_{s-j}q_{j+1} + A_{s-j-1}q_j} = \frac{A_{s-j}}{q_{s+1}},$$

but we want to make it more precise. For this purpose we define $\tilde{A}_{s-j}$ by the equality

$$\frac{\alpha_{j+2}}{\alpha_{j+2}q_{j+1} + q_j} = \frac{\tilde{A}_{s-j}}{A_{s-j}q_{j+1} + A_{s-j-1}q_j} = \frac{\tilde{A}_{s-j}}{q_{s+1}},$$

and our goal is to describe the the relation $\tilde{A}_{s-j} \approx A_{s-j}$ in quantitative form.

We claim

$$\frac{\tilde{A}_{s-j}}{A_{s-j}} = 1 + O\left(\frac{1}{a_{j+1}(a_{j+2}a_{j+3} + 1)(a_{j+3}a_{j+4} + 1)\cdots(a_s a_{s+1} + 1)} \right). \quad (1)$$

To prove (1) consider first the special case $\tilde{A}_0$. We have

$$\frac{\tilde{A}_0}{q_{s+1}} = \frac{\alpha_{s+2}}{\alpha_{s+2}q_{s+1} + q_s} = \frac{1}{q_{s+1} + \frac{1}{\alpha_{s+2}}q_s} = \frac{1}{q_{s+1}(1 + \frac{1}{\alpha_{s+2}}\frac{q_s}{q_{s+1}})},$$

and so

$$\frac{\tilde{A}_0}{A_0} = \tilde{A}_0 = 1 + O(\frac{1}{a_{s+1}a_{s+2}}),$$

as claimed.

In general we have

$$\frac{\tilde{A}_{s-j}}{q_{s+1}} = \frac{\alpha_{j+2}}{\alpha_{j+2}q_{j+1} + q_s} = \frac{1}{q_{j+1} + \frac{1}{\alpha_{j+2}}q_j},$$

and also

$$\frac{A_{s-j}}{q_{s+1}} = \frac{A_{s-j}}{A_{s-j}q_{j+1} + A_{s-j-1}q_s} = \frac{1}{q_{j+1} + \frac{A_{s-j-1}}{A_{s-j}}q_j},$$

and so

$$\frac{\tilde{A}_{s-j}}{A_{s-j}} = \frac{q_{j+1} + \frac{A_{s-j-1}}{A_{s-j}}q_j}{q_{j+1} + \frac{1}{\alpha_{j+2}}q_j} = 1 - \frac{\left(\frac{1}{\alpha_{j+2}} - \frac{A_{s-j-1}}{A_{s-j}}\right)q_j}{q_{j+1}\left(1 + \frac{q_j}{\alpha_{j+2}q_{j+1}}\right)}.$$

Note that $\frac{1}{\alpha_{j+2}} = [0; a_{j+2}, a_{j+3}, a_{j+4}, \cdots]$ and

$$\frac{A_{s-j-1}}{A_{s-j}} = [0; a_{j+2}, a_{j+3}, \cdots, a_{s+1}],$$

i.e. the second one is a long finite initial segment of the infinite continued fraction of the first one. It follows

$$\left|\frac{1}{\alpha_{j+2}} - \frac{A_{s-j-1}}{A_{s-j}}\right| \leq \frac{1}{(a_{j+2}a_{j+3} + 1)(a_{j+3}a_{j+4} + 1)\cdots(a_s a_{s+1} + 1)}.$$

(Indeed, here we used the well-known facts that $|\beta - \frac{p_i}{q_i}| < \frac{1}{q_i q_{i+1}}$ where $\frac{p_i}{q_i}$ is the i-th convergent of $\beta = [B_0; B_1, B_2, \cdots]$, and $q_0 = 1, q_1 = B_1, q_2 = B_1 B_2 + 1, \cdots, q_{i+2} = (B_{i+1}B_{i+2} + 1)q_i + B_{i+2}q_{i-2} \geq (B_{i+1}B_{i+2} + 1)q_i$, and so $q_i q_{i+1} \geq (B_1 B_2 + 1)(B_2 B_3 + 1) \cdots (B_{i+1}B_{i+2} + 1)$.)

From the inequalities above (1) easily follows.

Now we are ready to evaluate $\sum_2$. Just like in the case of $\sum_1$ (see the proof of Lemma 2) we have

$$\sum_2 = \sum_{j=0}^{s}(-1)^j \mathbf{E}\left(b_j|\varepsilon_j|\left(\sum_{0 \leq i < j}^{*} b_i q_i\right)\right) =$$

$$= \sum_{j=0}^{s}(-1)^j \frac{\tilde{A}_{s-j}}{q_{s+1}}\left(\frac{A_{s-j}q_{j+1}}{q_{s+1}}\frac{q_j}{q_{j+1}}(0 + 1 + 2 + \cdots + (a_{j+1} - 1))\frac{q_j}{2}\right) +$$

$$+ \sum_{j=0}^{s}(-1)^j \frac{\tilde{A}_{s-j}}{q_{s+1}}\left(\frac{A_{s-j}q_{j+1}}{q_{s+1}}\frac{q_{j-1}}{q_{j+1}}a_{j+1}\frac{q_{j-1}}{2}\right).$$

Here we used the geometrically obvious "decomposition law" that if the "random variable" b_j equals to any of the values $0, 1, 2, \cdots, (a_{j+1} - 1)$ then the other "random variable" $\sum_{0 \leq i < j}^{*} b_i q_i$ is uniformly distributed in the interval $\{0, 1, 2, \cdots q_j\}$, and if b_j equals to a_{j+1} then $b_{j-1} = 0$ and so $\sum_{0 \leq i < j}^{*} b_i q_i$ is uniformly distributed in the shorter interval $\{0, 1, 2, \cdots q_{j-1}\}$. Thus we have $\sum_2 =$

$$= \sum_{j=0}^{s}(-1)^j \frac{a_{j+1}}{4q_{s+1}^2}\left(\tilde{A}_{s-j}A_{s-j}q_j^2 a_{j+1} - \tilde{A}_{s-j}A_{s-j}q_j^2 + 2\tilde{A}_{s-j}A_{s-j}q_{j-1}^2\right).$$

Repeating the same argument for $\sum_3$ we have

$$\sum_3 = \sum_{j=0}^{s}(-1)^j \mathbf{E}\left(\frac{b_j^2|\varepsilon_j|q_j}{2}\right) =$$

$$= \sum_{j=0}^{s}(-1)^j \frac{\tilde{A}_{s-j}}{q_{s+1}}\left(\frac{A_{s-j}q_{j+1}}{q_{s+1}}\frac{q_j}{q_{j+1}}(0^2 + 1^2 + 2^2 + \cdots + (a_{j+1} - 1)^2)\frac{q_j}{2}\right) +$$

$$+ \sum_{j=0}^{s}(-1)^j \frac{\tilde{A}_{s-j}}{q_{s+1}}\left(\frac{A_{s-j}q_{j+1}}{q_{s+1}}\frac{q_{j-1}}{q_{j+1}}a_{j+1}^2\frac{q_{j-1}}{2}\right) =$$

$$= \sum_{j=0}^{s}(-1)^j \frac{a_{j+1}}{12q_{s+1}^2}\left(2\tilde{A}_{s-j}A_{s-j}q_j^2 a_{j+1}^2 - 3\tilde{A}_{s-j}A_{s-j}q_j^2 a_{j+1}\right) +$$

$$+ \sum_{j=0}^{s}(-1)^j \frac{a_{j+1}}{12q_{s+1}^2}\left(\tilde{A}_{s-j}A_{s-j}q_j^2 + 6\tilde{A}_{s-j}A_{s-j}q_j q_{j-1}a_{j+1}\right).$$

Therefore

$$\sum_2 + \sum_3 = \sum_{j=0}^{s}(-1)^j \frac{a_{j+1}}{6q_{s+1}^2}\left(\tilde{A}_{s-j}A_{s-j}q_j^2 a_{j+1}^2\right) +$$

$$+ \sum_{j=0}^{s}(-1)^j \frac{a_{j+1}}{6q_{s+1}^2}\left(3\tilde{A}_{s-j}A_{s-j}q_{j-1}^2 - \tilde{A}_{s-j}A_{s-j}q_j^2 + 3\tilde{A}_{s-j}A_{s-j}q_j q_{j-1}a_{j+1}\right).$$

$$(2)$$

In order to handle (2) we make use of the following trivial identities:

$$a_{j+1}q_j = (q_{j+1} - q_{j-1})^2 = q_{j+1}^2 + q_{j-1}^2 - 2q_{j+1}q_{j-1},$$

$$\tilde{A}_{s-j}A_{s-j}(a_{j+1}q_j)^2 = \tilde{A}_{s-j}A_{s-j}q_{j+1}^2 + \tilde{A}_{s-j}A_{s-j}q_{j-1}^2 - 2\tilde{A}_{s-j}A_{s-j}q_{j+1}q_{j-1}.$$

Since

$$\tilde{A}_{s-j}A_{s-j}q_{j+1}^2 = \frac{\tilde{A}_{s-j}}{A_{s-j}}\left(q_{s+1}^2 - A_{s-j-1}q_j\right)^2 =$$

$$= \frac{\tilde{A}_{s-j}}{A_{s-j}}\left(q_{s+1}^2 + A_{s-j-1}^2 q_j^2 - 2q_{s+1}A_{s-j-1}q_j\right),$$

we have

$$\tilde{A}_{s-j}A_{s-j}a_{j+1}^2 q_j^2 = \frac{\tilde{A}_{s-j}}{A_{s-j}}q_{s+1}^2 + \frac{\tilde{A}_{s-j}}{A_{s-j}}A_{s-j-1}^2 q_j^2 - 2q_{s+1}\frac{\tilde{A}_{s-j}}{A_{s-j}}A_{s-j-1}q_j +$$

$$+ \tilde{A}_{s-j}A_{s-j}q_{j-1}^2 - 2\tilde{A}_{s-j}A_{s-j}q_{j+1}q_{j-1}.$$

Going back to (2) we obtain

$$\sum_2 + \sum_3 = \sum_4 + \sum_{51} + \sum_{52} + \sum_{53} + \sum_{54} + \sum_{55} + \sum_{56}, \quad (3)$$

where

$$\sum_4 = \sum_{j=0}^{s} (-1)^j \frac{a_{j+1}}{6} \frac{\tilde{A}_{s-j}}{A_{s-j}}, \quad (4)$$

$$\sum_{51} = \sum_{j=0}^{s} (-1)^j \frac{a_{j+1}}{6q_{s+1}^2} \left(4\tilde{A}_{s-j} A_{s-j} q_{j-1}^2 \right), \quad (5)$$

$$\sum_{52} = \sum_{j=0}^{s} (-1)^j \frac{a_{j+1}}{6q_{s+1}^2} \left(3\tilde{A}_{s-j} A_{s-j} q_j q_{j-1} a_{j+1} \right), \quad (6)$$

$$\sum_{53} = \sum_{j=0}^{s} (-1)^j \frac{a_{j+1}}{6q_{s+1}^2} \left(-\tilde{A}_{s-j} A_{s-j} q_j^2 \right), \quad (7)$$

$$\sum_{54} = \sum_{j=0}^{s} (-1)^j \frac{a_{j+1}}{6q_{s+1}^2} \left(-2\tilde{A}_{s-j} A_{s-j} q_{j+1} q_{j-1} \right), \quad (8)$$

$$\sum_{55} = \sum_{j=0}^{s} (-1)^j \frac{a_{j+1}}{6q_{s+1}^2} \left(\frac{\tilde{A}_{s-j}}{A_{s-j}} A_{s-j-1}^2 q_j^2 \right), \quad (9)$$

$$\sum_{56} = \sum_{j=0}^{s} (-1)^j \frac{a_{j+1}}{6q_{s+1}^2} \left(-2q_{s+1} \frac{\tilde{A}_{s-j}}{A_{s-j}} A_{s-j-1} q_j \right). \quad (10)$$

In view of (1) we have

$$\sum_4 = \sum_{j=0}^{s} (-1)^j \frac{a_{j+1}}{6} \frac{\tilde{A}_{s-j}}{A_{s-j}} = \left(\sum_{j=0}^{s} (-1)^j \frac{a_{j+1}}{6} \right) + O(1). \quad (11)$$

In order to estimate $\sum_{51}$ we start with the identity $A_{s-j} a_{j+1} = A_{s-j+1} - A_{s-j-1}$ which implies

$$\tilde{A}_{s-j} A_{s-j} q_{j-1}^2 a_{j+1} = \tilde{A}_{s-j} A_{s-j+1} q_{j-1}^2 - \tilde{A}_{s-j} A_{s-j-1} q_{j-1}^2.$$

By using $A_{s-j} q_{j-1} = q_{s+1} - A_{s-j+1} q_j$, we conclude

$$\sum_{51} = \sum_{j=0}^{s} (-1)^j \frac{1}{6q_{s+1}^2} \left(4\tilde{A}_{s-j} A_{s-j} q_{j-1}^2 a_{j+1} \right) =$$

$$= \sum_{j=0}^{s} (-1)^j \frac{1}{6q_{s+1}^2} \left(4\tilde{A}_{s-j} A_{s-j+1} q_{j-1}^2 + 4\frac{\tilde{A}_{s-j}}{A_{s-j}} A_{s-j+1} A_{s-j-1} q_j q_{j-1} \right) -$$

$$\sum_{j=0}^{s}(-1)^j \frac{1}{6q_{s+1}^2}\left(4\frac{\tilde{A}_{s-j}}{A_{s-j}}q_{s+1}A_{s-j-1}q_{j-1}\right). \tag{12}$$

In order to estimate $\sum_{52}$ we start with the identity $a_{j+1}q_j = q_{j+1} - q_{j-1}$, which implies

$$\tilde{A}_{s-j}A_{s-j}q_jq_{j-1}a_{j+1}^2 = \tilde{A}_{s-j}A_{s-j}q_{j+1}q_{j-1}a_{j+1} - \tilde{A}_{s-j}A_{s-j}q_{j-1}^2a_{j+1}. \tag{13}$$

By (6),(8),(12) and (13), $\sum_{51} + \sum_{52} + \sum_{54} =$

$$= \sum_{j=0}^{s}\frac{(-1)^j}{6q_{s+1}^2}\left(\tilde{A}_{s-j}A_{s-j}q_{j+1}q_{j-1}a_{j+1} + \tilde{A}_{s-j}A_{s-j+1}q_{j+1}q_{j-1}^2\right) +$$

$$+ \sum_{j=0}^{s}\frac{(-1)^j}{6q_{s+1}^2}\left(\frac{\tilde{A}_{s-j}}{A_{s-j}}A_{s-j+1}A_{s-j-1}q_jq_{j-1} - \frac{\tilde{A}_{s-j}}{A_{s-j}}q_{s+1}A_{s-j-1}q_{j-1}\right). \tag{14}$$

To estimate the first term on the right-hand side of (14) we use $A_{s-j}a_{j+1} = A_{s-j+1} - A_{s-j-1}$, and so

$$\tilde{A}_{s-j}A_{s-j}q_{j+1}q_{j-1}a_{j+1} = \tilde{A}_{s-j}A_{s-j+1}q_{j+1}q_{j-1} + \tilde{A}_{s-j}A_{s-j-1}q_{j+1}q_{j-1}. \tag{15}$$

Moreover, by $A_{s-j}q_{j+1} = q_{s+1} - A_{s-j-1}q_j$ we get

$$\tilde{A}_{s-j}A_{s-j+1}q_{j+1}q_{j-1} = \frac{\tilde{A}_{s-j}}{A_{s-j}}A_{s-j+1}q_{s+1}q_{j-1} - \frac{\tilde{A}_{s-j}}{A_{s-j}}A_{s-j+1}q_jq_{j-1}, \tag{16}$$

and

$$-\tilde{A}_{s-j}A_{s-j-1}q_{j+1}q_{j-1} = -\frac{\tilde{A}_{s-j}}{A_{s-j}}A_{s-j-1}q_{s+1}q_{j-1} + \frac{\tilde{A}_{s-j}}{A_{s-j}}A_{s-j-1}^2q_jq_{j-1}, \tag{17}$$

Next consider $\sum_{53}$. By using $A_{s-j}a_{j+1} = A_{s-j+1} - A_{s-j-1}$ we get

$$-\tilde{A}_{s-j}A_{s-j}q_j^2a_{j+1} = \tilde{A}_{s-j}A_{s-j-1}q_j^2 - \tilde{A}_{s-j}A_{s-j+1}q_j^2. \tag{18}$$

Now we collect the second term on the right-hand side of (14) and the first term on the right-hand side of (18), and claim

$$\frac{1}{q_{s+1}^2}\sum_{j=0}^{s}(-1)^j\left(\tilde{A}_{s-j}A_{s-j+1}q_{j-1}^2 + \tilde{A}_{s-j}A_{s-j-1}q_j^2\right) = O(1). \tag{19}$$

Indeed, first observe the perfect cancellation

$$\sum_{j=0}^{s}(-1)^j\left(A_{s-j}A_{s-j+1}q_{j-1}^2 + A_{s-j}A_{s-j-1}q_j^2\right) = 0,$$

and then (19) easily follows from (1).

Next consider the second term in the right-hand side of (18). By using $A_{s-j+1}q_j = q_{s+1} - A_{s-j}q_{j-1}$, we get

$$-\tilde{A}_{s-j}A_{s-j+1}q_j^2 = \tilde{A}_{s-j}A_{s-j}q_jq_{j-1} - \tilde{A}_{s-j}q_{s+1}q_j. \tag{20}$$

Next we study $\sum_{55}$. By using $a_{j+1}q_j = q_{j+1} - q_{j-1}$ we get

$$\frac{\tilde{A}_{s-j}}{A_{s-j}}A_{s-j-1}^2 q_j^2 a_{j+1} = \frac{\tilde{A}_{s-j}}{A_{s-j}}A_{s-j-1}^2 q_{j+1}q_j - \frac{\tilde{A}_{s-j}}{A_{s-j}}A_{s-j-1}^2 q_{j-1}q_j. \tag{21}$$

In view of $A_{s-j-1}q_j = q_{s+1} - A_{s-j}q_{j+1}$, we obtain

$$\frac{\tilde{A}_{s-j}}{A_{s-j}}A_{s-j-1}^2 q_jq_{j+1} = \frac{\tilde{A}_{s-j}}{A_{s-j}}A_{s-j-1}q_{j+1}q_{s+1} - \frac{\tilde{A}_{s-j}}{A_{s-j}}A_{s-j-1}A_{s-j}q_{j+1}^2,$$
$$\tag{22}$$

and

$$-\frac{\tilde{A}_{s-j}}{A_{s-j}}A_{s-j-1}^2 q_jq_{j-1} = -\frac{\tilde{A}_{s-j}}{A_{s-j}}A_{s-j-1}q_{j-1}q_{s+1}$$
$$+ \frac{\tilde{A}_{s-j}}{A_{s-j}}A_{s-j-1}A_{s-j}q_{j+1}q_{j-1}. \tag{23}$$

Finally consider $\sum_{56}$. By using $a_{j+1}q_j = q_{j+1} - q_{j-1}$ we get

$$-\frac{\tilde{A}_{s-j}}{A_{s-j}}A_{s-j-1}q_{s+1}q_ja_{j+1} = \frac{\tilde{A}_{s-j}}{A_{s-j}}A_{s-j-1}q_{s+1}q_{j-1}$$
$$- \frac{\tilde{A}_{s-j}}{A_{s-j}}A_{s-j-1}q_{s+1}q_{j+1}. \tag{24}$$

Collecting all the terms we obtain (see (3)-(24))

$$\sum_2 + \sum_3 = \left(\sum_{j=0}^{s}(-1)^j\frac{a_{j+1}}{6}\right) + O(1) + \sum_6, \tag{25}$$

where

$$\sum_6 = \frac{1}{6q_{s+1}^2}\sum_{j=0}^{s}(-1)^j\left(\sum_{i=1}^{14}Z_i\right), \tag{26}$$

and

$$Z_1 = \frac{\tilde{A}_{s-j}}{A_{s-j}}A_{s-j+1}A_{s-j-1}q_jq_{j-1}, \quad Z_2 = -\frac{\tilde{A}_{s-j}}{A_{s-j}}A_{s-j+1}A_{s-j-1}q_jq_{j-1},$$
$$\tag{27}$$

$$Z_3 = \frac{\widetilde{A}_{s-j}}{A_{s-j}} A^2_{s-j-1} q_j q_{j-1}, \quad Z_4 = \widetilde{A}_{s-j} A_{s-j} q_j q_{j-1}, \tag{28}$$

$$Z_5 = -\frac{\widetilde{A}_{s-j}}{A_{s-j}} A_{s-j} A_{s-j-1} q^2_{j+1}, \quad Z_6 = \frac{\widetilde{A}_{s-j}}{A_{s-j}} A_{s-j} A_{s-j-1} q_{j+1} q_{j-1}, \tag{29}$$

$$Z_7 = -\frac{\widetilde{A}_{s-j}}{A_{s-j}} q_{s+1} A_{s-j-1} q_{j-1}, \quad Z_8 = \frac{\widetilde{A}_{s-j}}{A_{s-j}} q_{s+1} A_{s-j+1} q_{j-1}, \tag{30}$$

$$Z_9 = -\frac{\widetilde{A}_{s-j}}{A_{s-j}} q_{s+1} A_{s-j-1} q_{j-1}, \quad Z_{10} = -q_{s+1} A_{s-j} q_j, \tag{31}$$

$$Z_{11} = \frac{\widetilde{A}_{s-j}}{A_{s-j}} q_{s+1} A_{s-j-1} q_{j+1}, \quad Z_{12} = -\frac{\widetilde{A}_{s-j}}{A_{s-j}} q_{s+1} A_{s-j-1} q_{j-1}, \tag{32}$$

$$Z_{13} = 2\frac{\widetilde{A}_{s-j}}{A_{s-j}} q_{s+1} A_{s-j-1} q_{j-1}, \quad Z_{14} = -2\frac{\widetilde{A}_{s-j}}{A_{s-j}} q_{s+1} A_{s-j-1} q_{j+1}. \tag{33}$$

In what follows we simply detect the cancellations. Clearly

$$Z_1 + Z_2 = 0 \quad \text{and} \quad Z_9 + Z_{12} + Z_{13} = 0. \tag{34}$$

Furthermore,

$$Z_{11} + Z_{14} = -\frac{\widetilde{A}_{s-j}}{A_{s-j}} q_{s+1} A_{s-j-1} q_{j+1}. \tag{35}$$

For brevity let Z'_{14} denote the right-hand side of (35), so we can write

$$Z_{11} + Z_{14} = Z'_{14}. \tag{36}$$

By using $A_{s-j-1} q_j = q_{s+1} - A_{s-j} q_{j+1}$ we have

$$Z_3 = Z'_3 + Z''_3 \quad \text{where} \tag{37}$$

$$Z'_3 = \frac{\widetilde{A}_{s-j}}{A_{s-j}} q_{s+1} A_{s-j-1} q_{j-1}, \tag{38}$$

$$Z''_3 = -\frac{\widetilde{A}_{s-j}}{A_{s-j}} A_{s-j} A_{s-j-1} q_{j+1} q_{j-1}. \tag{39}$$

By using $A_{s-j} q_{j+1} = q_{s+1} - A_{s-j-1} q_j$ we obtain

$$Z_5 = Z'_5 + Z''_5 \quad \text{where} \tag{40}$$

$$Z'_5 = \frac{\widetilde{A}_{s-j}}{A_{s-j}} A^2_{s-j-1} q_{j+1} q_j, \tag{41}$$

$$Z''_5 = -\frac{\widetilde{A}_{s-j}}{A_{s-j}} q_{s+1} A_{s-j-1} q_{j+1}. \tag{42}$$

We claim

$$\frac{1}{q_{s+1}^2} \sum_{j=0}^{s} (-1)^j \, (Z_4 + Z_5') = O(1), \tag{43}$$

that is,

$$\frac{1}{q_{s+1}^2} \sum_{j=0}^{s} (-1)^j \left(\tilde{A}_{s-j} A_{s-j} q_j q_{j-1} + \frac{\tilde{A}_{s-j}}{A_{s-j}} A_{s-j-1}^2 q_{j+1} q_j \right) = O(1). \tag{44}$$

Indeed, first observe the perfect cancellation

$$\sum_{j=0}^{s} (-1)^j \left(A_{s-j}^2 q_j q_{j-1} + A_{s-j-1}^2 q_{j+1} q_j \right) = 0,$$

and then (44) easily follows from (1).

Next observe that

$$Z_6 + Z_3'' = 0 \quad \text{and} \quad Z_7 + Z_3' = 0. \tag{45}$$

So the remaining terms are Z_8, Z_{10}, Z_{14}' and Z_5''. We claim

$$\frac{1}{q_{s+1}^2} \sum_{j=0}^{s} (-1)^j \, (Z_8 + Z_{10} + Z_{14}' + Z_5'') = O(1). \tag{46}$$

Indeed,

$$\sum_{j=0}^{s} (-1)^j \left(A_{s-j-1} q_{j-1} - A_{s-j} q_j - 2 A_{s-j-1} q_{j+1} \right) = -2 A_s + (-1)^s q_s. \tag{47}$$

Since $q_{s+1} = A_s q_1 + A_{s-1} q_0 = A_s a_1 + A_{s-1}$, we have $q_{s+1} \geq A_s$, and so (47) is $O(q_{s+1})$. Now (46) easily follows from (47) by using (1). This completes the long proof of Lemma 3.

Finally Theorem 3.2 immediately follows from Lemmas 1-3.

6 REFERENCES

[Ba] Baker, A.: *Transcendental Number Theory*, Cambridge University Press 1975.

[Be1] Beck, J.: On a problem of W.M. Schmidt concerning one-sided irregularities of point distributions, Math. Ann. 285 (1989), 29-55.

[Be2] Beck, J.: A two-dimensional van Aardenne-Ehrenfest theorem in irregularities of distribution, Compositio Math. 72 (1989), 269-339.

[Be3] Beck, J.: A central limit theorem for quadratic irrational rotations, Preprint (1991), 90 pages.

[Be4] Beck, J.: Probabilistic Diophantine Approximation-Part I: Kronecker sequences, Annals of Math., 140 (1994), 451-502.

[Be5] Beck, J.: Probabilistic Diophantine Approximation-Part II: Local Results, Preprint (1997).

[Be6] Beck, J.: Probabilistic Diophantine Approximation-Part III: Global Results, Preprint (1997).

[Be7] Beck, J.: On the Cesaro Mean of $\sum\{n\alpha\}$, Preprint (1997).

[Be8] Beck, J.: Continued fractions and quadratic fields, Preprint (1997).

[BCDL] Bleher, P.M. and Cheng, Z. and Dyson, F.J. and Lebowitz, J.L.: Distribution for the error term for the number of lattice points inside a shifted circle, Preprint 1992, 40 pages

[Bo] Bohl, P.: Über ein in der Theorie der säkularen Störungen vorkommendes Problem, Journ. Reine u. Angew. Math. 135 (1909), 189-283.

[El] Elliott, P.: *Probabilistic Number Theory II, central Limit Theorems*, Springer Verlag 1980.

[EK] Erdős, P. and Kac, M.: The Gaussian law of errors in the theory of additive number-theoretic functions, Amer. Journ. Math. 62 (1940), 738-742.

[Ha] Halász, G.: On Roth's method in the theory of irregularities of point distributions, Recent Progress in Analytic Number Theory, Vol.2, pp. 79-94, London, Academic Press 1981.

[Ha-Li1] Hardy, G. and Littlewood, J.: The lattice-points of a right-angled triangle. I, Proc. London Math. Soc. 3 (1920), 15-36.

[Ha-Li2] Hardy, G. and Littlewood, J.: The lattice-points of a right-angled triangle. II, Abh. Math. Sem. Hamburg 1, 212-249 (1922).

[Ha-Li3] Hardy, G. and Littlewood, J.: Some problems of Diophantine approximation: A series of cosecants, Bull. of the Calcutta Math. Soc. 20, 251-66 (1930).

[He] Heath-Brown, D.R.: The distribution and moments of the error term in the Dirichlet's divisor problem, Acta Arithmetica 60 (1992), 389-415.

[Hi] Hirzebruch, F.: Hilbert modular surfaces, L'Einseignement Math., 19 (3-4) (1973), 183-281.

[Ka] Kac, M.: Probability methods in some problems of analysis and number theory, Bull. Amer. Math. Soc. 55 (1949), 641-665.

[Kem] Kemperman, J.H.B.: Probability methods in the theory of distributions modulo one, Compositio Math. 16 (1964), 106-137.

[Kes1] Kesten, H.: Uniform distribution mod 1, Annals of Math. 71 (1960), 445-471 and Part II in Acta Arith. 7 (1961), 355-360.

[Kes2] Kesten, H.: Some probabilistic theorems on Diophantine Approximations, Trans. Amer. Math. Soc. 103 (1962), 189-217.

[Kes3] Kesten, H.: On a conjecture of Erdös and Szüsz related to uniform distribution mod 1, Acta Arithm. 12 (1966), 193-212.

[Kh1] Khintchine, A.: Ein Satz über Kettenbrüche, mit arithmetischen Anwendungen, Math. Z. (1923), 289-306.

[Kh2] Khintchine, A.: *Continued Fractions*, English translation, P. Noordhoff, Groningen, The Netherlands 1963.

[La] Lang, S.: *Introduction to Diophantine Approximations*, Addison-Wesley 1966.

[Mo] Mollin, R. A.: An overview of the solution to the class number one problem for real quadratic fields of Richaud-Degert type, (Hungary) 1987, 871-88.

[Os] Ostrowski, A.: Bemerkungen zur Theorie der Diophantischen Approximationen. I, Abh. Hamburg Sem. 1 (1922), 77-98.

[Ro] Roth, K.F.: On irregularities of distribution, Mathematika 1 (1954), 73-79.

[S-Z] Salem, R. and Zygmund, A.: On lacunary trigonometric series, Proc. Nat. Acad. Sci. U.S.A. 34 (1947), 333-338.

[Sch1] Schmidt, W.M.: A metrical theorem in diophantine approximation, Canad. J. Math. 12 (1960), 619-631.

[Sch2] Schmidt, W.M.: Metrical theorems on fractional parts of sequences, Trans. Amer. Math. Soc. 110 (1964), pp. 493-518.

[Si1] Sinai, Ya.: Mathematical problems in the theory of quantum chaos, preprint (1990).

[Si2] Sinai, Ya.: Poisson distribution in a geometrical problem, preprint (1990).

[Sk] Skriganov, M.M.: Construction of uniform distributions in terms of geometry of numbers, Algebra and Analysis (Russian Academy of Sciences) 6 (3) (1994), 200-230.

[So] Sós, Vera: On the discrepancy of the sequence $\{n\alpha\}$, Coll. Math. Soc. János Bolyai 13 (1974), 359-367.

[T-W] Tijdeman, G. and Wagner, G.: A sequence has almost nowhere small discrepancy, Monatshefte Math. 90 (1980), 315-329.

[We] Weyl, H.: Über die Gleichverteilung von Zahlen mod Eins, Math. Ann. 77 (1916), 313-352.

[Za1] Zagier, D.: Nombres de classes et fractions continues, Journées Arithmétiques de Bordeaux, Astérisque No. 24-25 (1975), 81-97.

[Za2] Zagier, D.: A Kronecker limit formula for real quadratic fields, Math. Ann. 213 (1975), 153-184.

[Za3] Zagier, D.: *Zeta-Funktionen und quadratische Körper*, Springer-Verlag 1981, Hochschultext.

[Za4] Zagier, D.: On the values at negative integers of the zeta-function of a real quadratic field, Enseignement Math. (2) 22 (1976), no. 1-2, 55-95.

[Za5] Zagier, D.: Valeurs des functions zeta des corps quadratiques réels aux entiers négatifs, Journées Arithmétiques de Caen, Astérisque No. 41-42 (1977), 135-151.

József Beck
Mathematics Department, Busch Campus, Hill Center
Rutgers University, New Brunswick, NJ 08903 USA

e-mail: jbeck@math.rutgers.edu

On the Assessment of Random and Quasi-Random Point Sets

Peter Hellekalek

1 Introduction

The Monte Carlo and the quasi-Monte Carlo method are two of the most important techniques to solve multidimensional problems. In both cases, we employ deterministic algorithms to generate finite point sets in high dimensions. The quality of the equidistribution of these point sets may have a decisive influence on our numerical results. In this contribution, we will discuss the main concepts to assess equidistribution. In particular, we will present a new interpretation of the well-known spectral test.

If we choose the Monte Carlo method to solve our problem, then we should keep in mind Compagner's[Com95] warning message: "*Monte Carlo results are misleading when correlations hidden in the random numbers and in the simulated system interfere constructively*". The random numbers or points that we use in Monte Carlo simulations are produced by deterministic algorithms, so-called random number generators (RNGs). Correlations between those numbers or points are unavoidable. As a consequence, we will need methods to assess such correlations in order to prevent (computational or real-life) disaster. At present, all important concepts for theoretical correlation analysis of random numbers or random points lead to the assessment of the equidistribution of certain high-dimensional point sets.

If we choose the quasi-Monte Carlo approach, then we will have to generate deterministic point sets with some kind of "super-uniform" distribution. This leads to the search for so-called *low-discrepancy point sets* (LDPs). Again, we have to deal with figures of merit that measure the equidistribution of point sets in high dimensions.

In the Monte Carlo and in the quasi-Monte Carlo case, what we would like to achieve by the theoretical or a priori assessment is some kind of *reliable prediction* that the numbers or points we use in practice will yield the correct results.

In this contribution, we will discuss how to attain this objective. Our goal is to present a readable account of the main theoretical concepts in this field, to summarize the state of the art for practitioners, and to show that the spectral test is just the tip of a mathematical iceberg.

This paper is organized as follows. In Chapter 2 in our "Chapter for the Practitioner", we will give an overview of the theoretical safety-measures to avoid incorrect simulation results. We will discuss the practical relevance of this kind of a priori assessment of RNGs and LDPs. Chapter 3 contains the preliminaries to study the figures of merit for point sets that will be introduced in the more theory-oriented Chapters 4 to 7. The keywords there will be the spectral test, diaphony, and discrepancy.

We refer the reader to the comprehensive monographs of Fishman[Fis96] and Niederreiter[Nie92b] for a general introduction to the Monte Carlo and the quasi-Monte Carlo method.

2   Chapter for the Practitioner

In Monte Carlo and quasi-Monte Carlo simulation, one possible source of error are inappropriate RNGs or LDPs. I would like to discuss now if there are "safe" RNGs or LDPs and what safety-measures exist to keep the risk of incorrect simulation results as small as possible. This will lead us to a concise review of the state of the art.

Let us talk about the bad news first. Five decades of mathematical research have passed, hundreds of research papers have been published in this field, but there is no figure of merit for RNGs and LDPs available that is able to guarantee correct results in your simulation. Even reliable RNGs or LDPs will fail in certain circumstances. In more technical terms, no promising discrepancy estimate, no excellent spectral test values, or any other theoretical quantities can protect you against wrong results in practice. Guarantees will remain an impossible dream. There are no safe RNGs or LDPs.

The good news are that such unpleasant surprises will be rare events, provided that we respect some mathematical guidelines. Nowadays, there are several theoretical figures of merit available that keep their promises. They deliver very reliable predictions of the performance of RNGs and LDPs in numerical practice.

At this point, let us review the interplay between theory and practice. The best way to make sure that a RNG or a LDP will perform well in our simulation is to test it with a similar problem where we already know the answer in advance. In this (very lucky) case, we may compare our empirical to the expected results and gain confidence in the RNG or LDP we use.

Of course, this approach will improve our chances of success with the original problem. Alas, in most cases, such *application-specific* testing will be impossible, for a simple reason. If we are able to solve a closely related problem analytically, then, in most cases, we will be able to solve our original modelization problem by analytic means as well. Hence, there will be no need for simulation.

As the next best option, batteries of statistical tests for RNGs and collections of test functions for high-dimensional numerical integration with

LDPs have been designed. We refer the reader to the important examples of Knuth[Knu81], Marsaglia[Mar85][1], and L'Ecuyer[L'E92, L'E97c] for RNGs, and to Genz[Gen84, Gen87] for LDPs. The idea behind these batteries is to cover as many practical cases as possible. If well-designed, such tests will constitute *prototypes* of practical simulations. If a RNG or an LDP performs well in such a battery of tests, this fact will increase our confidence that we will get the correct results in our particular Monte Carlo or quasi-Monte Carlo problem. Of course, this does not constitute a guarantee.

Empirical testing consumes time and resources. We will have to carry out these tests with samples of similar size and in similar dimensions as in our original simulation. For this reason, *theoretical* figures of merit have been designed that allow to assess RNGs and LDPs without the need to generate a single random number or quasi-random point. The goal is to *predict* the practical performance of RNGs and LDPs. We will discuss this kind of a priori analysis first for RNGs.

2.1   Assessing RNGs

RNGs for Monte Carlo simulation have to be efficient and portable algorithms. Their output should be reproducible in order to verify simulation results and to debug the code. In cryptography the requirements for RNGs are different. There, Monte Carlo generators are not appropriate because of their low complexity and high predictability. We will not cover these aspects here but refer the reader to Lagarias[Lag93] and Schneier[Sch96].

The designer of an RNG faces a considerable problem. He does not know the user's simulation in advance. There will be practitioners that use very large samples compared to the period length of the generator, others will combine consecutive random numbers to generate random vectors. Even others will assign different substreams of an RNG's output to different processors on parallel machines. Clearly, it is impossible to foresee all these applications. No designer is able to check in advance by statistical tests if his RNG will perform well in all imaginable situations.

There is a partial solution to this dilemma. We may perform a *theoretical analysis* of RNGs. It consists of the study of the period length, the analysis of the intrinsic structures of the numbers and points we produce with a given type of RNG, and the assessment of correlations between these random numbers and points.

Well-designed algorithms for random number generation allow us to find conditions for their parameters that will guarantee a certain period length of the output sequence, see Niederreiter[Nie95] for a comprehensive survey of this subject. Further, it will be possible to detect in advance, by theoretical analysis, certain weaknesses of the algorithm. For example, one can

---

[1]see also the web-site http://stat.fsu.edu/~geo/diehard.html

show that the low-order bits produced by linear congruential generators (LCGs) with a power-of-two modulus cannot be trusted (see Eddy[Edd90] and Anderson[And90]).

From a practical viewpoint, any promising abstract algorithm will be useless if we are unable to compute parameters that yield long periods. For an illustration for this kind of theoretical research, we refer to the interesting inversive congruential generator (ICG) of Eichenauer and Lehn[EL86], see also the author's survey [Hel95c] for the background and for tables of parameters, Chou[Cho95] for the search strategy, and Eichenauer-Herrmann et al.[EHHW97] for numerous additional results.

The period length of a generator will impose bounds on the usable sample size. In practice, the sample size should be much smaller than the period length of the generator. Otherwise, the regularities of the RNG will show up in the simulation results. There are no general mathematical results available that tell us which sample sizes will be safe. All recommendations for the maximal usable sample size are based upon numerical experience. For example, in the case of linear RNGs, the square root of the period length seems to be an upper bound for the usable sample size. We refer the reader to the contribution of L'Ecuyer and the author in this volume for an extensive empirical analysis and to [Hel95c], Leeb and Wegenkittl[LW97], L'Ecuyer et al. [LCS96, LCC96, L'E97c], and MacLaren[Mac92] for further numerical evidence.

It is very important to know the intrinsic structures of RNGs. For example, linear types of generators produce grid structures (i.e. shifted lattices) in any dimension. They have been studied in great detail, see L'Ecuyer [L'E97b] for references. If we are aware of this fact, then we will not be surprised that "bad" linear congruential generators may perform better in a stochastic simulation like the nearest-pair test than "good" ones, see the contribution of L'Ecuyer and the author in this volume and L'Ecuyer[L'E92, LCS96]. Knowledge of intrinsic structures like the grid structure of LCGs helps to control such effects. In this context, the classification as "good" refers to the performance in the spectral test. This theoretical test will be discussed below and in extensive detail in Chapter 5.

This example illustrates another important fact. The nearest-pair test is certainly not an esoteric simulation. The spectral test is one of the most reliable figures of merit for RNGs. But even in this case, a RNG performing well in the spectral test may fail in the nearest-pair test. The correlations hidden in the random numbers interfere with the Monte Carlo simulation. In other words, properties of the grid exert a strong influence on the simulation results. This example underlines the significance of Compagner's statement.

2.2   Correlation Analysis for RNGs I

Correlation analysis is the central and most demanding part of the a priori assessment of RNGs. There are figures of merit for RNGs available that give us very reliable predictions of the performance in practical simulations. These quantities can only *forecast* the performance of the samples we are going to produce with our RNG. As it is the case with all forecasts, sometimes reality will be different from what we have predicted. Nevertheless, the reliability of these figures of merit is truely remarkable. Hence, the importance of theoretical correlation analysis for numerical practice is beyond question.

The basic concept to analyze correlations between random numbers is simple. Suppose we are given random numbers $x_0, x_1, \ldots$ in the unit interval $[0, 1[$. These numbers should behave like a realization of a sequence of independent, identically distributed ("i.i.d.") random variables $(X_n)_{n \geq 0}$ where each X_n is uniformly distributed in the unit interval. In particular, the random numbers should be uncorrelated, i.e. by knowing a string of some $d - 1$ consecutive random numbers $x_n, x_{n+1}, \ldots, x_{n+d-2}$, $d \geq 2$, it should be impossible to predict the next random number x_{n+d-1}. All possibilities for x_{n+d-1} should be equally likely. In other words, x_{n+d-1} should be able to range uniformly within the whole unit interval $[0, 1[$. Hence, in order to check for correlations between consecutive random numbers, we may construct either *overlapping* d-tuples,

$$\mathbf{x}_n := (x_n, x_{n+1}, \ldots, x_{n+d-1}),$$

or *non-overlapping* d-tuples

$$\mathbf{x}_n := (x_{nd}, x_{nd+1}, \ldots, x_{nd+d-1}),$$

and assess the empirical distribution of the finite sequences

$$\omega = (\mathbf{x}_n)_{n=0}^{N-1}$$

in the d-dimensional unit cube $[0, 1[^d$ that arise in this way. The task is to measure how "well" ω is uniformly distributed. Strong correlations between consecutive random numbers will lead to significant deviations from uniform distribution, at least in some dimensions d. Clearly, the restricted type of d-tuples being considered here cannot ensure against all kinds of correlations. In particular, this will be no guarantee against *long-range* correlations among the numbers x_n, as they may appear with the splitting technique in multitasking. For this topic, we refer the reader to DeMatteis and Pagnutti[DP90]. In rare cases, more general types of d-tuples have been studied in the literature. We refer the reader to Entacher[Ent98, Ent97a] and L'Ecuyer[L'E97a] for results on linear generators and to Eichenauer-Herrmann[EH93] and Niederreiter[Nie94a, Nie95] for explicit-inversive generators (EICGs).

It has turned out that the behavior of *full-period* sequences ω (i.e., N equals the period of the RNG) with respect to theoretical figures of merit allows very reliable predictions of the performance of the random numbers x_n themselves in empirical tests. If the full-period point set ω is well-behaved with respect to certain figures of merit in various dimensions d, then good empirical performance of the samples is highly probable. Practical evidence tells us that many target distributions will be simulated very well, see, for example, the empirical results in [FM82, L'E92, LW97]. This relation between properties of the full-period sequences in higher dimensions and the behavior of relatively small samples in low dimensions has not yet been studied in detail.

In correlation analysis there are two schools of thought. Both schools justify their approach by heuristic arguments, mainly derived from results of probability theory. In both cases, these concepts yield very good RNGs.

School I searches for reliable RNGs by choosing the parameters for a given type of generator in the following way. If the sets ω have a grid structure, then the *spectral test*, due to Coveyou and MacPherson[CM67], allows to measure the coarseness of this grid. Different parameters for the RNG will lead to different streams of random numbers in $[0, 1[$ and, hence, to different point sets $\omega = \omega(d)$ and grids in dimension d. Among all possible parameters, we select those with the best spectral test values for the lattices that underlie these grids. The selection criterion is based on the performance of the sets ω in a whole range of dimensions d. This procedure will yield some kind of "super-uniformity" for the sets ω in all dimensions d that have been scanned. At this point, it was not clear if such a procedure will really select those sets ω in $[0, 1[^d$ where the empirical distribution function of these sets approximates uniform distribution as closely as possible. In other words, it was unknown if the spectral test is a measure of uniform distribution like diaphony or discrepancy. We will answer this question in Chapter 5.

The spectral test quantity has a nice geometric interpretation, see Subsection 5.4. It computes the maximal distance between successive parallel hyperplanes covering all possible points $\mathbf{x}_n$ that the generator produces. This interpretation leads to very efficient algorithms to compute the value of the spectral test. We refer the reader to [Die75, Knu81, Rip87, Fis96, LC97, L'E97b, Tez95] for details.

As an illustration of this approach, we plot overlapping pairs (x_n, x_{n+1}) and triples (x_n, x_{n+1}, x_{n+2}) of LCG(2^{31}, 65539, 0), called RANDU in the literature. We observe that in dimension $d = 2$ this RNG produces a relatively fine grid (in the plot, note the magnification), whereas in dimension $d = 3$ it exhibits its well-known catastrophic correlations. All possible points happen to lie on only 15 planes in $[0, 1[^3$, which is unacceptably low. RANDU should not be used in serious stochastic simulations.

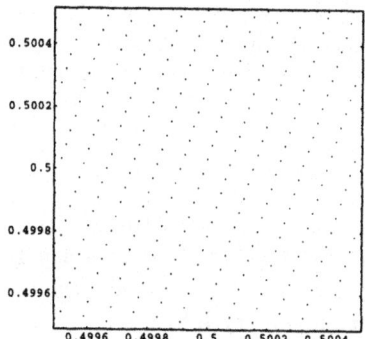

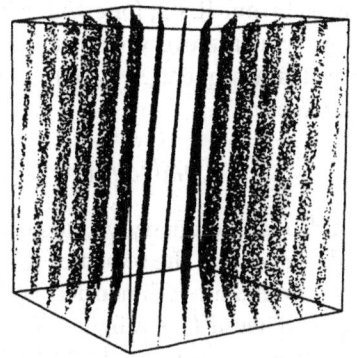

Figure 1.1
Zoom into the Unit Square

Figure 1.2
The 15 Planes in the Unit Cube

This example shows that, for a given RNG, it is not enough to compute spectral test values in only one dimension d. In order to check the correlations, we have to scan the lattices being involved here for a whole range of dimensions. Typical values are $2 \leq d \leq 30$, where the upper bound will depend upon the time required for the computations, see L'Ecuyer and Couture[LC97] and L'Ecuyer[L'E98] for recent results.

If we use samples of a linear RNG where the generator has been selected with the spectral test and if the sample size happens to be considerably larger than the square root of the generator's period, then the samples will be too regular for many simulations. The RNG might loose its ability to randomize the test statistic. For convincing numerical evidence, we refer to the author's paper [Hel95c], to Leeb and Wegenkittl[LW97] and Wegenkittl[Weg95] for the overlapping serial test, and to L'Ecuyer[LCS96] for the nearest-pair test. In the latter test, LCGs with worse spectral test values may perform better, depending on the size of the samples that we use. This relation between the period length and the maximal usable sample size is studied in some detail in the joint contribution of L'Ecuyer and the author in this volume.

Sometimes *Beyer quotients* are used instead of the spectral test to assess lattice structures. This quantity is well-defined only in dimension $d \leq 6$. Above this dimension, there might be two different Minkowski-reduced lattice bases generating the same lattice but having different Beyer quotients. Hence, at the present state of knowledge, any computations of Beyer quotients in dimensions beyond $d = 6$ will be difficult to justify, see Leeb[Lee95, Section 4.3.1] for further information.

In the case of RNGs that produce a sequence of binary words, like generalized feedback shift-register generators (GFSR), see Niederreiter[Nie92b], and [Nie95] for surveys, there is another interesting figure of merit, the notion of *d-distribution to l-bit accuracy*.

Suppose that $(w_n)_{n \geq 0}$ is the sequence of binary words of length L that is produced by our RNG, for example a GFSR. This sequence will be periodic. We now truncate every word w_n and take only the leading l bits, $l \leq L$. This yields a sequence $(w_n(l))_{n \geq 0}$ of l-bit words. Next, we construct the $d \cdot l$-bit vectors

$$(w_n(l), w_{n+1}(l), \ldots, w_{n+d-1}(l)), \quad n \geq 0.$$

If all 2^{dl} possible binary vectors occur the same number of times within a period, the we call the sequence $(w_n)_{n \geq 0}$ d-*distributed to* l-*bit accuracy*. Let $d(l)$ be the maximal dimension with l-bit accuracy for the given sequence $(w_n)_{n \geq 0}$. Then $d(l)$ is a figure of merit for our binary RNG.

For example, the twisted GFSR (tGFSR) generator called TT800 of Matsumoto and Kurita[MK92, MK94] has $d(16) = 50$ and $d(32) = 25$, and the Mersenne Twister MT19937[MN98], a tGFSR with the incredible period length $2^{19937} - 1$, has $d(32) = 623$.

The concept of d-distribution to l-bit accuracy is closely related to the notion of (t, m, s)-nets, which are a fundamental notion of LDPs in quasi-Monte Carlo. We refer the reader to Niederreiter[Nie92b] for background information. In both concepts, the goal is to construct super-uniform point sets in the highest possible dimension.

The super-uniformity approach of School I leads to fast and reliable RNGs. If the full-period point sets ω in $[0, 1[^d$ have a grid structure, then we should not consider this as a deficiency of the RNG. This property allows us to compute a powerful figure of merit, the spectral test. It is a measure for the coarseness of the lattice that lies behind the grid structure. Of course, there will be cases where these linear structures will interfere with the simulation and bias the results, like in the nearest-pair test. In such a situation, we may switch to nonlinear RNGs without a lattice structure like ICGs with prime modulus (see below).

2.3   Correlation Analysis for RNGs II

School II of correlation analysis proposes RNGs where the empirical distribution function associated with ω does not approximate uniform distribution on $[0, 1[^d$ "too well". The two-sided Kolmogorov-Smirnov test statistic is used to measure the distance between these two distribution functions. The law of the iterated logarithm tells us that, in the case of N realizations of independent, identically uniformly on $[0, 1]^d$ distributed random variables, this statistic will have an order of magnitude close to $1/\sqrt{N}$, see [Kie61, Nie95]. We will call this kind of uniformity of point sets $\omega = (\mathbf{x}_n)_{n=0}^{N-1}$ "random-uniformity". Here, an asymptotic result (i.e. N tends to infinity) is employed to propose conditions for the finite point sets ω (i.e. N is fixed) that we deal with in practice. Nevertheless, this heuristic approach works very well. We refer the reader to the author's survey [Hel95c] of particular types of nonlinear generators. In this paper, inversive RNGs that have been constructed with respect to this criterion are compared to the best LCGs with the same period length.

In metric number theory, the two-sided Kolmogorov-Smirnov test statistic is known under the name "star-discrepancy". A powerful number-theoretic method to estimate discrepancy by exponential sums is available due to Niederreiter[Nie78, Nie92b, Nie95], see Chapter 7 for details. Niederreiter's technique has allowed to assess most types of RNGs by discrepancy.

The random-uniformity approach of School II leads to nonlinear RNGs where even very large samples are sufficiently irregular to simulate many test statistics whereas linear generators with the same period length will usually fail. The usable sample sizes are considerably larger than for linear methods. Of course, the choice of a linear RNG with a much longer period will have the same effect. A more important advantage of many types of nonlinear RNGs is the absence of any lattice structure that might interfere with the simulation. As an example, let us consider $ICG(2^{31} - 1, 9102, 36884165, 1)$. The zoom into all possible points (x_n, x_{n+1}), $0 \leq n < 2^{31} - 1$, in the unit interval shows the typical behavior of inversive RNGs.

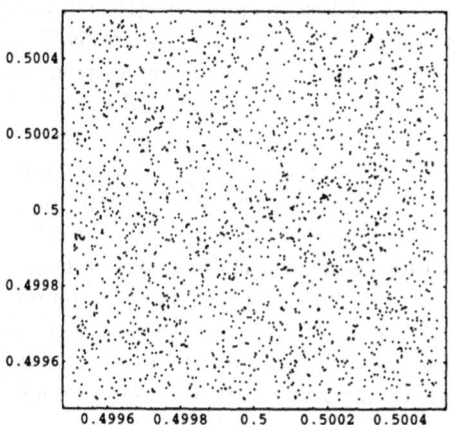

Figure 2
The Point Structure of $ICG(2^{31} - 1, 9102, 36884165, 1)$

ICGs also possess their regularities. They are of a nonlinear type and less prone to interfere with the simulation. There is one disadvantage of inversive generators in simulation practice. They are considerably slower than linear RNGs. We refer the reader to the author's surveys [Hel95c, Hel98], to Leeb and Wegenkittl[LW97] for empirical results, in comparison to LCGs, and to Niederreiter[Nie92b, Nie95] and Eichenauer-Herrmann et al.[EHHW97] for comprehensive reviews of the theoretical background and the intrinsic structure of nonlinear, in particular inversive algorithms.

Both discrepancy and the spectral test have their advantages and deficiencies. The spectral test can be computed efficiently, provided that the point set ω has a lattice structure. This will be the case for all linear methods for random number generation, where ω is a full-period point set. Discrepancy is a more general concept. It is not limited to point sets with a lattice structure. It is well-defined for arbitrary point sets ω in $[0, 1[^d$. However, in practice it is impossible to compute its value. The computational complexity for N points in dimension d is $\mathcal{O}(N^d)$. There are upper and lower bounds for discrepancy available, due to Niederreiter[Nie92b, Nie95]. Recently, a probabilistic algorithm to compute discrepancy has been presented by Winker and Fang[WF95], but empirical evidence is not yet available.

For both figures of merit, their distribution in dimension $d \geq 2$ was not known. Therefore, we could not design an empirical test for random number generators from this quantities. In a forthcoming paper, Leeb and the author[LH98] will present an extension of the concept of the spectral test and the asymptotic distribution of this quantity in arbitrary dimensions.

The *weighted spectral test* introduced by the author in [Hel95a, Hel97] is a another concept to assess the equidistribution of arbitrary point sets ω in $[0, 1[^d$. At present, there are two numerical quantities that realize this concept. They are called *diaphony*, see Hellekalek and Niederreiter[HN98], and *dyadic diaphony*, see Hellekalek and Leeb[HL97]. Diaphony was originally introduced into number-theoretical numerics by Zinterhof[Zin76], who used this quantity to estimate the integration error in certain quasi-Monte Carlo integration methods. Both versions of diaphony are closely related to discrepancy. As the latter, they can be estimated from above and below, but it takes only $\mathcal{O}(d \cdot N^2)$ steps to compute their values for an arbitrary set $\omega = (\mathbf{x}_n)_{n=0}^{N-1}$ in dimension d. The weighted spectral test has an interesting interpretation in numerical integration. It was shown that it is the mean square integration error for certain classes of integrands. Further, the asymptotic distribution of diaphony has also been found. Diaphony allows to carry out the same kind of theoretical research as discrepancy, which is pure number theory. In addition, it also allows to assess practical point sets and to compare the empirical distribution of samples to the theoretical distribution. We refer to Chapter 6 for details.

In Chapter 5 we present a very general form of the spectral test and indicate how this concept and the weighted spectral test are related.

2.4   Theory vs. Practice I: Leap-Frog Streams

Random number generation on parallel processors requires great care. We will show by examples how theoretical analysis of RNGs is able to provide valuable assistance for practice.

There are two widely used parallelization strategies. Strategy I assigns different RNGs to different processors. Strategy II uses different substreams of one large RNG on different processors. The first strategy carries the risk that there might be unknown correlations between the different RNGs. Using the same type of RNG but with different parameters might lead to unpleasant surprises. For certain nonlinear generators like the explicit-inversive congruential generator strong theoretical support on uncorrelatedness is available, see Niederreiter[Nie92a, Nie94a] and Eichenauer-Herrmann and Niederreiter[EHN94].

Strategy II is more common. It can be controlled better, but its risks should not be neglected. It comes in two variations. Method II.a, the "leapfrog" technique with parameter $L \in \mathbb{N}$, assigns the substream $(x_{nL+j})_{n\geq 0}$ to the j-th processor, $0 \leq j \leq L - 1$. In other words, we use substreams of lag L of the original sequence $(x_n)_{n\geq 0}$.

Method II.b, the "splitting technique", partitions the original sequence into L (very long) consecutive blocks. To each processor, we assign a different block. These blocks are defined by unique seeds. Splitting is particularly easy for linear types of RNGs, see L'Ecuyer[LC91, LA97]. It has to be used with caution, as there might be strong long-range correlations.

As a first example of what may happen with these parallelization techniques, we study lagged subsequences of the good linear congruential generator $LCG(2^{48}, 55151000561141, 0)$ which was proposed by Fishman[Fis96]. We will analyze the leap-frog subsequence $(x_{23n})_{n\geq 0}$ with lag $L = 23$ by a theoretical and an empirical test and compare the results to those of the original sequence. This subsequence is produced by an LCG with the same modulus and the same period length as the original generator.

The *normalized* spectral test values σ of the two RNGs are given below. Values close to one mean good performance in the spectral test. In the two tables below, d denotes the dimension and σ the normalized spectral test value.

Table 1
Spectral Test Values, Original Sequence

d	2	3	4	5	6	7	8
σ	0.9246	0.8170	0.9240	0.8278	0.8394	0.6274	0.4471

Table 2
Spectral Test Values, Subsequence with Lag $L = 23$

d	2	3	4	5	6	7	8
σ	0.2562	0.0600	0.0114	0.0462	0.1275	0.2031	0.2077

We observe a marked decrease in the lattices' quality in dimension $d \geq 3$. The values in dimension 3, 4, and 5 are unacceptably low. We will now compare these theoretical findings to some empirical results. It will be interesting to see how the empirical performance of the samples of this RNG is related to the theoretical prediction.

The *load test* is a particular version of the overlapping serial test or m-tuple test of Marsaglia[Mar85]. We use the setup of Wegenkittl[Weg95]. The test statistic is the following. We consider overlapping d-tuples $(x_n, x_{n+1}, \ldots, x_{n+d-1})$ of random numbers, $n \geq 0$, where the dimension d varies, $d \in \{1, \ldots, 5\}$. For every component of this vector, we consider the first four digits. Every such digit block represents an integer in the range $\{0, \ldots, 15\}$. We apply Marsaglia's[Mar85] m-tuple test to these d-tuples of integers and obtain one value of the test statistic for a given sample size N and a given dimension d. In our setup, the sample size varies between 2^{18} and 2^{26}, and, for each pair (N, d) of parameters, we compute 32 independent samples. In the Figures 3.1 and 4.1 below, the bar chart shows the value of the two-sided KS-test statistic applied to the empirical distribution of the upper-tail probabilities of these 32 values. On the axes, we indicate the dimension d, $d \in \{1, \ldots, 5\}$, and the dyadic logarithm of the sample size N, $N = 2^{18}$, $\ldots, 2^{26}$. For the 1% level of significance that we use here, the critical value is equal to 1.59. Dark grey bars signify rejection, i.e. values of the test statistic larger than the critical value. Our target distribution is $U[0, 1]$, i.e. uniform distribution on $[0, 1]$. The pattern plots in Figures 3.2 and 4.2 show the 32 values of the upper-tail probabilities in a grey scale. Irregular patterns should appear. If a box is white, then the sample distribution approximates the target distribution too evenly. This will be a consequence of super-uniformity. If a box is all black, this will indicate that the distance between the empirical distribution function and the cumulative distribution function of the target distribution is too large. In both of these extreme cases, the RNG under study fails the test. Its samples do not randomize the test statistic correctly.

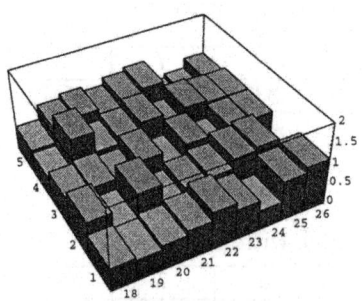

Figure 3.1
KS Values

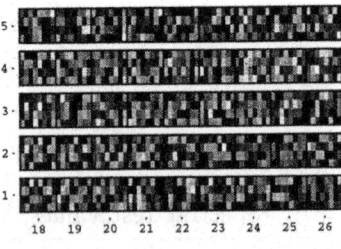

Figure 3.2
Upper Tail Probabilities

The original sequence of random numbers performs considerably better than the lagged sequence. The theoretical figures of merit, i.e. the spectral test values, have indicated that the lagged random numbers might fail in empirical tests, from dimension 3 upwards.

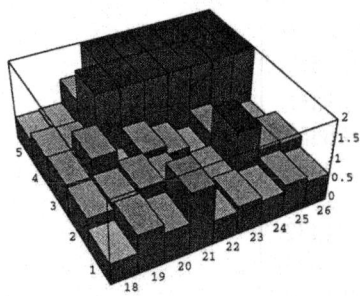

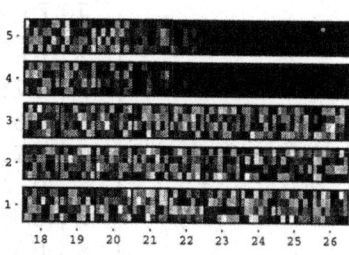

Figure 4.1
Lag 23: KS Values

Figure 4.2
Lag 23: Upper Tail Probabilities

We observe that, indeed, in dimensions 4 and 5, the prediction given by the spectral test was correct, the lagged sequence fails the empirical test. In dimension 3, the empirical test was unable to detect the bad lattice structure.

The situation we have encountered in this example is quite common. Theoretical figures of merit have a tendency to be much more stringent than empirical tests. If the figure of merit is bad, then this does not necessarily imply that the RNG should not be used at all. For small sample sizes, many simulations will not detect the deficiencies that the theoretical test has found. As we observe in dimensions 4 and 5, if the sample size increases, then the empirical test will begin to reject the samples. The bad distribution of the full-period point set begins to shine through and to affect the statistical test.

2.5 Theory vs. Practice II: Parallel Monte Carlo Integration

In our next example, we will study the effects of a recommended parallelization strategy in a Monte Carlo integration. We will use the CRAY system generator Ranf, which is LCG(2^{48}, 44485709377909, 0, 1). CRAY systems use the lags $L = 64$ or $L = 128$ for vectorization and parallelization.

First, we compare the normalized spectral test values of the CRAY generator to those of the leap-frog subsequence with lag 128. We observe rather small values for the leap-frog substream in higher dimensions. This means that the maximal distances between adjacent parallel hyperplanes are large. As a consequence, the spectral test predicts bad empirical behavior in dimensions $d \geq 6$.

Table 3
Spectral Test for CRAY RNG and Lag 128

d	CRAY	Lag 128
2	0.8269	0.2917
3	0.7416	0.8355
4	0.3983	0.6459
5	0.7307	0.2439
6	0.6177	0.1359
7	0.6670	0.1432
8	0.5642	0.1524
9	0.6952	0.1590
10	0.6730	0.1986

For a given test function $f : [0,1]^d \to \mathbb{R}$ and a given sequence $\omega = (x_n)_{n \geq 0}$ of integration nodes in the d-dimensional unit cube, we will study the integration error

$$R_N(f, \omega) := \int_{[0,1]^d} f(\mathbf{x})\, d\mathbf{x} - \frac{1}{N} \sum_{n=0}^{N-1} f(\mathbf{x}_n) \tag{1.1}$$

In the Monte Carlo method, $R_N(f, \omega)$ is treated as a random variable that depends on i.i.d. random variables $\mathbf{X}_n$, $n \geq 0$, where each $\mathbf{X}_n$ is supposed to be uniformly distributed on $[0,1[^d$. The numerical vectors $\mathbf{x}_n$ that are used in (1.1) are interpreted as realizations of the random variables $\mathbf{X}_n$. We derive from the Strong Law of Large Numbers that the integration error $R_N(f, \omega)$ will tend to zero almost surely if the sample size N increases to infinity. Of course, this probabilistic result will tell us nothing about the integration error that is produced by a particular node set ω.

Our test function will be the polynomial

$$f(\mathbf{x}) := \prod_{i=1}^{d} g(x_i), \quad \mathbf{x} = (x_1, \ldots, x_d) \in [0,1[^d,$$

where

$$g(x) := x^5 - 1/6, \quad x \in [0,1[,$$

and the dimension d will be equal to eight. The functions g and f have zero integral.

In a statistical setup that follows Entacher, Uhl and Wegenkittl[EUW98], we generate 64 independent samples $\omega_k = (\mathbf{x}_n^{(k)})_{n=0}^{N-1}$, $1 \leq k \leq 64$, in $[0,1[^d$ and compute the 64 associated integration errors $R_N(f, \omega_k)$, $1 \leq k \leq 64$. Let

$$\hat{\mu} := \frac{1}{64} \sum_{k=1}^{64} R_N(f, \omega_k) \tag{1.2}$$

denote the mean error and

$$\hat{\sigma}^2 := \frac{1}{63} \sum_{k=1}^{64} \left(R_N(f,\omega_k) - \hat{\mu} \right)^2 \tag{1.3}$$

the sample variance. We then compute the 99% confidence interval for these values by Student's t-distribution. This interval is given by

$$]\hat{\mu} - 0.332\hat{\sigma}, \hat{\mu} + 0.332\hat{\sigma}[,$$

where $0.332 = 1/\sqrt{64}\, t_{63,0.995}$. Here, the number $t_{63,0.995}$ denotes the 0.005-quantile of Student's t-distribution with 63 degrees of freedom.

We will now describe how the node sets are constructed. With the CRAY generator, we produce the first consecutive $d \cdot N$ random numbers x_n in $[0,1[$ and construct N nonoverlapping d-tuples $\mathbf{x}_n = (x_{nd}, \ldots, x_{nd+d-1})$ in $[0,1[^d$, $0 \le n < N$. This yields the first set $\omega_1 := (\mathbf{x}_n^{(1)})_{n=0}^{N-1}$ of integration nodes and the first value $R_N(f,\omega_1)$ of the integration error. Then, we take the next $d \cdot N$ random numbers x_n, construct the associated N points in the d-dimensional unit cube and obtain the second integration error $R_N(f,\omega_2)$. Altogether, we repeat this procedure 64 times to get the integration errors $R_N(f,\omega_k)$, $1 \le k \le 64$. From this sample of 64 values, the mean error $\hat{\mu}_N$, the sample variance $\hat{\sigma}_N^2$, and the bounds for the confidence interval are computed, see (1.2) and (1.3).

In Figure 5 below, the ticks on the abscissa represent the dual logarithm of N. The sample size N is in the range between 2^{20} and 2^{27}. The associated value of $\hat{\mu}_N$ is given on the ordinate by the dotted line. The shaded area indicates the 99% confidence interval. The horizontal line at level 0 represents the expected value in this experiment.

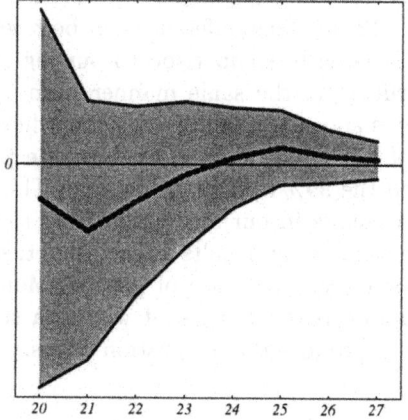

Figure 5
Integration Error, CRAY Generator

We observe that the mean error is simulated correctly. In all cases, the 99% confidence interval contains the true value 0. The CRAY generator performs as it may be expected in such a harmless Monte Carlo integration.

Even in such a simple application, leap-frog must be done with care. The CRAY system generator is very sensible to the lag $L = 128$, as we will show now. In the case of the original CRAY generator, we have used 64 samples ω_k, $1 \leq k \leq 64$, each of size N. Every sample has consumed $d \cdot N$ consecutive random numbers to produce N points in dimension d. Now, we will produce 64 samples ω_k whose size $\tilde{N}$ is comparable to N, but each sample will be the union of 128 point sets that stem from 128 leap-frog subsequences with lag $L = 128$. We will observe that these 64 samples perform much worse than the original CRAY samples although they do not stem from one single leap-frog sequence but are just unions of such subsequences. Even in this "mild" leap-frog case, we get incorrect results in our simulation. The situation would have been much worse if we had compared the samples of the first setup to node sets of similar size that were constructed from *just one* leap-frog sequence and not from a union of 128 such subsequences, see Figure 7 below.

We construct the node set in the following way. We produce the first block of $d \cdot N$ random numbers x_n in $[0, 1[$ with the CRAY generator as before. Now, in the first block, we take the leap-frog random numbers with lag $L = 128$ and index $j = 0$. This yields the sequence x_0, x_{128}, x_{256}, $\ldots$, i.e. the numbers $x_{n \cdot 128}$, $0 \leq n < N'$, $N' := \lfloor d \cdot N/128 \rfloor$. From these N' numbers, we construct $N'' := \lfloor N'/d \rfloor$ nonoverlapping d-tuples $\mathbf{x}_n$ in $[0, 1[^d$. Then, in the next step, we use the leap-frog sequence $x_{n \cdot 128+1}$, $0 \leq n < N'$, in the same manner to construct another N'' points in $[0, 1[^d$, then the subsequence $x_{n \cdot 128+2}$, $0 \leq n < N'$ and so on. Finally, we unite all these 128 point sets of size N'' in one big point set ω_1. The set ω_1 will have $\tilde{N} = 128 \cdot N''$ elements with $\tilde{N} \leq N$. It is elementary to see that $N - \tilde{N} < 128 - 128/d$. This difference can be neglected in numerical computations if N is large in comparison to our lag $L = 128$. The next sample ω_2 is constructed in the same manner from the second block of $d \cdot N$ random numbers and so on, until we have produced 64 node sets ω_k, $1 \leq k \leq 64$, each of size $\tilde{N}$. As before, we compute the mean error, the sample variance, and the 99% confidence interval. This time, the samples give strongly biased results in our simulation from sample size $N = 2^{25}$ onwards. Visibly, the simulation results are off target, see Figure 6.

This example was a very mild case of parallel Monte Carlo. Figure 7 shows that the situation is much worse if we use a single leap-frog subsequence of lag 128 to produce the integration nodes.

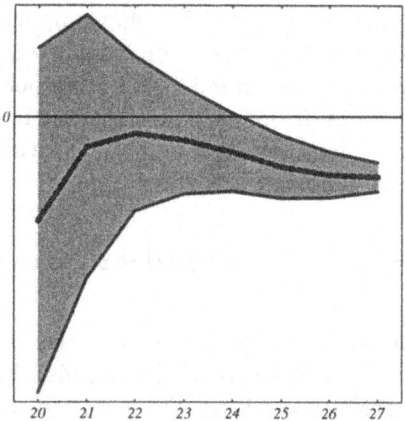

Figure 6
Integration Error, CRAY Generator, Lag 128, Leap-Frog Union

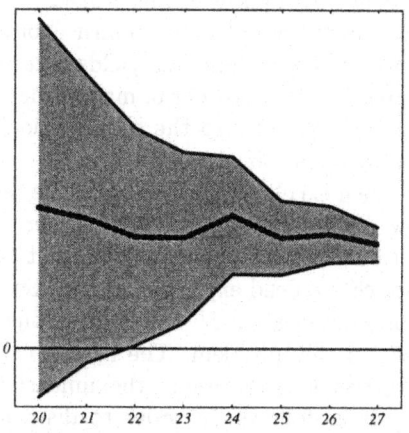

Figure 7
Integration Error, CRAY Generator, Lag 128, Single Leap-Frog

2.6 Assessing LDPs

Numerical practice like option pricing leads to questions like the following
(see the contribution of Tezuka in this volume). Let $f : [0,1]^d \to \mathbb{R}$ be an
integrable function where we are unable to compute the integral directly.
Suppose further that the Monte Carlo approximations

$$S_N(f,\omega) = \frac{1}{N} \sum_{n=0}^{N-1} f(\mathbf{x}_n), \quad \omega = (\mathbf{x}_n)_{n=0}^{N-1}, \tag{1.4}$$

with their rate $\mathcal{O}(N^{-1/2})$ converge too slowly for our purposes.

How can we solve this problem without changing our simulation model represented by the function f? One solution would be to buy a faster computer. A better solution is to switch in (1.4) to other point sets ω. Number-theoretic numerics provides the following background information. Let $\omega = (\mathbf{x}_n)_{n=0}^{N-1}$ be an arbitrary finite sequence in $[0, 1[^d$. The absolute value of the integration error can be bounded as follows:

$$\left| S_N(f, \omega) - \int_{[0,1]^d} f(\mathbf{x}) \, d\mathbf{x} \right| \leq v(f) \, \delta_N(\omega), \qquad (1.5)$$

where $v(f)$ is a quantity that depends only on the "smoothness" of the integrand f and $\delta_N(\omega)$ is a measure of the equidistribution of ω. In this inequality, the associated keywords are the *variation* of f and the *discrepancy* of ω. The classical result in this context is the inequality of Koksma-Hlawka, see Kuipers and Niederreiter[KN74] and Sloan and Joe[SJ94] for further reading.

At this point the concept of quasi-Monte Carlo comes into play. We construct finite point sets ω with figures of merit $\delta_N(\omega)$ as small as possible, yielding super-uniform point sets. The standard choice for $\delta_N(\omega)$ is discrepancy. The quasi-Monte Carlo approach yields *deterministic* bounds for the integration error $R_N(f, \omega)$ with order of magnitude $N^{-1}(\log N)^{d-1}$, for large classes of functions. We refer to the contribution of Larcher in this volume for additional information.

In order to achieve such excellent error bounds, the sequence ω will have to be an LDP, a low discrepancy point set. In other words, ω has to be super-uniformly distributed in the d-dimensional unit cube $[0, 1[^d$.

For numerical practice, we need explicit construction methods for LDPs, for a broad range of sample sizes N and dimensions d. There are two important approaches to this problem. The first is called the concept of (t, m, s)-nets. In this notation, s represents the dimension, which we denote by d. The basic idea is as follows. In order to distribute the points $\mathbf{x}_n$, $0 \leq n < N$, of the finite sequence ω as evenly as possible, we try to put the same number of points in every "cell" of $[0, 1[^d$ of the same size. There are only finitely many points available here, hence this procedure of distributing points uniformly will have to stop at a certain cell size. For smaller cells, irregularities in the distribution will be impossible to avoid. Sobol' [Sob67, Sob69] was the first to mold this heuristic concept into mathematical form. He used cells that were elementary dyadic intervals in the d-dimensional unit cube (for this notion see Chapter 3). Sobol's method has been extended considerably by Niederreiter[Nie92b]. Niederreiter was able to find explicit construction methods for what he called (t, m, s)-nets, see [Nie87, Nie88, Nie92b] and the contributions of Niederreiter and Xing and Larcher in this volume. Further, by deep number-theoretic methods, he was able to give upper bounds for the maximal difference between the relative number of points $\mathbf{x}_n$ in a cell and the cell volume. This quantity is called the discrepancy of the sequence ω, see Chapter 7.

2.7 Good Lattice Points

The second important concept for LDPs is called the concept of "good lattice points (GLPs)". It has been introduced by Korobov[Kor59, Kor63]. For a given integer $N \geq 2$ and a given dimension d, an integer point $\mathbf{g} = (g_1, \ldots, g_d) \in \mathbb{Z}^d$ is called a *good lattice point modulo N* if the finite sequence

$$w_{\mathbf{g}} = \left(\left\{ \frac{n}{N} \mathbf{g} \right\} \right)_{n=0}^{N-1}$$

in the d-dimensional unit cube $[0, 1[^d$ is "very uniformly" distributed in the sense that its discrepancy is low. Here, $\{y\}$ denotes the fractional part of the number (or vector) y.

There exist several theoretical figures of merit to prove the existence of good lattice points $\mathbf{g}$ for prime moduli N and to search for them. Some of these quality measures are only suitable for existence proofs, while others happen to be just the quadrature error for a particular function.

For practical reasons, the search for good lattice points in dimension d is restricted to lattice points of the *Korobov type*:

$$\mathbf{g} = (1, a, a^2, \ldots, a^{d-1}),$$

where a is an integer, $1 < a < N$.

The common method to obtain tables of good lattice points $\mathbf{g}$ is to minimize the odd-looking quantity $P_\alpha(\mathbf{g}, N)$, $\alpha > 1$,

$$P_\alpha(\mathbf{g}, N) := \sum_{\substack{\mathbf{k} \neq 0: \\ \mathbf{k} \cdot \mathbf{g} \equiv 0 \bmod N}} \frac{1}{r(\mathbf{k})^\alpha}, \tag{1.6}$$

where summation is over all $\mathbf{k} \in \mathbb{Z}^d \setminus \{0\}$, and

$$r(\mathbf{k}) := \prod_{i=1}^{d} \max\{1, |k_i|\}, \quad \mathbf{k} = (k_1, \ldots, k_d) \in \mathbb{Z}^d,$$

see [Nie92b, page 103] and [SJ94, page 72]. In fact, this quantity has a practical background. For even integers α, $P_\alpha(\mathbf{g}, N)$ can be computed easily. It is the quadrature error $S_N(f_\alpha, \omega_{\mathbf{g}}) - I(f_\alpha)$, where $I(f_\alpha)$ denotes the integral of f_α, for a certain function f_α constructed from the α-th Bernoulli polynomial B_α:

$$f_\alpha(\mathbf{x}) := \prod_{i=1}^{d} \left(1 - (-1)^{\alpha/2} \frac{(2\pi)^\alpha B_\alpha(x_i)}{\alpha!} \right), \quad \mathbf{x} = (x_1, \ldots, x_d) \in [0, 1[^d.$$

Available tables of GLPs were computed by minimizing the "cost function" $P_2(\mathbf{g}, N)$, see [Nie92b, p.145] and [SJ94, p.73] for details and references. Certainly, it would be preferable to search for good lattice points with figures of merit independent of such regularity parameters α, like the quantity

$$\rho(\mathbf{g}, N) := \min \left\{ r(\mathbf{k}) : \mathbf{k} \in \mathbb{Z}^d \setminus \{\mathbf{0}\}, \ \mathbf{k} \cdot \mathbf{g} \equiv 0 \bmod N \right\}. \tag{1.7}$$

This is one of the most important figures of merit in the theory of GLPs, but it is to complex to be computed in practice. We refer the reader to Entacher, Hellekalek and L'Ecuyer[EHL98] for a new approach to this problem and, in particular, to Chapter 5.

As we will see in Chapter 6, Remark 6.2, $P_2(\mathbf{g}, N)$ is just the diaphony of the sequence $\omega = (\{(n/N)\mathbf{g}\})_{n=0}^{N-1}$. Minimizing $P_2(\mathbf{g}, N)$ means minimizing the diaphony of the point set $\omega_{\mathbf{g}}$. Diaphony is a measure of the uniform distribution of sequences, similar to discrepancy. It is much more appropriate for empirical studies than the latter. We will discuss diaphony in Chapter 6. In view of this rich theoretical background, the choice of $P_2(\mathbf{g}, N)$ as a figure of merit for good lattice points appears much less arbitrary than before.

2.8   GLPs versus (t, m, s)-Nets

It is an interesting question to compare some of the best available (t, m, s)-nets to the classical type of node sets generated by GLPs. These nets have been designed with considerable mathematical effort. The aim was to keep the related discrepancy estimates as small as possible. In comparison, for GLPs much less theory is involved. We simply search for integer lattice points $(1, a, a^2, \ldots, a^{d-1})$ where the spectral test

$$\sigma(\mathbf{g}, N) := \min \left\{ \|\mathbf{k}\|_2 : \mathbf{k} \in \mathbb{Z}^d \setminus \{\mathbf{0}\}, \ \mathbf{k} \cdot \mathbf{g} \equiv 0 \bmod N \right\}, \tag{1.8}$$

is large, see Entacher, Hellekalek, and L'Ecuyer[EHL98].

Larcher et al.[LT94, LSW94, LSW96] have shown that integrands with fast-converging Walsh series are particularly suited for quasi-Monte Carlo integration with (t, m, s)-nets in an appropriate basis. For this reason, we have compared the integration error for the best implemented (t, m, s)-nets and for GLPs in Entacher, Hellekalek, and L'Ecuyer[EHL98]. The nets were taken from Bratley, Fox, and Niederreiter[BFN92] and Niederreiter[Nie88]. We have used the spectral test to find good lattice points modulo N for various sample sizes N, see the tables published in L'Ecuyer[L'E98]. Our integrands were of the form

$$f(\mathbf{x}) = \prod_{i=1}^{d} g(x_i), \quad \mathbf{x} = (x_1, \ldots, x_d) \in [0, 1[^d,$$

where
$$g(x) = x^r - \frac{1}{r+1}.$$

The regularity parameter r was varied between 1,2, and 3. For larger values of r, the peaks of the function f become very prominent and suggest the use of adaptive methods instead of LDPs.

Our empirical results did not detect any superiority of (t, m, s)-nets, quite to the contrary. This fact came somewhat unexpected because the integrands were supposed to be particularly suitable for numerical quadrature with such nets, see the contribution of Larcher in this volume. Further, our integrands were extremely well-behaved with respect to their Haar series development, which also suggests the use of (t, m, s)-nets for numerical integration (see Entacher[Ent97b]).

2.9   Conclusion

The main difference between RNG and LDP algorithms is the fact that an LDP algorithm is used for much more specific purposes than an RNG. The designer of an LDP algorithm knows (from his theoretical studies) for which dimensions d and which sample sizes N his method will work. The practitioner will be unable to bypass these limits unknowingly. With RNGs, the situation is completely different. An RNG has to be a *general purpose* algorithm. The designer does not know what operations the practitioner will perform with his random numbers or points. He does not know how these numbers or points will be mixed up, which sample sizes will be used and in which dimensions the practitioner will work. He cannot foresee all possible cases of applications. Contrary to the case of LDPs, much less theoretical support like diaphony or discrepancy estimates is available. There are simply too many cases that would have to be treated, many of them still inaccessible to theoretical analysis. Further, empirical testing of RNGs can cover only a rather limited number of prototypical applications.

The assessments of RNGs and that of LDPs have several important points in common. In both cases, we measure the equidistribution of point sets in high dimension. Further, we design point sets with super-uniform distribution. We have several figures of merit for equidistribution at our disposition, the most prominent being diaphony, discrepancy, and the spectral test. In the case of good lattice points, discrepancy and the quantities ρ and P_2 are related by inequalities, see Niederreiter[Nie92b, Ch.5]. We shall exhibit in the latter sections that ρ and P_2 are special cases of general measures of equidistribution.

The spectral test σ has a similar meaning for RNGs as the quantities ρ and P_2 for LDPs. It allows to assess particular point sets, those with grid structure, but it was unknown if it also represents some measure of uniform distribution. We shall present our general concept of the spectral test in Chapter 5.

For the practitioner, the most important aspect of all these considerations is the fact that RNGs and LDPs that are designed with respect to these figures of merit perform very well in practice. The empirical evidence for the reliability of these a priori tests is convincing. RNGs and LDPs without such a theoretical background should be used with great caution.

The well-known generators that we have studied in our examples illustrate the interplay between a priori testing of RNGs (with the spectral test) and the empirical performance of samples. These were neither isolated nor pathological examples. The phenomena we have described in this section occur with many other RNGs and tests.

3   Mathematical Preliminaries

Throughout this paper, we shall identify the d-dimensional torus $(\mathbb{R}/\mathbb{Z})^d$, $d \geq 1$, with the half-open unit cube $[0,1[^d$. Normalized Haar measure on $(\mathbb{R}/\mathbb{Z})^d$, respectively Lebesgue measure on $[0,1[^d$, will be denoted by λ_d.

Let $\omega = (\mathbf{x}_n)_{n \geq 0}$ be a sequence in $[0,1[^d$ and let $f : [0,1[^d \to \mathbb{C}$. We define

$$S_N(f,\omega) := \frac{1}{N} \sum_{n=0}^{N-1} f(\mathbf{x}_n).$$

The inner product of two vectors $\mathbf{x}$ and $\mathbf{y}$ in $\mathbb{R}^d$, $\mathbf{x} = (x_1, \ldots, x_d)$, $\mathbf{y} = (y_1, \ldots, y_d)$, will be denoted by

$$\mathbf{x} \cdot \mathbf{y} = \sum_{i=1}^{d} x_i y_i.$$

In this section, we will introduce generalized Haar and Walsh function systems and some basic notions from the geometry of numbers.

3.1   Haar and Walsh Series

We will define Haar and Walsh function systems to arbitrary integer bases $q \geq 2$. In the classical case q equals 2. We refer the reader to the monograph of Schipp, Wade, Simon and Pál[SWS90] and to Golubov, Efimov, and Skvortsov[GES91] for further reading.

It is possible to extend these classes of functions to other systems of numeration, like the so-called Cantor series expansion of real numbers. We refer to [Hel84, Hel95b] for examples and references.

Notation (1) Let $q \geq 2$ be a fixed integer. For a nonnegative integer k, let

$$k = \sum_{j=0}^{\infty} k_j q^j, \quad k_j \in \{0, 1, \ldots, q-1\},$$

be the unique q-adic expansion of k in base q. Every number $x \in [0, 1[$ has a unique q-adic expansion

$$x = \sum_{j=0}^{\infty} x_j q^{-j-1}, \quad x_j \in \{0, 1, \ldots, q-1\},$$

under the condition that $x_j \neq q-1$ for infinitely many j. In the following, this uniqueness condition will be assumed without further notice. For $g \in \mathbb{N}$ we define

$$x(g) := 0.x_0 x_1 \ldots x_{g-1},$$

and

$$k(g) := \sum_{j=0}^{g-1} k_j q^j.$$

Then $x(g) \in \{aq^{-g} : 0 \leq a < q^g\}$ and $k(g) \in \{0, 1, \ldots, q^g - 1\}$. Further, put

$$x(0) := 0, \ k(0) := 0.$$

(2) An *elementary q-adic interval of length* q^{-g} is a subinterval of $[0, 1[$ of the form $[aq^{-g}, (a+1)q^{-g}[, \ 0 \leq a < q^g, \ g \geq 0$, a and g integers.

(3) Let $b_0, b_1, \ldots, b_{g-1}$ be arbitrary digits in $\{0, 1, \ldots, q-1\}$. The *cylinder set* $I(b_0, b_1, \ldots, b_{g-1})$ *of order g defined by* the digits $b_0, b_1, \ldots, b_{g-1}$ denotes the following set:

$$I(b_0, b_1, \ldots, b_{g-1}) := \{x \in [0, 1[: \ x_j = b_j, \quad \forall j : 0 \leq j < g\} .$$

Note that, for every elementary q-adic interval $I = [aq^{-g}, (a+1)q^{-g}[$ of length q^{-g}, $g \in \mathbb{N}$, there is a unique cylinder set $I(b_0, b_1, \ldots, b_{g-1})$ such that

$$I = I(b_0, b_1, \ldots, b_{g-1}).$$

We only have to observe that $aq^{-g} = 0.b_0 b_1 \ldots b_{g-1}$ with suitable digits b_j. The class of cylinder sets of order g is identical with the class of elementary q-adic intervals of length q^{-g}.

Definition 3.1 *Let* $\phi_0 : \mathbb{Z}_q \to \mathbf{K}$, $\mathbb{Z}_q = \{0, 1, \ldots, q-1\}$ *the least residue system modulo q*, $\mathbf{K} = \{z \in \mathbb{C} : |z| = 1\}$, *denote the function*

$$\phi_0(a) := e^{2\pi i a/q}, \quad a \in \mathbb{Z}_q.$$

The k-th Walsh function w_k, $k \geq 0$, *to the base q is defined as*

$$w_k(x) := \prod_{j=0}^{\infty} \phi_0(x_j k_j), \tag{1.9}$$

where $x = 0.x_0 x_1 \ldots$ *is the q-adic expansion of* $x \in [0, 1[$ *and the index k is represented by* $k = \sum_{j=0}^{\infty} k_j q^j$.

Definition 3.2 *The k-th Haar function h_k, $k \geq 0$, to the base q is defined as follows: If $k = 0$, then*

$$h_0(x) := 1 \quad \forall x \in [0, 1[.$$

If $k \geq 1$, $q^g \leq k < q^{g+1}$, with $g \geq 0$, then

$$h_k(x) := \begin{cases} q^{\frac{g}{2}} \phi_0(ak_g) & : \quad \text{if } x \in \left[\frac{k(g)}{q^g} + \frac{a}{q^{g+1}}, \frac{k(g)}{q^g} + \frac{a+1}{q^{g+1}} \right[, \\ 0 & : \quad \text{otherwise.} \end{cases}$$

The k-th normalized Haar function H_k on $[0, 1[$ is defined as $H_0 := h_0$, and, if $k \geq 1$, $q^g \leq k < q^{g+1}$, with $g \geq 0$, then

$$H_k := q^{-\frac{g}{2}} h_k.$$

The notion below of a "fundamental domain" will help us to illustrate the behavior of Haar functions.

Definition 3.3 *The fundamental domain D_k of the k-th Haar function h_k is defined as the following elementary q-adic interval :*
If $k = 0$, then
$$D_0 := [0, 1[.$$

If $k \geq 1$, $q^g \leq k < q^{g+1}$, with $g \geq 0$, then

$$D_k := \left[\frac{k(g)}{q^g}, \frac{k(g) + 1}{q^g} \right[.$$

Remark 3.1 *(Properties of the functions h_k and w_k)*
   (1) If k has the q-adic expansion $k = \sum_{j=0}^{g} k_j q^j$, with $k_g \neq 0$, then D_k is simply the cylinder set $I(k_{g-1}, \ldots, k_0)$.
   (2) The functions h_k and H_k are "concentrated" on D_k. They vanish outside this interval.
   (3) For all $k < q^g$, the functions h_k and w_k are constant on the elementary q-adic intervals of length q^{-g}.
   (4) There are exactly $q - 1$ Haar functions h_k with $q^g \leq k < q^{g+1}$, that share the same fundamental domain.
   (5) The Haar system to base q is an orthonormal base of $L^2([0, 1[, \lambda)$ (see, for example, [SWS90]).
   (6) If $0 \leq k < q$, then $h_k = w_k$.
   (7) $|H_k(x)| = 1 \quad \forall x \in D_k$.

Definition 3.4 *Let $\mathcal{H}(q) := \{ h_{\mathbf{k}} : \mathbf{k} = (k_1, k_2, \ldots, k_d), k_i \geq 0 \}$ denote the system of Haar functions to the base q on the d-dimensional torus $[0, 1[^d$. The $\mathbf{k}$-th Haar function $h_{\mathbf{k}}$ is defined as*

$$h_{\mathbf{k}}(\mathbf{x}) := \prod_{i=1}^{d} h_{k_i}(x_i) , \quad \mathbf{x} = (x_1, \ldots, x_d) \in [0, 1[^d.$$

In analogy to Definition 3.2, let

$$H_{\mathbf{k}}(\mathbf{x}) := \prod_{i=1}^{d} H_{k_i}(x_i) , \quad \mathbf{x} = (x_1, \ldots, x_d) \in [0, 1[^d,$$

denote the $\mathbf{k}$-th normalized Haar function on $[0, 1[^d$ and $\mathcal{H}_0(q) := \{H_{\mathbf{k}}\}$ the system of normalized Haar functions.

The fundamental domain $D_{\mathbf{k}}$ of the $\mathbf{k}$-th Haar function $h_{\mathbf{k}}$ on $[0, 1[^d$, where $\mathbf{k} = (k_1, \ldots, k_d)$, is defined as the elementary q-adic interval

$$D_{\mathbf{k}} := \prod_{i=1}^{d} D_{k_i}.$$

Definition 3.5 *Let $\mathcal{W}(q) := \{w_{\mathbf{k}} : \mathbf{k} = (k_1, k_2, \ldots, k_d), k_i \geq 0\}$ denote the system of Walsh functions to the base q on the d-dimensional torus $[0, 1[^d$. The $\mathbf{k}$-th Walsh function $w_{\mathbf{k}}$ is defined as*

$$w_{\mathbf{k}}(\mathbf{x}) := \prod_{i=1}^{d} w_{k_i}(x_i), \quad \mathbf{x} = (x_1, \ldots, x_d) \in [0, 1[^d.$$

Remark 3.2 *(Properties of the functions $h_{\mathbf{k}}$ and $w_{\mathbf{k}}$)*
It is obvious how to rewrite Remark 3.1 for the multidimensional case. Properties (2), (5) and (7) remain unchanged.

Notation In this paper, the symbol $\mathcal{F}$ will represent one of the following function systems on the d-dimensional torus $[0, 1[^d$:

1. $\mathcal{T} = \{e_{\mathbf{k}} : \mathbf{k} \in \mathbb{Z}^d\}$, the system of trigonometric functions,

$$e_{\mathbf{k}} := e^{2\pi\sqrt{-1}\mathbf{k}\cdot\mathbf{x}}, \quad \mathbf{x} \in [0, 1[^d,$$

2. $\mathcal{H}(q) = \{h_{\mathbf{k}} : \mathbf{k} = (k_1, \ldots, k_d), k_i \geq 0\}$, the system of Haar functions to the base q, $q \geq 2$,

3. $\mathcal{H}_0(q) = \{H_{\mathbf{k}} : \mathbf{k} = (k_1, \ldots, k_d), k_i \geq 0\}$, the system of normalized Haar functions to the base q, $q \geq 2$,

4. $\mathcal{W}(q) = \{w_{\mathbf{k}} : \mathbf{k} = (k_1, \ldots, k_d), k_i \geq 0\}$ the system of Walsh functions to the base q, $q \geq 2$.

Let $\mathcal{F} = \{\chi_{\mathbf{k}}\}$. The Fourier coefficients of an integrable function f defined on $[0, 1[^d$ will be denoted by $\hat{f}(\cdot)$,

$$\hat{f}(\mathbf{k}) := \int_{[0,1[^d} f(\mathbf{x}) \, \overline{\chi_{\mathbf{k}}(\mathbf{x})} \, d\mathbf{x} .$$

It will be clear from the context which function system we use.

One interesting and very helpful property of q-adic intervals in $[0,1[^d$ is the fact that their Haar and Walsh series are finite. Further, it is possible to give close estimates of the related Haar and Walsh coefficients. The following results were proved by the author in [Hel94b, Hel95b].

Lemma 3.3 *Let* $f(x) := 1_I(x) - \lambda(I)$, *where* I *is a subinterval of the interval* $[0,1[$. *Then the Haar coefficients of* f *may be estimated as follows:*

1. *If* I *is an elementary* q-adic interval of length $q^{-\alpha}$, then

$$\hat{f}(k) = 0 \quad \forall k \geq q^\alpha .$$

2. *If* $I = [0, \beta[$, $0 < \beta < 1$, then $\hat{f}(0) = 0$ and, for all k with $q^g \leq k < q^{g+1}$, $g \geq 0$,*

$$\overline{\hat{f}(k)} = \begin{cases} q^{g/2} \cdot \left(\dfrac{1}{q^{g+1}} \dfrac{\phi_0(k_g \cdot \beta_g) - 1}{\phi_0(k_g) - 1} + \right. \\ \left. \phi_0(k_g \cdot \beta_g) \, (\beta - \beta(g+1)) \right) & \text{if } k(g) = q^g \cdot \beta(g), \\ 0 & \text{otherwise.} \end{cases}$$

Lemma 3.4 *Let* $f(x) := 1_I(x) - \lambda(I)$, *where* $I = [0, \beta[, 0 < \beta < 1$. *Suppose that* $q^g \leq k < q^{g+1}$, *where* $g \geq 0$. *Then the Walsh coefficients of* f *may be estimated as follows.*

1. *If* I *is an elementary* q-adic interval of length $q^{-\alpha}$, then

$$\hat{f}(k) = 0 \quad \forall k \geq q^\alpha .$$

2. *The Walsh coefficient* $\hat{f}(k)$ *of* f *has the following value:*

$$\overline{\hat{f}(k)} = w_{k(g)}(\beta(g)) \left(\frac{1}{q^{g+1}} \frac{e^{2\pi i \frac{k_g}{q} \beta_g} - 1}{e^{2\pi i \frac{k_g}{q}} - 1} + e^{2\pi i \frac{k_g}{q} \beta_g} (\beta - \beta(g+1)) \right),$$

3. *The following estimate holds:*

$$|\hat{f}(k)| \leq \frac{1}{q^{g+1} \sin \pi \frac{k_g}{q}}$$

3.2   Integration Lattices

If we consider overlapping d-tuples $(x_n, x_{n+1}, \ldots, x_{n+d-1})$ of random numbers that are produced by linear algorithms like the LCG, then, in most cases, these point sets will have a grid structure. We refer to Figures 1.1 and 1.2 for an illustration. Similar geometric structures appear when we choose an LDP produced by a good lattice point for quasi-Monte Carlo integration.

Figures of merit to assess such structures are not defined for the point sets themselves but for the lattices that are generated by these grids. We refer the reader to the monographs of Niederreiter[Nie92b] and Sloan and Joe[SJ94] for details.

Definition 3.6 *Let* $g_1, \ldots, g_d$ *be d independent points in* $\mathbb{R}^d$. *The set L of all integer linear combinations of these vectors,*

$$L := \left\{ \sum_{i=1}^{d} t_i g_i : t_i \in \mathbb{Z} \right\}$$

is called the d-dimensional lattice generated by $g_1, \ldots, g_d$. *The set of points* $\{g_1, \ldots, g_d\}$ *is called a basis of the lattice L.*

A d-dimensional lattice that contains the set $\mathbb{Z}^d$ *as a subset is called an integration lattice.*

The basis of a lattice L need not be unique. Any unimodular transformation of $\{g_1, \ldots, g_d\}$ will yield another basis of L. Every discrete additive subgroup of $\mathbb{R}^d$ not contained in any proper linear subspace of $\mathbb{R}^d$ is a d-dimensional lattice and vice versa.

If we want to complete a subset of $e < d$ independent points of a d-dimensional lattice L to a basis, this will lead us to the notion of a *primitive* set, see Gruber and Lekkerkerker[GL87]. In particular, a single point $g \in L$, $g \neq 0$, is called a *primitive lattice point of L* if the line segment joining 0 and g does not contain any other lattice points than those two.

The dual of a lattice L gives us information about the hyperplanes that cover L. It is exactly this kind of information we are interested in when comparing the grid structures of generators like LCGs. Further, the dual lattice plays an important role in error estimates for lattice rules in quasi-Monte Carlo integration, see Sloan and Joe[SJ94].

Definition 3.7 *The dual lattice* $L^\perp$ *of a lattice L in* $\mathbb{R}^d$ *is the set*

$$L^\perp := \left\{ z \in \mathbb{R}^d : z \cdot x \in \mathbb{Z}, \ \forall x \in L \right\}.$$

Clearly, the dual lattice of an integration lattice will be a sublattice of $\mathbb{Z}^d$.

Definition 3.8 *A d-dimensional N-point lattice rule for the integral*

$$\int_{[0,1]^d} f(x) \, dx,$$

where the integrand f is continuous on the closed unit cube and periodic with period one in every component, is a mean of the form

$$S_N(f, \omega) = \frac{1}{N} \sum_{n=0}^{N-1} f(x_n),$$

where the finite sequence $\omega = (\mathbf{x}_n)_{n=0}^{N-1}$ consists of N distinct points in $[0, 1[^d$ and

$$\{\mathbf{x}_0, \ldots, \mathbf{x}_{N-1}\} = [0, 1[^d \cap L,$$

L some integration lattice. We will call ω the node set of this lattice rule.

Suppose that $\mathbf{g} = (g_1, \ldots, g_d)$ is a good lattice point modulo N (see Subsection 2.7). If we assume that

$$(g_i, N) = 1, \quad \forall i, 1 \leq i \leq d,$$

then it is easy to show that the set $\omega_{\mathbf{g}}$ is the node set of a d-dimensional N-point lattice rule. Further, it is elementary to prove that

$$L^{\perp} = \{\mathbf{k} \in \mathbb{Z}^d, \mathbf{k} \neq \mathbf{0} : \mathbf{k} \cdot \mathbf{g} \equiv 0 \bmod N\}. \tag{1.10}$$

Remark 3.5 If the points $\mathbf{x}_0, \ldots, \mathbf{x}_{N-1}$ are the node set of a d-dimensional N-point lattice rule L, then we have the relation

$$S_N(e_{\mathbf{k}}, \omega) = \begin{cases} 1 & : & \mathbf{k} \in L^{\perp} \\ 0 & : & \text{otherwise.} \end{cases} \tag{1.11}$$

For a proof of this result we refer to Niederreiter[Nie92b, Lemma 5.21].

4   Uniform Distribution Modulo One

4.1   The Definition of Uniformly Distributed Sequences

In random number generation, we are dealing with finite sequences $\omega = (\mathbf{x}_n)_{n=0}^{N-1}$ of numbers or points in the standard domain $[0, 1[^d, d \geq 1$. For reasons that we have discussed in Chapter 2, the sequence ω should simulate the realization of a sequence of i.i.d. random variables X_n, distributed uniformly in the d-dimensional unit cube. Hence, to a certain extent, ω should be equidistributed in $[0, 1[^d$. We employ the theory of uniform distribution of sequences modulo one to measure the uniform distribution of ω. The notion "modulo one" means that we work on the torus, which we have identified with the half-open interval.

The usual definition of uniform distribution modulo one of a sequence is the following.

Definition 4.1 The sequence $\omega = (\mathbf{x}_n)_{n \geq 0}$ is uniformly distributed modulo one (u.d. mod 1) in $[0, 1[^d$ if

$$\lim_{N \to \infty} S_N(f, \omega) = 0$$

for all functions $f = 1_I(\cdot) - \lambda_d(I)$, where $I = \prod_{i=1}^d [u_i, v_i[$ is a subinterval of $[0, 1[^d, 0 \leq u_i < v_i \leq 1, 1 \leq i \leq d,$ and $\lambda_d(I)$ denotes its volume.

The q-adic analog to this definition is obvious:

Definition 4.2 *(q-adic version) The sequence $\omega = (\mathbf{x}_n)_{n\geq 0}$ is uniformly distributed modulo one (u.d.mod 1) in $[0, 1[^d$ if*

$$\lim_{N\to\infty} S_N(f, \omega) = 0$$

for all functions $f = 1_I(\cdot) - \lambda_d(I)$, where $I = \prod_{i=1}^d [a_i q^{-g_i}, (a_i + 1)q^{-g_i}[$ is an elementary q-adic subinterval of $[0, 1[^d$, $g_i \geq 0$ and $0 \leq a_i < q^{g_i}$, for all i, $1 \leq i \leq d$.

By using approximation arguments, it is easily seen that these two definitions are equivalent. The q-adic version has certain advantages over the standard definition:

Remark 4.1 (1) *A technical advantage.* Let $\mathcal{F} \in \{\mathcal{H}(q), \mathcal{W}(q)\}$, $\mathcal{F} = \{\chi_\mathbf{k}\}$. Then the Fourier series of f, $f(\mathbf{x}) := 1_I(\mathbf{x}) - \lambda_d(I)$, I as in Definition 4.2, with respect to $\mathcal{F}$ is finite, see Lemma 3.3 and Lemma 3.4.

(2) *A conceptual advantage.* Discretization of the definition is easy. It simply means to work with *finite precision*, i.e. with α digits: Let $M = q^\alpha$. A sequence ω is called uniformly distributed in $1/M \; \mathbb{Z}^d$ mod 1 if

$$\lim_{N\to\infty} S_N(f, \omega) = 0$$

for all functions f, $f(\mathbf{x}) = 1_I(\mathbf{x}) - \lambda_d(I)$, $I = \prod_{i=1}^d [a_i/M, b_i/M[$, $0 \leq a_i < b_i \leq M$ for all i, $1 \leq i \leq d$.

4.2 Weyl Sums and Weyl's Criterion

We introduce the notion of Weyl sums relative to the function system $\mathcal{F}$. Weyl sums play a central role in the qualitative as well as in the quantitative analysis of the equidistribution of sequences.

Definition 4.3 *Let $\omega = (\mathbf{x}_n)_{n\geq 0}$ be a sequence in $[0, 1[^d$ and let $\mathcal{F} = \{\chi_\mathbf{k}\}$ be a function system, $\mathcal{F} \in \{\mathcal{T}, \mathcal{H}(q), \mathcal{H}_0(q), \mathcal{W}(q)\}$. A Weyl sum with respect to ω and $\mathcal{F}$ is a sum of the form*

$$S_N(\chi_\mathbf{k}, \omega), \quad \chi_\mathbf{k} \in \mathcal{F}.$$

Weyl's Criterion is the fundamental qualitative result of the theory of uniform distribution of sequences. It establishes a relation between the uniform distribution of a sequence ω and the asymptotic behavior of the Weyl sums $S_N(\chi_\mathbf{k}, \omega)$.

Theorem 4.2 *(Weyl's Criterion) Let $\omega = (\mathbf{x}_n)_{n\geq 0}$ be a sequence in $[0, 1[^d$ and let $\mathcal{F} \in \{\mathcal{T}, \mathcal{H}(q), \mathcal{H}_0(q), \mathcal{W}(q)\}$. The sequence ω is uniformly distributed modulo one if and only if*

$$\lim_{N\to\infty} S_N(\chi_\mathbf{k}, \omega) = 0 \tag{1.12}$$

for all nontrivial functions $\chi_{\mathbf{k}} \in \mathcal{F}$ (i.e. $\mathbf{k} \neq \mathbf{0}$).

Proof In the case $\mathcal{F} = \mathcal{T}$, this is the classical Weyl Criterion. For a proof, we refer to the monograph of Kuipers and Niederreiter[KN74]. From now on, let $\mathcal{F} \in \{\mathcal{H}(q), \mathcal{W}(q)\}$.

(1) Suppose that Relation (1.12) holds. Remark 4.1(1) implies that

$$\lim_{N \to \infty} S_N(f, \omega) = 0,$$

for all functions $f = 1_I(\cdot) - \lambda_d(I)$, where I is an elementary q-adic interval in $[0, 1[^d$. Hence Definition 4.2 applies.

(2) Suppose that ω is u.d. mod 1 in $[0, 1[^d$. Let $\mathbf{k}$ be such that $k_i < q^{g_i}$ for all i, $1 \leq i \leq d$, with $g_i \geq 0$. Then the function $\chi_{\mathbf{k}}$ is constant on all elementary q-adic intervals $I = \prod_{i=1}^d [a_i q^{-g_i}, (a_i + 1)q^{-g_i}[$, $0 \leq a_i < q^{g_i}$. Hence it is a step function. In other words, it is a finite linear combination of functions $f = 1_I(\cdot) - \lambda_d(I)$, I an elementary q-adic subinterval of $[0, 1[^d$. For these functions we know, from the definition of uniform distribution, that

$$\lim_{N \to \infty} S_N(f, \omega) = 0.$$

This implies Relation (1.12). The case $\mathcal{F} = \mathcal{H}_0(q)$ is now trivial. $\quad\square$

The notion of (t, m, s)-nets allows to construct excellent quadrature methods in high dimension. In this notation, the parameter s stands for the dimension, which we have denoted by the symbol d. Weyl sums with respect to such nets behave as follows.

Proposition 4.3 *Let $\mathcal{P} = (\mathbf{x}_n)_{n=0}^{q^m-1}$ be a (t, m, s)-net to the base q and suppose that $\mathcal{F} \in \{\mathcal{H}(q), \mathcal{H}_0(q), \mathcal{W}(q)\}$. Then, with $s = d$,*

$$S_N(\chi_{\mathbf{k}}, \mathcal{P}) = 0$$

for all $\mathbf{k} \neq \mathbf{0}$ such that there exist integers $g_i \geq 0$ with $k_i < q^{g_i}$ for all i, $1 \leq i \leq d$, and $\sum_{i=1}^d g_i = m - t$.

Proof $\mathcal{P}$ is a (t, m, s)-net. This implies that every elementary q-adic interval $I = \prod_{i=1}^d [a_i q^{-g_i}, (a_i + 1)q^{-g_i}[$, with $\sum_{i=1}^d g_i \leq m-t$, $g_i \geq 0$, $1 \leq i \leq s$, contains the same number of points of $\mathcal{P}$. Hence, for the system $\mathcal{H}(q)$,

$$S_N(h_{\mathbf{k}}, \mathcal{P}) = 0$$

for all $\mathbf{k} \neq \mathbf{0}$ with $k_i < q^{g_i}$ $\forall i$, $1 \leq i \leq s$, which also proves the proposition for $\mathcal{H}_0(q)$. The analogous result for the system $\mathcal{W}(q)$ was proved by Larcher and Traunfellner[LT94]. $\quad\square$

4.3   Remarks

The concepts of good lattice points and of (t, m, s)-nets have structural similarities which are not sufficiently known. In both cases we are searching for parameters such that the associated point sets ω in $[0, 1[^d$ have the following property: For an appropriate function system $\mathcal{F} = \{\chi_\mathbf{k}\}$, the Weyl sums $S_N(\chi_\mathbf{k}, \omega)$ are either zero or the norm of those indices $\mathbf{k}$ where these sums do not vanish is large.

Sobol's concept of (t, m, s)-nets has its roots in this idea. We take $\mathcal{F} = \mathcal{H}(2)$ and construct point sets ω where $S_N(h_\mathbf{k}, \omega) = 0$ for as many Haar functions $h_\mathbf{k}$ as possible. In particular, this should happen for all "small" indices $\mathbf{k}$, that is to say, for all $\mathbf{k}$ close to the origin $\mathbf{0}$. From Lemma 3.3 and its multidimensional analog in [Hel94a] we deduce that every elementary dyadic interval I will contain the right number of points, i.e. the local discrepancy $S_N(1_I(\cdot) - \lambda_d(I), \omega)$ will be equal to zero. This means that ω hits all d-dimensional elementary dyadic intervals with the right frequency, provided that the intervals are not too small. This property is very close to that of being a (t, m, s)-net.

The notion of good lattice points $\mathbf{g}$ modulo N of Korobov[Kor59, Kor63] follows the very same idea, with $\mathcal{F} = \mathcal{T}$ and $\omega = (\{(n/N)\mathbf{g}\})_{n=0}^{N-1}$.

The observation that we are trying to find point sets ω such that the Weyl sums $S_N(\chi_\mathbf{k}, \omega)$ are zero for as many $\mathbf{k}$ as possible, and such that the norm of those indices $\mathbf{k}$ where these sums do not vanish is large, is a fundamental ingredient of the concept of the *spectral test* in Chapter 5. From this viewpoint, we may say that both Sobol' and Korobov tried to find point sets ω with a spectral test value as small as possible. The main difference between these two approaches lies in the function system that is employed.

5   The Spectral Test

We have observed that the equidistribution properties of a sequence $\omega = (\mathbf{x}_n)_{n \geq 0}$ in $[0, 1[^d$ are closely related to the behavior of the Weyl sums $S_N(\chi_\mathbf{k}, \omega)$, where the functions $\chi_\mathbf{k}$ belong to some appropriate system $\mathcal{F}$. In this contribution, we limit our discussion to the case where $\mathcal{F}$ is one of the systems $\mathcal{T}$, $\mathcal{W}(q)$, or $\mathcal{H}(q)$. Analogous results will hold for other function systems $\mathcal{F}$ as long as they are sufficiently large. In this context, we refer the reader to the notion of a convergence-determining class of functions on $[0, 1[^d$ and to results on the uniform distribution of sequences in topological groups, see Lemma 1.2 and Corollary 1.1 in Kuipers and Niederreiter[KN74, Ch. 4].

Coveyou and MacPherson[CM67] introduced the spectral test to measure the equidistribution of point sets ω in $[0, 1[^d$ that were generated by overlapping d-tuples of consecutive random numbers. Their approach may be summarized as follows (see [CM67, p. 105]).

A uniform RNG with output sequence $\omega = (\mathbf{x}_n)_{n \geq 0}$ is satisfactory only if, for each $\mathbf{k} \neq 0$, $\mathbf{k} = (k_1, \ldots, k_d)$, one of the numbers

$S_N(e_{\mathbf{k}}, \omega)$ *and* $1/(k_1^2 + \ldots + k_d^2)$ *is very small.*

This heuristic concept led to a numerical quantity to assess the coarseness lattices L in $\mathbb{R}^d$, the so-called spectral test. It is defined as the quantity

$$\sigma(L) := \frac{1}{\min\{\|\mathbf{k}\|_2 : \mathbf{k} \in L^\perp, \mathbf{k} \neq \mathbf{0}\}} \tag{1.13}$$

where $\|\mathbf{k}\|_2 = \sqrt{k_1^2 + \ldots + k_d^2}$ denotes the Euclidean norm of the vector $\mathbf{k}$ and $L^\perp$ stands for the dual lattice (see Subsection 3.2 for this notion). At first sight, this quantity seems to be of a rather theoretical nature. This impression is wrong. The spectral test is the most powerful figure of merit for linear RNGs. For example, all good LCGs pass this test and all LCGs known to be bad (like RANDU) fail it.

The spectral test has an important geometric interpretation. It is the maximal distance between two adjacent hyperplanes, taken over all families of parallel hyperplanes that cover the lattice L. We refer the reader to Subsection 5.4 for an exposition of the details. Based on results of Dieter[Die75] and Knuth[Knu81], a very efficient algorithm to compute the spectral test for linear generators like LCGs and MRGs is available due to L'Ecuyer and Couture[LC97]. The idea to assess RNGs by the spectral test is the following. The d-tuples of consecutive random numbers or random vectors produced with linear RNGs will have a grid structure. We refer to Figure 1 in Chapter 2 as an illustration and to Niederreiter[Nie92b, Nie95] for the theoretical results. If we translate the points of ω (and, if not already an element of ω, also the origin) by $\mathbb{Z}^d$, then, in many cases, we will obtain an integration lattice L in $\mathbb{R}^d$. Assessing an RNG with the spectral test in dimension d means that we assess the lattice L by the figure of merit $\sigma(L)$.

The "classical" form of the spectral test is an *indirect* approach to measure the equidistribution of finite point sets $\omega = (\mathbf{x}_n)_{n=0}^{N-1}$ in the d-dimensional unit cube $[0, 1[^d$. We asses the lattice that is generated by the point set and not the set ω itself. We have to restrict our study to linear structures ω. For arbitrary point sets ω, the classical spectral test is not defined.

In Subsection 3.2, Relation (1.11) we have pointed out that for any node set ω of a d-dimensional N point lattice rule L we have the relation

$$S_N(e_{\mathbf{k}}, \omega) = \begin{cases} 1 & : \quad \mathbf{k} \in L^\perp \\ 0 & : \quad \text{otherwise.} \end{cases}$$

This simple observation proves to be very useful. It allowed Leeb and Hellekalek[LH98] to answer an open question of Niederreiter[Nie92b, p.168], who remarked in connection with the classical spectral test: *"The difficulty here is to find a convincing quantitative formulation of this idea."* We will define a version of the spectral test that applies to arbitrary sequences ω. It generalizes the classical figure of merit $\sigma(L)$.

Definition 5.1 *Let $\omega = (x_n)_{n\geq 0}$ be an arbitrary sequence of points in the d-dimensional unit cube $[0, 1[^d$, not necessarily with a grid structure.*

The spectral test $\sigma_N(\omega)$ of the first N points of ω is defined as the quantity

$$\sigma_N(\omega) := \sup \left\{ \frac{|S_N(e_{\mathbf{k}}, \omega)|}{\|\mathbf{k}\|_2} : \quad \mathbf{k} \in \mathbb{Z}^d, \mathbf{k} \neq \mathbf{0} \right\}. \qquad (1.14)$$

Relation (1.11) implies that this quantity coincides with the classical spectral test in the case where ω generates an integration lattice. We have the important property

$$\omega \text{ is uniformly distributed in } [0, 1[^d \Leftrightarrow \lim_{N\to\infty} \sigma_N(\omega) = 0.$$

In other words, $\sigma_N(\omega)$ is a measure of uniform distribution, like discrepancy or diaphony. This result will be shown in Theorem 5.2 below in a more general setting.

5.1 Definition

We will extend our concept of the spectral test to other function systems $\mathcal{F}$ and to other weight functions than the simple case $\mathbf{k} \mapsto 1/\|\mathbf{k}\|_2$. As special cases, we will encounter the quantities ρ and σ of the theory of good lattice points, see Chapter 2 for their definition.

Definition 5.2 *Let $\omega = (x_n)_{n\geq 0}$ be an arbitrary sequence of points in the d-dimensional unit cube $[0, 1[^d$, not necessarily with a grid structure. Let $\mathcal{F} = \{\chi_{\mathbf{k}}\}$ be one of the function systems $\mathcal{T}$, $\mathcal{H}_0(q)$, or $\mathcal{W}(q)$, and let $\|\cdot\|$ denote an arbitrary norm on $\mathbb{R}^d$. Let $r : \mathbf{k} \mapsto r(\mathbf{k})$ be a real-valued function with the properties*

(i) $r(\mathbf{k}) > 0 \quad \forall \mathbf{k}$,

(ii) $r(0) = 1$,

(iii) $\lim_{\|\mathbf{k}\|\to\infty} 1/r(\mathbf{k}) = 0$.

The spectral test $\sigma_N(\mathcal{F}, r; \omega)$ of the first N elements of ω relative to $\mathcal{F}$ and r is defined as the quantity

$$\sigma_N(\mathcal{F}, r; \omega) := \sup \left\{ \frac{|S_N(\chi_{\mathbf{k}}, \omega)|}{r(\mathbf{k})} : \mathbf{k} \neq \mathbf{0} \right\}. \qquad (1.15)$$

We will call r the weight function. For reasons of simplicity, we will write $\sigma_N(\omega)$ for the spectral test quantity, provided that it is clear from the context which function system and which weight function we use.

5.2   Properties

In the classical spectral test for lattices, we have to compute a maximum. This property holds also in the general case.

Proposition 5.1 Let $\mathcal{F}$, r, and ω be as in Definition 5.2. Then $\sigma_N(\omega) = \sigma_N(\mathcal{F}, r; \omega)$ is a maximum.

Proof We may assume $\sigma_N(\omega) > 0$. Otherwise, all Weyl sums $S_N(\chi_{\mathbf{k}}, \omega)$ would be equal to zero, which is impossible.

As a consequence, there exists $\epsilon > 0$ such that

$$\epsilon < \sigma_N(\omega),$$

and $K_0 = K_0(\epsilon) \in \mathbb{N}$ such that

$$\frac{1}{r(\mathbf{k})} < \epsilon \quad \forall \, \mathbf{k} : \|\mathbf{k}\| > K_0.$$

This implies

$$\sup\left\{ \frac{|S_N(\chi_{\mathbf{k}}, \omega)|}{r(\mathbf{k})} : \|\mathbf{k}\| > K_0 \right\} \leq \sup\left\{ \frac{1}{r(\mathbf{k})} : \|\mathbf{k}\| > K_0 \right\} \leq \epsilon < \sigma_N(\omega).$$

Hence,

$$\sigma_N(\omega) = \sup\left\{ \frac{|S_N(\chi_{\mathbf{k}}, \omega)|}{r(\mathbf{k})} : \|\mathbf{k}\| \leq K_0 \right\}.$$

We observe that the set $\{\mathbf{k} : \|\mathbf{k}\| \leq K_0\}$ is finite. $\square$

Theorem 5.2 Let $\mathcal{F}$, r, and ω be as in Definition 5.2. The sequence ω is uniformly distributed modulo one if and only if

$$\lim_{N \to \infty} \sigma_N(\omega) = 0.$$

Proof Suppose first that $\lim_{N \to \infty} \sigma_N(\omega) = 0$. This implies the relation $\lim_{N \to \infty} S_N(\chi_{\mathbf{k}}, \omega) = 0$ for all $\mathbf{k} \neq \mathbf{0}$. Weyl's Criterion then yields the uniform distribution of ω.

In order to prove the remaining part of the theorem, assume that ω is uniformly distributed. For an arbitrary $\epsilon > 0$, there exists a positive integer $K_0 = K_0(\epsilon)$ such that $r(\mathbf{k})^{-1} < \epsilon$ for all $\mathbf{k}$ with $\|\mathbf{k}\| > K_0$. We have the relation

$$\sigma_N(\omega) = \max\{A, B\},$$

where

$$A := \sup\left\{ \frac{|S_N(\chi_{\mathbf{k}}, \omega)|}{r(\mathbf{k})} : \|\mathbf{k}\| \leq K_0 \right\}$$

and

$$B := \sup\left\{ \frac{|S_N(\chi_{\mathbf{k}}, \omega)|}{r(\mathbf{k})} : \|\mathbf{k}\| > K_0 \right\}$$

Clearly, $B \leq \epsilon$, and, hence,

$$\sigma_N(\omega) \leq A + \epsilon.$$

A is the supremum over a finite set of indices $\mathbf{k}$. As a consequence, it is a maximum. Weyl's Criterion implies that

$$A = \max \left\{ \frac{|S_N(\chi_{\mathbf{k}}, \omega)|}{r(\mathbf{k})} : \|\mathbf{k}\| \leq K_0 \right\} < \epsilon \quad \forall N \geq N_0(\epsilon).$$

We deduce the relation

$$\sigma_N(\omega) < 2\epsilon \quad \forall N \geq N_0(\epsilon).$$

This proves the theorem.   □

Theorem 5.2 generalizes Theorem 1 in Leeb and Hellekalek [LH98].

5.3 Examples

Example 1 Let $\mathcal{F} = \mathcal{T}$ and $r(\mathbf{k}) = \|\mathbf{k}\|_2$. Then $\sigma_N(\mathcal{F}, r; \omega)$ is the spectral test introduced in Definition 5.1. Further, if ω is as in Remark 3.5, in particular, if $\omega = (\{(n/N)\mathbf{g}\})_{n=0}^{N-1}$, with $(g_i, N) = 1$, $\forall i$, $1 \leq i \leq d$, then $\sigma_N(\mathcal{F}, r; \omega)$ equals the classical spectral test, as it is well-known in random number generation.

Example 2 Let $\mathcal{F} = \mathcal{T}$, $r(\mathbf{k}) = \prod_{i=1}^{d} \max\{1, |k_i|\}$, and suppose that $\omega = (\{(n/N)\mathbf{g}\})_{n=0}^{N-1}$, $(g_i, N) = 1$, $1 \leq i \leq d$. Then

$$\sigma_N(\mathcal{F}, r; \omega) = \frac{1}{\rho(\mathbf{g}, N)},$$

where the figure of merit $\rho(\mathbf{g}, N)$ has been defined in Subsection 2.7.

Example 3 Let $\mathcal{F} = \mathcal{W}(2)$ and $r(\mathbf{k}) := \prod_{i=1}^{d} r(k_i)$, where

$$r(k) := \begin{cases} 2^{2g} & \text{if } k: \ 2^g \leq k < 2^{g+1}, \text{with } g \geq 0, \ g \in \mathbb{Z}, \\ 1 & \text{if } k = 0. \end{cases}$$

Then $\sigma_N(\mathcal{F}, r; \omega)$ is closely related to the dyadic diaphony of Hellekalek and Leeb[HL97].

In addition to the examples above, we may construct many other realizations of our general version of the spectral test. Themes of ongoing research are the asymptotic distribution of these quantities, laws of the iterated logarithm, and the relation to other figures of merit like diaphony and discrepancy.

5.4   Geometric Interpretation

We have already mentioned the important geometric interpretation of the spectral test. This test computes the maximal distance between two adjacent parallel hyperplanes covering all possible points x_n that the generator can produce. From this interpretation, very efficient algorithms to compute the value of the spectral test and to choose suitable parameters for linear RNGs are derived, see the results of Dieter[Die75], Knuth[Knu81], and L'Ecuyer and Couture[LC97], and their application in the computations by Entacher[Ent98, Ent97a], Fishman[Fis96], and L'Ecuyer et al.[LBC93, L'E96, L'E97a].

The following presentation of the geometric background of the spectral test was inspired by the discussion in Fishman[Fis96] and Leeb[Lee95]. The main result in this subsection will be Corollary 5.7. The next lemma will be needed in its proof.

Lemma 5.3 *Let L be a d-dimensional lattice and let A be a nonempty subset of L.*

Then A contains a lattice point with minimal norm, i.e.

$$\exists \mathbf{a} \in A : \ ||\mathbf{a}||_2 = \inf\{||\mathbf{b}||_2 : \mathbf{b} \in A\}.$$

Proof Let $\alpha := \inf\{||\mathbf{b}||_2 : \mathbf{b} \in A\}$. If A has only finitely many elements, then the statement of the lemma is trivial.

From now on, we will consider the case where A contains infinitely many elements.

Let $\delta > 0$ be arbitrary. The intersection of A with the δ-neighborhood $U_\delta(\mathbf{0})$ of $\mathbf{0}$,

$$U_\delta(\mathbf{0}) = \{\mathbf{x} \in \mathbb{R}^d : ||\mathbf{x}||_2 < \delta\},$$

will be a finite set,

$$\sharp \, A \cap U_\delta(\mathbf{0}) < \infty.$$

This statement is proved as follows. The set A is discrete. In other words, there exists a real number $\rho > 0$ such that

$$||\mathbf{b} - \mathbf{a}||_2 \geq \rho \quad \forall \mathbf{a} \neq \mathbf{b}, \mathbf{a} \text{ and } \mathbf{b} \text{ in } A.$$

We have

$$\bigcup_{\mathbf{a} \in A \cap U_\delta(\mathbf{0})} U_\rho(\mathbf{a}) \subseteq U_{\delta+\rho}(\mathbf{0}).$$

The ρ-neighborhoods $U_\rho(\mathbf{a})$ in the union on the left side are pairwise disjoint, hence the d-dimensional Lebesgue measure λ_d of this union is the sum of the individual measures:

$$\lambda_d\left(\bigcup_{\mathbf{a} \in A \cap U_\delta(\mathbf{0})} U_\rho(\mathbf{a})\right) = \sum_{\mathbf{a} \in A \cap U_\delta(\mathbf{0})} \lambda_d(U_\rho(\mathbf{a})).$$

Clearly,

$$\sum_{\mathbf{a}\in A\cap U_\delta(\mathbf{0})} \lambda_d(U_\rho(\mathbf{a})) \le \lambda_d\left(U_{\delta+\rho}(\mathbf{0})\right) < \infty. \tag{1.16}$$

λ_d is invariant under translations, hence

$$\lambda_d\left(U_\rho(\mathbf{a})\right) = \lambda_d\left(U_\rho(\mathbf{0})\right) \quad \forall \mathbf{a}.$$

Relation (1.16) implies that the set $A \cap U_\delta(\mathbf{0})$ is finite. This proves our statement above.

Now, let $\epsilon > 0$ be arbitrary. Then

$$A \cap U_{\alpha+\epsilon}(\mathbf{0}) \ne \emptyset$$

and we have

$$\alpha = \inf\left\{\|\mathbf{b}\|_2 : \mathbf{b} \in A \cap U_{\alpha+\epsilon}(\mathbf{0})\right\}.$$

The set $A \cap U_{\alpha+\epsilon}(\mathbf{0})$ is finite. The results follows.   □

A hyperplane in $\mathbb{R}^d$ is uniquely determined by a vector $\mathbf{n} = (n_1, \ldots, n_d) \ne \mathbf{0}$ and a real number c,

$$H_{\mathbf{n},c} = \{\mathbf{x} \in \mathbb{R}^d : \mathbf{n} \cdot \mathbf{x} = c\}.$$

A family of parallel hyperplanes in $\mathbb{R}^d$ is defined by a nonempty set C of real numbers,

$$\mathcal{H}_{\mathbf{n},C} = \{H_{\mathbf{n},c} : c \in C\}.$$

The distance between two hyperplanes $H_{\mathbf{n},c}$ and $H_{\mathbf{n},d}$ is given by the number

$$d(H_{\mathbf{n},c}, H_{\mathbf{n},d}) = \frac{|c - d|}{\|\mathbf{n}\|_2}.$$

Definition 5.3 *(1) Let L be a d-dimensional lattice in $\mathbb{R}^d$. A family $\mathcal{H}_{\mathbf{n},C}$ of parallel hyperplanes in $\mathbb{R}^d$ is called a cover of L if*

(i) $L \subseteq \bigcup_{c \in C} H_{\mathbf{n},c}$,

(ii) C is the smallest set (in the sense of set-inclusion) with this property.

(2) Let $\mathcal{H}_{\mathbf{n},C}$ denote an arbitrary family of parallel hyperplanes. The spacing $d(\mathcal{H}_{\mathbf{n},C})$ of $\mathcal{H}_{\mathbf{n},C}$ will denote the minimal distance between adjacent hyperplanes in this family,

$$d(\mathcal{H}_{\mathbf{n},C}) := \inf\{d(H_{\mathbf{n},c}, H_{\mathbf{n},d}) : c \ne d, \ c, d \in C\}$$

(3) The spectral test of L is defined as the number

$$\sigma(L) := \sup\{d(\mathcal{H}_{\mathbf{n},C}) : \mathcal{H}_{\mathbf{n},C} \text{ is a cover of } L\}. \tag{1.17}$$

The distance between adjacent hyperplanes within some family $\mathcal{H}_{\mathbf{n},C}$ is independent of the particular choice of hyperplanes. In order to prove this statement we show the following.

Lemma 5.4 *Let $\mathcal{H}_{\mathbf{n},C}$ by a cover of the lattice L.*
Then $(C,+)$ is an abelian group.

Proof C is a subset of the additive group $\mathbb{R}$. Hence, it suffices to show that, for arbitrary c and d in C, also $c + d$ and $-c$ belong to C. Let c and d be elements of C. Then there exists two lattice points $\mathbf{x}$ in $L \cap H_{\mathbf{n},c}$ and $\mathbf{y}$ in $L \cap H_{\mathbf{n},d}$ such that $\mathbf{n} \cdot \mathbf{x} = c$ and $\mathbf{n} \cdot \mathbf{y} = d$. L is a subgroup of $\mathbb{R}^d$, hence $\mathbf{x} + \mathbf{y}$ is in L. Further,

$$\mathbf{n} \cdot (\mathbf{x} \pm \mathbf{y}) = c \pm d,$$

which shows that the number $c \pm d$ belongs to C. $\square$

Corollary 5.5 Let $\mathcal{H}_{\mathbf{n},C}$ be a cover of the lattice L. Then

$$d(\mathcal{H}_{\mathbf{n},C}) = \frac{1}{||\mathbf{n}||_2} \inf \{c \in C : \ c > 0, \}. \tag{1.18}$$

In general, the spacing of a cover $\mathcal{H}_{\mathbf{n},C}$ of a lattice L need not be strictly positive, see Leeb[Lee95, page 122] for an example. As we will show now, if $d(\mathcal{H}_{\mathbf{n},C}) > 0$, then we are able to compute the spectral test with the help of the dual lattice $L^{\perp}$.

Proposition 5.6 *Let $\mathcal{H}_{\mathbf{n},C}$ be a cover of the lattice L.*
Then the following are equivalent:

> (i) $d(\mathcal{H}_{\mathbf{n},C}) > 0$,
> (ii) $\exists \lambda > 0 : \lambda \mathbf{n} \in L^{\perp}$ *and this lattice point is primitive.* (1.19)

Proof Let $\gamma := \inf\{c \in C : c > 0\}$. Then

$$d(\mathcal{H}_{\mathbf{n},C}) = \frac{\gamma}{||\mathbf{n}||_2}.$$

Suppose that $d(\mathcal{H}_{\mathbf{n},C}) > 0$. Then $\gamma > 0$.
We will now prove that

$$C = \gamma \mathbb{Z}. \tag{1.20}$$

For, every interval of the form $[k\gamma, (k + 1)\gamma[$, $k \in \mathbb{N}$, contains at most one element of C:

$$k\gamma \le d < c < (k + 1)\gamma, \quad c, d \in C,$$

implies $0 < c - d < \gamma$. Because of $c - d \in C$, we obtain the inequality

$$\gamma \le c - d < \gamma,$$

which is a contradiction.

γ is the infimum over C. Hence, for all $\epsilon > 0$, there is some $c = c(\epsilon)$ in C such that

$$\gamma \le c < \gamma + \epsilon.$$

Because of the above reasoning, for $\epsilon < \gamma$, this will imply $c = \gamma$ and, hence, $C = \gamma \mathbb{Z}$.

Let us write $\lambda := 1/\gamma$. Then we have the relations

$$\mathcal{H}_{\mathbf{n},C} = \mathcal{H}_{\mathbf{n},\gamma \mathbb{Z}} = \mathcal{H}_{\lambda \mathbf{n},\mathbb{Z}},$$

$$d(\mathcal{H}_{\mathbf{n},C}) = \frac{1}{\|\lambda \mathbf{n}\|_2},$$

and, further,

$$\lambda \mathbf{n} \in L^{\perp}.$$

We will show now that $\lambda \mathbf{n}$ is a primitive lattice point. Let $\mu \lambda \mathbf{n}$ be an arbitrary element of the dual lattice $L^{\perp}$ that belongs to the line segment joining the points $\mathbf{0}$ and $\lambda \mathbf{n}$, $0 < \mu \le 1$. This will imply $\mu = 1$. For, $\mathcal{H}_{\lambda \mathbf{n},\mathbb{Z}}$ is a cover of L, hence there exists an element $\mathbf{x} \in L$ such that

$$\lambda \mathbf{n} \cdot \mathbf{x} = 1.$$

This implies $(\mu \lambda \mathbf{n}) \cdot \mathbf{x} = \mu$. The lattice point $\mu \lambda \mathbf{n}$ belongs to the dual lattice, hence μ must be an integer. This yields $\mu = 1$ and shows that $\lambda \mathbf{n}$ is primitive.

Conversely, let $\lambda \mathbf{n} \in L^{\perp}$ be primitive, $\lambda > 0$. Let $\mathcal{H}_{\mathbf{n},C}$ be a cover of the lattice L. Because of the relation

$$\mathcal{H}_{\mathbf{n},C} = \mathcal{H}_{\lambda \mathbf{n},\lambda C},$$

we will have

$$\lambda C \subseteq \mathbb{Z}.$$

This implies

$$d(\mathcal{H}_{\mathbf{n},C}) = d(\mathcal{H}_{\lambda \mathbf{n},\lambda C}) = \frac{1}{\|\mathbf{n}\|_2} \inf\{\lambda c : c \in C, c \ne 0\} \ge \frac{1}{\|\mathbf{n}\|_2} > 0.$$

$\square$

Proposition 5.6 allows us to compute the spectral test for a lattice L by searching for the shortest vector in the dual lattice $L^{\perp}$.

Corollary 5.7 *Let L be a d-dimensional lattice in $\mathbb{R}^d$. Then*

$$\sigma(L) = \frac{1}{\min\{\|\mathbf{g}\|_2 : \mathbf{g} \in L^{\perp},\ \mathbf{g}\ \text{primitive}\}} \qquad (1.21)$$

Proof

$$\sigma(L) = \sup \{d(\mathcal{H}_{\mathbf{n},C}) : \mathcal{H}_{\mathbf{n},C} \text{ a cover of } L \text{ and } d(\mathcal{H}_{\mathbf{n},C}) > 0\}.$$

From Proposition 5.6 it follows that $\sigma(L)$ equals the number

$$\sup \{d(\mathcal{H}_{\mathbf{n},C}) : \mathcal{H}_{\mathbf{n},C} \text{ a cover of } L \text{ and } \exists \lambda > 0 : \lambda \mathbf{n} \in L^{\perp} \text{primitive}\}.$$

Hence,

$$\sigma(L) \le \sup \left\{\frac{1}{\|\mathbf{g}\|_2} : \mathbf{g} \in L^{\perp}, \ \mathbf{g} \text{ primitive}\right\}.$$

On the other hand, let $\mathbf{g} \in L^{\perp}$ be primitive. Then $\mathbf{g} \cdot \mathbf{x} \in \mathbb{Z}$ for all $\mathbf{x} \in L$. Hence, there exists a nonempty subset C of $\mathbb{Z}$ such that $\mathcal{H}_{\mathbf{g},C}$ is a cover of the lattice L. Because of Corollary 5.5, we have $d(\mathcal{H}_{\mathbf{g},C}) > 0$. Proposition 5.6 implies that $C = \gamma \mathbb{Z}$ with some positive number γ in C and, hence, in $\mathbb{Z}$. Further, the lattice point $\gamma^{-1}\mathbf{g}$ will be a primitive element of the dual lattice $L^{\perp}$. The element $\gamma^{-1}\mathbf{g}$ belongs to the line segment joining the origin and the lattice point $\mathbf{g}$. The latter is primitive, hence $\gamma = 1$. This yields

$$d(\mathcal{H}_{\mathbf{g},C}) = \frac{1}{\|\mathbf{g}\|_2},$$

and, as a consequence,

$$\sup \left\{\frac{1}{\|\mathbf{g}\|_2} : \mathbf{g} \in L^{\perp}, \ \mathbf{g} \text{ primitive}\right\} \le \sigma(L).$$

Lemma 5.3 finishes the proof of Corollary 5.7. □

Dieter[Die75] has found a very efficient method to compute the shortest vector in the dual lattice. His algorithm is used for computing the spectral test in practice, see for example L'Ecuyer et al.[L'E96, LC97].

5.5   Remarks

We have already discussed the forerunners of our concept of the spectral test, the numerical quantities $\rho(\mathbf{g}, N)$ and $\sigma(\mathbf{g}, N)$, see Subsection 2.7.

Klaus Schmidt[Sch71] investigated the question how to measure the equidistribution of sequences in locally compact groups. He introduced the following metric d_r on the set $M(X)$ of positive, normed Radon measures on a locally compact abelian group X,

$$d_r(\mu_1, \mu_2) := \sup_{y \in Y} \frac{|\hat{\mu}_1(y) - \hat{\mu}_2(y)|}{r(y)}. \tag{1.22}$$

Here Y denotes the group of characters of X, $\hat{\mu}(\cdot)$ stands for the Fourier transform of $\mu \in M(X)$, and r is a fixed real-valued continuous function on Y with the properties $r(0) = 1$, $1 \le r(y) < \infty$, and $\lim_{r \to \infty} 1/r(y) = 0$.

The metric d_r leads to the notion of discrepancy of a sequence of measures $(\mu_n)_{n \geq 0}$ in $M(X)$ relative to $\mu \in M(X)$:

$$d_r(N, \mu) := d_r\left(\frac{1}{N} \sum_{n=0}^{N-1} \mu_n, \ \mu\right). \tag{1.23}$$

It turns out that, in the special case of the d-dimensional torus $X = [0, 1[^d$, with μ the Haar measure (i.e. Lebesgue measure) on X, and μ_n the point measure at $\mathbf{x}_n$, this quantity is equal to the spectral test $\sigma_N(\mathcal{T}, r; \omega)$, where $\omega = (\mathbf{x}_n)_{n \geq 0}$.

In the literature, other variants of the classical spectral test have been introduced, based on the Walsh system $\mathcal{W}(2)$. In the context of generalized feedback-shift register generators, Tezuka[Tez87] defined the figure of merit

$$\max\{\alpha : \ S_N(w_{\mathbf{k}}, \omega) = 0 \ \ \forall \mathbf{k} = (k_1, \ldots, k_s) \neq \mathbf{0}, \ 0 \leq k_i < 2^\alpha\},$$

which is very similar in spirit to the parameter t of (t, m, s)-nets. We also refer the reader to Yuen[Yue77].

In order to establish the spectral test as a measure of equidistribution, we have to answer several questions:

1. Is σ_N a measure of uniform distribution? This means, do we have the relation

   $$\omega \text{ is uniformly distributed mod } 1 \Leftrightarrow \lim_{N \to \infty} \sigma_N(\omega) = 0?$$

2. What is the asymptotic distribution of the random variable $\sigma_N(\omega)$, where ω is a sequence of independent, identically uniformly distributed random variables on $[0, 1[^d$?

3. How is $\sigma_N(\omega)$ related to other equidistribution measures like diaphony or discrepancy?

We have already answered the first question in this contribution. The second question is solved in Leeb and Hellekalek[LH98] for the cases $\mathcal{F} = \mathcal{T}$, $r(\mathbf{k}) = \|\mathbf{k}\|_2$ and $r(\mathbf{k}) = \prod_{i=1}^{d} \max\{1, |k_i|\}$. The general case as well as the third question are the subject of ongoing research.

Hickernell[Hic98] has postulated a set of conditions that should be satisfied by a measure of equidistribution. Even at this early state of research, it can be shown that the spectral test $\sigma_N(\omega)$ fulfills most of these requirements.

6   The Weighted Spectral Test

Coveyou and MacPherson were striving for finite sequences or point sets $\omega = (\mathbf{x}_n)_{n=0}^{N-1}$ in $[0,1[^d$ such that at least one of the numbers $S_N(\mathbf{e_k},\omega)$ or $\|\mathbf{k}\|_2^{-1}$ were small. We may use weighted means to mold this condition into a general quantitative concept, *the weighted spectral test*. This notion is closely related to the spectral test.

6.1   Definition

For a function system $\mathcal{F} = \{\chi_\mathbf{k}\}$, a norm $\|\cdot\|$ on $\mathbb{R}^d$, and a "weight" function $r : \mathbf{k} \rightarrow r(\mathbf{k})$ with

(i) $r(\mathbf{k}) > 0 \quad \forall \mathbf{k}$,

(ii) $r(\mathbf{0}) = 1$,

(iii) $\sum_\mathbf{k} 1/r(\mathbf{k})^2 < \infty$ (which implies $\lim_{\|\mathbf{k}\| \to \infty} 1/r(\mathbf{k}) = 0$)

we may define the weighted mean below.

Definition 6.1 *Let $\omega = (\mathbf{x}_n)_{n\geq 0}$ be an arbitrary sequence of points in the d-dimensional unit cube $[0,1[^d$. Let $\mathcal{F} = \{\chi_\mathbf{k}\}$ be one of the function systems $\mathcal{T}$ or $\mathcal{W}(q)$, and let $\|\cdot\|$ denote an arbitrary norm on $\mathbb{R}^d$. The quantity*

$$F_N(\mathcal{F}, r; \omega) := \left(\sum_{\mathbf{k} \neq \mathbf{0}} \frac{1}{r(\mathbf{k})^2} |S_N(\chi_\mathbf{k}, \omega)|^2 \right)^{1/2} \tag{1.24}$$

will be called the weighted spectral test of the first N elements of ω relative to $\mathcal{F}$ and r. In case there is no ambiguity, we will shorten this notation to $F_N(\omega)$.

Remark 6.1 (i) With every finite or infinite sequence ω in $[0,1[^d$, each system $\mathcal{F}$, and each weight function r, we may associate the array

$$T(\omega) := \left(\frac{S_N(\chi_\mathbf{k}, \omega)}{r(\mathbf{k})} \right)_{\mathbf{k} \neq \mathbf{0}} \tag{1.25}$$

We observe that

$$\sigma_N(\omega) = \|T(\omega)\|_\infty,$$

and

$$F_N(\omega) = \|T(\omega)\|_2.$$

Hence, the spectral test and the weighted spectral test are just two different norms of the same mathematical object.

(ii) If we are able to find a function $f : [0,1[^d \to \mathbb{R}$ such that its Fourier series with respect to $\mathcal{F}$ fulfills the following conditions,

$$\hat{f}(\mathbf{k}) = r(\mathbf{k})^{-2} \quad \forall \mathbf{k} \neq 0,$$

$$f = \sum_{\mathbf{k}} \hat{f}(\mathbf{k}) \chi_{\mathbf{k}} \quad \text{(pointwise)},$$

and, further, a difference operation $\ominus$ on the torus $[0,1[^s$ that is compatible with the function system $\mathcal{F}$, then

$$F_N^2(\mathcal{F}, r; \omega) = \frac{1}{N^2} \sum_{n=0}^{N-1} \sum_{m=0}^{N-1} f(\mathbf{x}_n \ominus \mathbf{x}_m).$$

In this context, compatibility of $\ominus$ with $\mathcal{F}$ means the property

$$\chi_{\mathbf{k}}(x \ominus y) = \chi_{\mathbf{k}}(x)\overline{\chi_{\mathbf{k}}(y)}.$$

In the case $\mathcal{F} = \mathcal{T}$, the operation $\ominus$ will be the inverse of addition modulo one. If $\mathcal{F} = \mathcal{W}(2)$, then it will be dyadic addition, which is the same as dyadic subtraction (or the XOR operation for finite blocks of bits).

6.2 Examples and Properties

At present, we are able to present two realizations of this general concept. Both are ready to be employed in theoretical and practical computations.

Example 6.1 (Classical Diaphony, Zinterhof[Zin76]) We choose $\mathcal{F} = \mathcal{T}$ and the weight function $r(\mathbf{k}) = \prod_{i=1}^d \max\{1, |k_i|\}^2$, $\mathbf{k} = (k_1, \ldots, k_d) \in \mathbb{Z}^d$. An appropriate function f is given by

$$f(\mathbf{x}) = -1 + \prod_{i=1}^d g(x_i), \quad \mathbf{x} = (x_1, \ldots, x_d) \in [0,1[^d, \quad \text{where}$$

$$g(x) = 1 - \frac{\pi^2}{6} + \frac{\pi^2}{2}(1 - 2x)^2, \quad x \in [0,1[. \tag{1.26}$$

Zinterhof[Zin76] has not only introduced the notion of diaphony, but he has already proved the fundamental identity

$$F_N^2(\mathcal{T}, r; \omega) = \frac{1}{N^2} \sum_{n=0}^{N-1} \sum_{m=0}^{N-1} f(\mathbf{x}_n - \mathbf{x}_m \bmod 1). \tag{1.27}$$

Remark 6.2 (1) Zinterhof[Zin76] introduced the notion of "diaphony" in order to estimate the integration error for a special number-theoretic integration method adapted to integrands with rapidly converging Fourier series. The basic properties of diaphony and its relation to discrepancy were studied in Zinterhof[Zin76] and Zinterhof and Stegbuchner[ZS78]. We refer to Drmota and Tichy[DT97] for a concise review of these results.

(2) In the case of a d-dimensional N point lattice rule node set $\omega = (\{(n/N)\mathbf{g}\})_{n=0}^{N-1}$ with a good lattice point $\mathbf{g} = (g_1,\ldots,g_d)$ modulo N, $(g_i, N) = 1$ for all i, $1 \leq i \leq d$, we derive the following identity from Example 6.1,

$$F_N(\omega) = P_2(\mathbf{g}, N).$$

This relation shows that the figure of merit P_2 for lattice points $\mathbf{g}$ is a much more natural quantity than it might appear at the first glance.

(3) If we would like to have the relation $0 \leq F_N(\omega) \leq 1$, then we will have to multiply $F_N(\omega)$ by the factor $((1 + \pi^2/3)^d - 1)^{-1/2}$.

(4) The interpretation of diaphony as an example of the weighted spectral test is due to Niederreiter and the author [HN98, Hel97].

Example 6.2 (Dyadic Diaphony, Hellekalek and Leeb[HL97]) We choose $\mathcal{F} = \mathcal{W}$, and the weight function

$$r(\mathbf{k}) = \prod_{i=1}^{d} r(k_i),$$

where

$$r(k) := \begin{cases} 2^g & \text{if } k: 2^g \leq k < 2^{g+1}, \text{with } g \geq 0, \ g \in \mathbb{Z}, \\ 1 & \text{if } k = 0, \end{cases}$$

for an integer vector $\mathbf{k} = (k_1,\ldots,k_d)$ with nonnegative coordinates k_i. Hellekalek and Leeb[HL97] have proved that the weighted spectral test $F_N(\mathcal{W},r;\omega)$ of ω can be realized by the function

$$f(\mathbf{x}) = -1 + \prod_{i=1}^{d} g(x_i), \quad \mathbf{x} = (x_1,\ldots,x_d),$$

$$g(x) = \begin{cases} 3 - 3 \cdot 2^{1+\lfloor \log_2 x \rfloor} & \text{if } x \in]0,1[, \\ 3 & \text{if } x = 0, \end{cases} \tag{1.28}$$

and dyadic addition $\dotplus$,

$$F_N^2(\mathcal{W},r;\omega) = \frac{1}{N^2} \sum_{n=0}^{N-1} \sum_{m=0}^{N-1} f(\mathbf{x}_n \dotplus \mathbf{x}_m). \tag{1.29}$$

Remark 6.3 (1) Dyadic diaphony may be viewed as a quantitative and continuous version of Tezuka's Walsh spectral test in [Tez87].

(2) Both examples above of a weighted spectral test define measures of uniform distribution of sequences. This means that the diaphony converges to

zero for N to infinity if and only if the sequence ω is uniformly distributed. This fact is derived easily from Weyl's Criterion.

(3) We also have appropriate variants of the Erdös-Turán-Koksma inequality for diaphony and dyadic diaphony at our disposition. Hence, discrepancy estimates that have been obtained with Niederreiter's technique (see Chapter 7) can be rewritten to yield estimates of diaphony. We refer the reader to [HN98, HL97] for examples and details.

6.3 Remarks

Although the weighted spectral test is a recent concept, the asymptotic distribution of diaphony and dyadic diaphony is already known, due to the work of James, Hoogland, and Kleiss[JHK96], van Hameren, Kleiss, and Hoogland[vHKH97], Leeb[Lee97b, Lee97a], and Leeb and Hellekalek[LH98]. This is a marked difference to discrepancy, where such results exist only for dimension one. On the other hand, a law of the iterated logarithm is known for discrepancy, see Kiefer[Kie61], but not (yet) for diaphony.

We have seen that the computational complexity of diaphony and dyadic diaphony is $\mathcal{O}(dN^2)$ for any set of N points in the d-dimensional unit cube. This compares favorably to the computational complexity of discrepancy, which is $\mathcal{O}(N^d)$ and allows the use of diaphony for practical computations.

James, Hoogland, and Kleiss[JHK96] have been able to provide an important interpretation for diaphony. They have shown that diaphony and dyadic diaphony are the mean squared integration error for particular classes of integrands. In this context, it is important to mention the work of Frank and Heinrich [FH96, FH97], who have investigated the question of computing discrepancies related to spaces of smooth periodic functions, in particular for Smolyak quadrature rules.

Strauch[Str94] has introduced a generalized version of L^2-discrepancy in the form

$$D_F(\omega, N) := \frac{1}{N^2} \sum_{m,n=0}^{N-1} F(\mathbf{x}_m, \mathbf{x}_n),$$

where $F : [0,1[^d \times [0,1[^d \to \mathbb{R}$ and $\omega = (\mathbf{x}_n)_{n \geq 0}$ is a sequence in the d-dimensional unit cube. He was able to show that this concept generalizes the classical L^2-discrepancy as well as diaphony.

An interesting interpretation of diaphony was given by Amstler[Ams95] who has used discrepancy operators to generalize the concepts of diaphony and L^2-discrepancy to compact groups. The same author has shown how to define diaphony in the abstract setting of (not necessarily abelian) compact groups in [Ams97].

An independent interpretation of diaphony is due to Lev[Lev95]. For a given finite sequence $\omega = (\mathbf{x}_n)_{n=0}^{N-1}$ in $[0,1[^d$, Lev assigns weights ρ_n to each point $\mathbf{x}_n$. The *local discrepancy* $R_N(I)$ of a d-dimensional interval $I =$

$[\mathbf{u}, \mathbf{v}[= \prod_{i=1}^{d}[u_i, v_i[,\ \mathbf{u} = (u_1, \ldots, u_d),\ \mathbf{v} = (v_1, \ldots, v_d),\ 0 \le u_i < v_i \le 1,$ $1 \le i \le d$, is defined as the number

$$R_N(I) := \sum_{n : \mathbf{x}_n \in I} \rho_n - \lambda_d(I).$$

The *weighted L^p-discrepancy* $D_N(\omega)$ is defined by Lev as

$$D_N(\omega) := \|R_N([\mathbf{0}, \cdot[)\|_p.$$

Further, a so-called Weyl-L^p-discrepancy is introduced. It is equivalent to the L^p-average of the weighted L^p-discrepancy $D_N(\omega + \alpha)$ of all shifted sets $\omega + \alpha$, where α runs through $[0, 1[^d$. Lev[Lev95, Th.1] showed that there is a direct relation between the Weyl-L^2-discrepancy and diaphony.

The contribution of Hickernell in this volume contains an extension of the concept of the L^2-discrepancy which is similar to the generalizations in [JHK96, Str94, Lev95] cited above.

7 Discrepancy

7.1 Definition

Discrepancy is the best-known numerical quantity to measure the equidistribution of (finite) sequences in $[0, 1[^d$. We will introduce this figure of merit by its statistical background and not in the usual number-theoretic terminology.

The two-sided Kolmogorov-Smirnov test statistic (KS statistic) is one of the most important goodness-of-fit tests in statistics. We are given a sample $\mathbf{x}_0, \ldots, \mathbf{x}_{N-1}$ of size N and want to test the hypothesis that this sample stems from a particular continuous probability distribution. In order to answer this question, we compute the empirical distribution function F_N of this sample and compare it to the cumulative distribution function F that underlies our hypothesis. The L^∞-norm $\|F_N - F\|_\infty$ of the difference of these two functions multiplied by $\sqrt{N}$ is called the KS-statistic.

In our case, we want to assess the equidistribution of a sample $\omega = (\mathbf{x}_n)_{n=0}^{N-1}$ on the d-dimensional torus. The theoretical distribution is uniform distribution on $[0, 1[^d$, hence the cumulative distribution function F equals $F(\mathbf{x}) = \prod_{i=1}^{d} x_i$, $\mathbf{x} = (x_1, \ldots, x_d) \in [0, 1[^d$. We now compare this target distribution to the empirical distribution of ω. This yields the following value of the KS-statistic:

$$D_N^*(\omega) = \sup_{\mathbf{x} \in [0,1[^d} \left| \frac{1}{N} \sharp \left\{ n, 0 \le n < N : \mathbf{x}_n \in \prod_{i=1}^{d}[0, x_i[\right\} - \prod_{i=1}^{d} x_i \right|.$$

In other words, we have molded the question of assessing the equidistribution of ω into a quantitative form by studying the distance between two distribution functions. Of course, we might have used the L^2-norm $\|F_N - F\|_2$

as well. This would have given rise to the so-called L^2-discrepancy. The associated goodness-of-fit test is known as the Cramér-von Mises test statistic, see Subsection 7.3 for further information on this interesting variant of discrepancy.

In the theory of uniform distribution of sequences modulo one, the quantity $D_N^*(\omega)$ is called the *star discrepancy*. It is common in this field to use a closely related notion of discrepancy to assess RNGs and LDPs, the *extreme discrepancy*.

Definition 7.1 *Let* $\omega = (x_n)_{n \geq 0}$ *be a sequence in* $[0,1[^s$. *The* extreme discrepancy $D_N(\omega)$ *of* ω *is defined as*

$$D_N(\omega) := \sup_{I \in \mathcal{J}} \left| \frac{1}{N} \cdot \sharp \{n, 0 \leq n < N : x_n \in I\} - \lambda_d(I) \right| ,$$

where $\mathcal{J}$ *denotes the class of all subintervals* I *of* $[0,1[^d$ *of the form* $I = \prod_{i=1}^{d} [u_i, v_i[, \ 0 \leq u_i < v_i \leq 1, \ 1 \leq i \leq d$.

We would like to point out that in statistics one uses closed intervals in the definition of the KS statistic. The choice of half-open intervals does not change the numerical value of discrepancy. The (number-theoretic) tradition of taking half-open intervals stems from interpreting $[0,1[^d$ as the d-dimensional torus.

7.2   The Inequality of Erdös-Turán-Koksma

In number-theoretical practice, we are unable to compute discrepancy, but we can give bounds for it in many cases. In addition to elementary counting arguments, as they are employed in the study of sequences of the van-der-Corput family, there is a powerful analytic technique, the *inequality of Erdös-Turán-Koksma*. The general setting for this inequality is the following.

(1) *The question.* We want to derive an upper bound for the discrepancy $D_N^*(\omega)$ or $D_N(\omega)$ of some given sequence ω in the d-dimensional unit cube $[0,1[^d$.

(2) *The prerequisite.* For a function system $\mathcal{F} = \{\chi_\mathbf{k}\}$ of our choice, we have to know upper bounds for the absolute value of the Weyl sums

$$S_N(\chi_\mathbf{k}, \omega),$$

for all indices $\mathbf{k} = (k_1, \ldots, k_d) \neq \mathbf{0}$ in some finite range $\Delta^*(M)$. The parameter M denotes a positive integer that defines this range.

(3) *The result.* The general inequality of Erdös-Turán-Koksma has the form

$$D_N(\omega) \leq C_d \cdot \left(\epsilon_{\mathcal{F}}(M) + \sum_{\mathbf{k} \in \Delta^*(M)} \rho_{\mathcal{F}}(\mathbf{k}) \cdot |S_N(\chi_\mathbf{k}, \omega)| \right) . \tag{1.30}$$

C_d is a constant that depends only on the dimension d. The functions $\epsilon_{\mathcal{F}}(M)$ and $\rho_{\mathcal{F}}(\mathbf{k})$ depend on the system $\mathcal{F}$, $\epsilon_{\mathcal{F}}(M)$ stands for an approximation error. For example, if $\mathcal{F} = \mathcal{W}(q)$ or $\mathcal{H}(q)$, then $\epsilon_{\mathcal{F}}(M)$ represents the error we commit when we approximate arbitrary intervals in $[0,1[^d$ by particular q-adic cubes (see [Hel95b] for details).

This inequality relates the behaviour of Weyl sums to discrepancy. We refer the reader to Drmota and Tichy[DT97] for a thorough treatment of this topic. We will now discuss several examples.

Example 7.1 The classical inequality of Erdös-Turán-Koksma (see Grabner and Tichy[GT90]).
$\mathcal{F} = \mathcal{T}$,
$\Delta^*(M) = \{\mathbf{k} = (k_1, \ldots, k_d) \in \mathbb{Z}^d \setminus \{\mathbf{0}\} : \max_{1 \le i \le d} |k_i| < M\}$,
$C_d = 2 \cdot (3/2)^d$,
$\epsilon_{\mathcal{F}}(M) = 1/M$,
$\rho_{\mathcal{F}}(\mathbf{k}) = \prod_{i=1}^{d} (\max\{1, |k_i|\})^{-1}$.

Example 7.2 The inequality of Erdös-Turán-Koksma for the Walsh system to the base q (see [Hel94b]).
$\mathcal{F} = \mathcal{W}(q)$,
$\Delta^*(M) = \{\mathbf{k} = (k_1, \ldots, k_d) \in \mathbb{Z}^d \setminus \{\mathbf{0}\} : 0 \le k_i < M \quad \forall\, i, 1 \le i \le d\}$,
$C_d = 1$,
$\epsilon_{\mathcal{F}}(M) = 1 - (1 - 2/M)^d$,
$\rho_{\mathcal{F}}(\mathbf{k}) = \prod_{i=1}^{d} \rho_{Walsh}(k_i)$,

$$\rho_{Walsh}(k) := \begin{cases} 1 & : \quad \text{if } k=0, \\ q^{-g-1} \cdot 2\left(\sin \pi \frac{k_g}{q}\right)^{-1} & : \quad \text{if } q^g \le k < q^{g+1}, g \in \mathbb{Z}, g \ge 0. \end{cases}$$

Example 7.3 The inequality of Erdös-Turán-Koksma for the Haar system to the base q (see [Hel95b]).
$\mathcal{F} = \mathcal{H}(q)$,
$\Delta^*(M)$ as in Example 7.2,
$C_d = 1$,
$\epsilon_{\mathcal{F}}(M) = 1 - (1 - 2/M)^d$.
Due to the "local" nature of the Haar functions, we present the inequality of Erdös-Turán-Koksma for $\mathcal{H}(q)$ in its "raw" form

$$D_N(\omega) \le 1 - (1 - 2/M)^d + \sup_{I \in \mathcal{J}} \sum_{\mathbf{k} \in \Delta^*(M)} |\hat{\mathbf{1}}_{G(I)}(\mathbf{k})|\, |S_N(h_{\mathbf{k}}, \omega)|.$$

The reason for this is as follows. For every interval $I \in \mathcal{J}$, $\mathcal{J}$ as in Definition 7.1, we construct an approximating set $G(I)$. It will turn out that the function $\mathbf{1}_{G(I)}$ has a finite Haar series and that $\hat{\mathbf{1}}_{G(I)}(\mathbf{k}) = 0$ for all $\mathbf{k} \notin \Delta^*(M)$. Further, most of the numbers $\hat{\mathbf{1}}_{G(I)}(\mathbf{k})$, $\mathbf{k} \in \Delta^*(M)$, will be zero, see [Hel95b] for the details. Hence it does not make sense to give a global estimate for $|\hat{\mathbf{1}}_{G(I)}(\mathbf{k})|$, in contrast to the case of Walsh functions. The inequality

$$D_N(\omega) \le 1 - (1 - 2/M)^d + \sum_{\mathbf{k} \in \Delta^*(M)} \rho(\mathbf{k}) \, |S_N(h_{\mathbf{k}}, \omega)|.$$

with

$$\rho(\mathbf{k}) := \sup_{I \in \mathcal{J}} |\hat{\mathbf{1}}_{G(I)}(\mathbf{k})|$$

would be too weak. We can do much better. It is possible to estimate the number of indices $\mathbf{k}$ where the Haar coefficient $\hat{\mathbf{1}}_{G(I)}(\mathbf{k})$ is different from zero. Further, if $\hat{\mathbf{1}}_{G(I)}(\mathbf{k}) \ne 0$, then

$$|\hat{\mathbf{1}}_{G(I)}(\mathbf{k})| \le \rho_{\mathcal{H}(q)}(\mathbf{k}) := \prod_{i=1}^{d} \rho_{Haar}(k_i),$$

where

$$\rho_{Haar}(k) := \begin{cases} 1 & : \quad \text{if } k = 0, \\ q^{-g/2-1} \cdot \left(\sin \pi \frac{k_g}{q} \right)^{-1} & : \quad \text{if } q^g \le k < q^{g+1}, g \in \mathbb{Z}, g \ge 0. \end{cases}$$

For this reason we have chosen to present the inequality of Erdös-Turán-Koksma in its strongest but "unpolished" form. We would like to call it the "raw form" of the inequality. It allows the best estimates.

In order to estimate the discrepancy of a finite sequence ω, where this sequence stems from an RNG or is an LDP itself, the common approach is the following. Suppose that we are able to establish a bound B for the Weyl sums,

$$B_{\mathcal{F}} := \max_{\mathbf{k} \in \Delta^*(M)} |S_N(\chi_{\mathbf{k}}, \omega)|$$

in the case where $\mathcal{F} \in \{\mathcal{T}, \mathcal{W}(q), \mathcal{H}_0(q)\}$, $\mathcal{F} = \{\chi_{\mathbf{k}}\}$. Then the following results hold.
   (1) $\mathcal{F} = \mathcal{T}$:

$$D_N(\omega) \le C_d \cdot \left(\frac{1}{M} + B_{\mathcal{T}} \cdot \sum_{\mathbf{k} \in \Delta^*(M)} \rho_{\mathcal{T}}(\mathbf{k}) \right).$$

The sum on the right side is well-known: see Niederreiter[Nie77, Ineq. (3.4), p. 112, and Lemma 2.3, p. 109].
   (2.1) $\mathcal{F} = \mathcal{W}(q), M = q$:

$$D_N(\omega) \le 1 - \left(1 - \frac{2}{M} \right)^d + B_{\mathcal{W}(q)} \cdot \left(\frac{4}{\pi^2} \log M + 1.72 \right)^d.$$

For this result, see Niederreiter[Nie92b, Cor. 3.11] and [Hel94b, Cor. 2]. Note that our discretization error is slightly larger. This is due to the fact that we allow *arbitrary* sequences ω. We get the same bounds for the discretization error if we restrict ourselves to the special point sets $\mathcal{P}$ that have been considered in the papers [Nie92b, Hel94b] cited above.

(2.2) $\mathcal{F} = \mathcal{W}(q), M = q^{\alpha}, \alpha \geq 2$:

$$D_N(\omega) \leq 1 - \left(1 - \frac{2}{M}\right)^d + B_{\mathcal{W}(q)} \cdot \left(2 \cdot C(q) \cdot \frac{\log M}{\log q} + 1\right)^d,$$

see [Hel94b, Cor. 4]. The quantity $C(q)$ is defined as

$$C(q) = \frac{1}{q} \sum_{a=1}^{q-1} \frac{1}{\sin \pi \frac{a}{q}}.$$

(3.1) $\mathcal{F} = \mathcal{H}(q), M = q$: In this case we get the same estimate as in (2.1), see [Hel95b].

(3.2) $\mathcal{F} = \mathcal{H}(q), M = q^{\alpha}, \alpha \geq 2$:

$$D_N(\omega) \leq 1 - \left(1 - \frac{2}{M}\right)^d + B_{\mathcal{H}_0(q)} \cdot \left(2 \cdot C(q) \cdot \left(\frac{\log M}{\log q} - \frac{1}{2}\right) + 1\right)^d,$$

see [Hel95b].

For the star discrepancy $D_N^*(\omega)$ with respect to $\mathcal{F}$, the inequality of Erdös-Turàn-Koksma has the same form, with the following adjustments. In Example 7.3, the function ρ_{Haar} remains unchanged. In Example 7.2, in the definition of the function ρ_{Walsh}, the factor $2 \cdot (\sin \pi(k_g/q))^{-1}$ has to be replaced by $(\sin \pi(k_g/q))^{-1}$ (see [Hel94b, Theorem 1]).

Niederreiter[Nie92b, p.215] has introduced the concept of *discrete discrepancy*. In [Nie94b, Lem. 1, Cor. 3], the same author has shown estimates of the discrete discrepancy. With our approach it is possible to prove these two results without difficulty (see [Hel95b, p. 39]).

7.3 Remarks

It is not (yet) known how to compute the star-discrepancy nor the extreme discrepancy in practice. Both quantities have a computational complexity of $\mathcal{O}(N^d)$ for N points in $[0, 1[^d$. There is an interesting probabilistic approach to compute discrepancy, see Winker and Fang[WF95] for first results. A limit law for discrepancy in dimensions higher than one is not available. Holt[Hol96] has obtained an interesting variant of the inequality of Erdös-Turán-Koksma for the *discrepancy over balls*, on the torus $[0, 1[^d$ (see also Holt and Vaaler[HV96]). Amstler[Ams97] has extended the inequality of Erdös-Turán-Koksma to compact abelian groups.

From a practical point of view, the L^2-discrepancy is much more attractive. As we have indicated in Chapter 6, it is closely related to diaphony, see

Strauch[Str94], Lev[Lev95], Frank and Heinrich[FH97], and the contribution of Hickernell in this volume. Due to Heinrich[Hei96], we know an efficient algorithm to compute its value (see also Frank and Heinrich[FH96]).

We refer the reader to Drmota and Tichy[DT97] for many further references on discrepancy.

8 Summary

We have seen that RNGs and LDPs have much more in common than is reflected by the published literature. Both in Monte Carlo and quasi-Monte Carlo methods, we employ *finite deterministic point sets*. There is nothing random about these numbers or points.

The fundamental property of the low-discrepancy point sets that are used in quasi-Monte Carlo is super-uniformity. School I of correlation analysis proposes the same criterion for the *full-period* point sets produced by RNGs. In addition to this condition, after a suitable transformation, the *samples* produced by RNGs should be able to simulate as many probability distributions as possible, not just uniform distribution.

The figures of merit to assess LDPs have been discrepancy or quantities like $\rho(\mathbf{g}, N)$ or $P_2(\mathbf{g}, N)$. In the case of RNGs, it has been mainly the spectral test $\sigma(L)$ (in its classical, geometric version) or, again, discrepancy. We have seen in Chapter 5 that P_2 is a special value of diaphony and that diaphony is the L^2-version of the general spectral test. Hence, tables of good lattice points based upon the figure of merit P_2 are nothing else than numerical results based on the L^2-version of the spectral test. On the other hand, anyone who uses the classical spectral test (i.e. the quantity $\sigma(L)$, L a lattice) to choose good parameters for RNGs is trying to optimize the equidistribution of his point sets with the L^∞-version of the good lattice point criterion P_2. For example, linear congruential generators like those of L'Ecuyer[L'E98] yield extremely good LDPs based on the good lattice point method up to dimension $d = 32$. Numerical results like those presented in Entacher, Hellekalek, and L'Ecuyer[EHL98] provide empirical evidence for these theoretical findings, by comparing the performance of these LDPs to those of excellent nets.

Hence, for the assessment of the equidistribution of LDPs as well as for the assessment of correlations within the output stream of an RNG, we use the same basic concepts. Many of the numerical quantities involved in this kind of research are contained in our concept of the general spectral test as particular examples.

Our future research in this field will deal with the relations between these figures of merit and empirical studies like the distribution of the spectral test over sets of parameters.

9   Acknowledgements

This work has been supported by the grants P9285, P11143, and P12654 of the Austrian Science Fund (FWF).

Harald Niederreiter and Pierre L'Ecuyer have provided valuable references and hints for this contribution. The author would also like to thank Françoise Axel and Jean-Pierre Gazeau for their hospitality during a stay at the Centre de Physique des Houches, France, in February 1998, where parts of this paper were written. Finally, the author would like to express his gratitude to his students Karl Entacher, Hannes Leeb, Otmar Lendl, Karin Schaber, and Stefan Wegenkittl, for very inspiring questions and comments and also for the assistance with computing.

10 REFERENCES

[Ams95]   C. Amstler. Discrepancy operators and numerical integration on compact groups. *Mh. Math.*, **119**:177–186, 1995.

[Ams97]   C. Amstler. Some remarks on a discrepancy in compact groups. *Arch. Math.*, **68**:274–284, 1997.

[And90]   S.L. Anderson. Random number generation on vector super-computers and other advanced architectures. *SIAM Review*, **32**:221–251, 1990.

[BFN92]   P. Bratley, B.L. Fox, and H. Niederreiter. Implementation and tests of low-discrepancy sequences. *ACM Trans. Modeling and Computer Simulation*, **2**:195–213, 1992.

[Cho95]   Wun-Seng Chou. On inversive maximal period polynomials over finite fields. *Appl. Algebra Engrg. Comm. Comput.*, **6**:245–250, 1995.

[CM67]   R.R. Coveyou and R.D. MacPherson. Fourier analysis of uniform random number generators. *J. Assoc. Comput. Mach.*, **14**:100–119, 1967.

[Com95]   A. Compagner. Operational conditions for random-number generation. *Phys. Review E*, **52**:5634–5645, 1995.

[Die75]   U. Dieter. How to calculate shortest vectors in a lattice. *Math. Comp.*, **29**:827–833, 1975.

[DP90]   A. DeMatteis and S. Pagnutti. Long-range correlations in linear and non-linear random number generators. *Parallel Computing*, **14**:207–210, 1990.

[DT97]   M. Drmota and R.F. Tichy. *Sequences, Discrepancies and Applications*, volume 1651 of *Lecture Notes in Mathematics*. Springer, Berlin, 1997.

[Edd90]   W.F. Eddy. Random number generators for parallel processors. *J. Comp. Appl. Math.*, **31**:63–71, 1990.

[EH93]   J. Eichenauer-Herrmann. Statistical independence of a new class of inversive congruential pseudorandom numbers. *Math. Comp.*, **60**:375–384, 1993.

[EHHW97]   J. Eichenauer-Herrmann, E. Herrmann, and S. Wegenkittl. A survey of quadratic and inversive congruential pseudorandom numbers. In Niederreiter et al. [NHLZ97], pages 66–97.

[EHL98]   K. Entacher, P. Hellekalek, and P. L'Ecuyer. Quasi-Monte Carlo integration with linear congruential generators. Submitted, 1998.

[EHN94]   J. Eichenauer-Herrmann and H. Niederreiter. Bounds for exponential sums and their applications to pseudorandom numbers. *Acta Arith.*, **67**:269–281, 1994.

[EL86]   J. Eichenauer and J. Lehn. A non-linear congruential pseudo random number generator. *Statist. Papers*, **27**:315–326, 1986.

[Ent97a] K. Entacher. A collection of selected pseudorandom number generators with linear structures. Technical report series, ACPC - Austrian Center for Parallel Computation, 1997.

[Ent97b] K. Entacher. Quasi-Monte Carlo methods for numerical integration of multivariate Haar series. *BIT*, **37**:846–861, 1997.

[Ent98]   K. Entacher. Bad subsequences of well-known linear congruential pseudorandom number generators. *ACM Trans. Modeling and Computer Simulation*, **8**:61–70, 1998.

[EUW98]   K. Entacher, A. Uhl, and S. Wegenkittl. Linear congruential generators for parallel Monte-Carlo: the leap-frog case. *Monte Carlo Methods and Appl.*, **4**:1–16, 1998.

[FH96]   K. Frank and S. Heinrich. Computing discrepancies of Smolyak quadrature rules. *J. Complexity*, **12**:287–314, 1996.

[FH97]   K. Frank and S. Heinrich. Computing discrepancies related to spaces of smooth periodic functions. In Niederreiter et al. [NHLZ97], pages 238–250.

[Fis96]   G.S. Fishman. *Monte Carlo: Concepts, Algorithms, and Applications.* Springer-Verlag, New York, 1996.

[FM82]   G.S. Fishman and L.R. Moore. A statistical evaluation of multiplicative congruential random number generators with modulus $2^{31} - 1$. *J. Amer. Statist. Assoc.*, **77**:129–136, 1982.

[Gen84]   A.C. Genz. Testing multidimensional integration routines. In B. Ford, J.C. Rault, and F. Thomasset, editors, *Tools, Methods, and Languages for Scientific and Engineering Computation*, pages 81–94. North-Holland, Amsterdam, 1984.

[Gen87]   A.C. Genz. A package for testing multiple integration subroutines. In P. Keast and G. Fairweather, editors, *Numerical Integration: Recent Developments, Software and Applications*, pages 337–340. D. Reidel, Dordrecht, 1987.

[GES91]   B. Golubov, A. Efimov, and V. Skvortsov. *Walsh Series and Transforms*. Kluwer, Doordrecht, 1991.

[GL87]   P.M. Gruber and C.G. Lekkerkerker. *Geometry of Numbers*. North-Holland, Amsterdam, second edition, 1987.

[GT90]   P.J. Grabner and R.F. Tichy. Remark on an inequality of Erdös-Turán-Koksma. *Anz. Österr. Akad. Wiss. Math.-Natur. Kl.*, **127**:15–22, 1990.

[Hei96]   Heinrich, S. Efficient algorithms for computing the L_2-discrepancy. *Math. Comp.*, **65**:1621–1633, 1996.

[Hel84]   P. Hellekalek. Regularities in the distribution of special sequences. *J. Number Th.*, **18**:41–55, 1984.

[Hel94a]   P. Hellekalek. General discrepancy estimates II: the Haar function system. *Acta Arith.*, **67**:313–322, 1994.

[Hel94b]   P. Hellekalek. General discrepancy estimates: the Walsh function system. *Acta Arith.*, **67**:209–218, 1994.

[Hel95a]   P. Hellekalek. Correlations between pseudorandom numbers: theory and numerical practice. In P. Hellekalek, G. Larcher, and P. Zinterhof, editors, *Proceedings of the 1st Salzburg Minisymposium on Pseudorandom Number Generation and Quasi-Monte Carlo Methods, Salzburg, 1994*, volume ACPC/TR 95-4 of *Technical Report Series*, pages 43–73. ACPC – Austrian Center for Parallel Computation, University of Vienna, 1995.

[Hel95b]   P. Hellekalek. General discrepancy estimates III: the Erdös-Turán-Koksma inequality for the Haar function system. *Monatsh. Math.*, **120**:25–45, 1995.

[Hel95c]   P. Hellekalek. Inversive pseudorandom number generators: concepts, results, and links. In C. Alexopoulos, K. Kang, W.R. Lilegdon, and D. Goldsman, editors, *Proceedings of the 1995 Winter Simulation Conference*, pages 255–262, 1995.

[Hel97]   P. Hellekalek. On correlation analysis of pseudorandom numbers. In Niederreiter et al. [NHLZ97], pages 251–265.

[Hel98]   P. Hellekalek. Good random number generators are (not so) easy to find. *Mathematics and Computers in Simulation*, 1998.

[Hic98]   F. J. Hickernell. A generalized discrepancy and quadrature error bound. *Math. Comp.*, **67**:299–322, 1998.

[HL97]   P. Hellekalek and H. Leeb. Dyadic diaphony. *Acta Arith.*, 80:187–196, 1997.

[HN98]   P. Hellekalek and H. Niederreiter. The weighted spectral test: diaphony. *ACM Trans. Modeling and Computer Simulation*, 8:43–60, 1998.

[Hol96]   J. J. Holt. On a form of the Erdös-Turán inequality. *Acta Arith.*, 74:61–66, 1996.

[HV96]   J. J. Holt and J. D. Vaaler. The Beurling-Selberg extremal functions for a ball in Euclidean space. Preprint, submitted, 1996.

[JHK96]   F. James, J. Hoogland, and R. Kleiss. Multidimensional sampling for simulation and integration: measures, discrepancies, and quasi-random numbers. Preprint submitted to Computer Physics Communications, 1996.

[Kie61]   J. Kiefer. On large deviations of the empiric d.f. of vector chance variables and a law of the iterated logarithm. *Pacific J. Math.*, 11:649–660, 1961.

[KN74]   L. Kuipers and H. Niederreiter. *Uniform Distribution of Sequences*. John Wiley, New York, 1974.

[Knu81]   D.E. Knuth. *The Art of Computer Programming, Vol. 2*. Addison-Wesley, Reading, Mass., second edition, 1981.

[Kor59]   N.M. Korobov. The approximate computation of multiple integrals. *Dokl. Akad. Nauk SSSR*, 124:1207–1210, 1959. (In Russian).

[Kor63]   N.M. Korobov. *Number-Theoretic Methods in Approximate Analysis*. Fizmatgiz, Moscow, 1963. (In Russian).

[LA97]   P. L'Ecuyer and T.H. Andres. A random number generator based on the combination of four LCGs. *Mathematics and Computers in Simulation*, 44:99–107, 1997.

[Lag93]   J. C. Lagarias. Pseudorandom numbers. *Statistical Science*, 8:31–39, 1993.

[LBC93]   P. L'Ecuyer, F. Blouin, and R. Couture. A search for good multiple recursive random number generators. *ACM Trans. Modeling and Computer Simulation*, 3:87–98, 1993.

[LC91]   P. L'Ecuyer and S. Coté. Implementing a random number package with splitting facilities. *ACM Trans. Math. Software*, 17:98–111, 1991.

[LC97]   P. L'Ecuyer and R. Couture. An implementation of the lattice
            and spectral tests for multiple recursive linear random number
            generators. *INFORMS J. on Comput.*, 9:206–217, 1997.

[LCC96]   P. L'Ecuyer, A. Compagner, and J.-F. Cordeau. Entropy tests
            for random number generators. Manuscript, 1996.

[LCS96]   P. L'Ecuyer, J.-F. Cordeau, and R. Simard. Close-point spa-
            tial tests for random number generators. Preprint, submitted,
            1996.

[L'E92]   P. L'Ecuyer. Testing random number generators. In J.J. Swain
            et al., editor, *Proc. 1992 Winter Simulation Conference (Ar-
            lington, Va., 1992)*, pages 305–313. IEEE Press, Piscataway,
            N.J., 1992.

[L'E96]   P. L'Ecuyer. Combined multiple-recursive random number
            generators. *Operations Res.*, 44:816–822, 1996.

[L'E97a]   P. L'Ecuyer. Bad lattice structures for vectors of non-
            successive values produced by some linear recurrences. *IN-
            FORMS J. on Computing*, 9:57–60, 1997.

[L'E97b]   P. L'Ecuyer. Random number generation. In Jerry Banks,
            editor, *Handbook on Simulation*. Wiley, New York, 1997.

[L'E97c]   P. L'Ecuyer. Random number generators and empirical tests.
            In Niederreiter et al. [NHLZ97], pages 124–138.

[L'E98]   P. L'Ecuyer. Tables of linear congruential generators of differ-
            ent sizes and good lattice structure. *Math. Comp.*, 1998. To
            appear.

[Lee95]   H. Leeb. Random numbers for computer simulation. Master's
            thesis, Institut für Mathematik, Universität Salzburg, Austria,
            1995. Available from http://random.mat.sbg.ac.at/.

[Lee97a]   H. Leeb. *Stochastic properties of diaphony*. PhD thesis,
            University of Salzburg, Dept. of Mathematics, University of
            Salzburg, Austria, 1997.

[Lee97b]   H. Leeb. Weak limits for diaphony. In Niederreiter et al.
            [NHLZ97], pages 330–339.

[Lev95]   V. F. Lev. On two versions of L^2-discrepancy and geometrical
            interpretation of diaphony. *Acta Math. Hungar.*, 69:281–300,
            1995.

[LH98]   H. Leeb and P. Hellekalek. Strong and weak laws for the spec-
            tral test and related quantities. Preprint, submitted, 1998.

[LSW94]   G. Larcher, W.Ch. Schmid, and R. Wolf. Representation of functions as Walsh series to different bases and an application to the numerical integration of high-dimensional Walsh series. *Math. Comp.*, **63**:701–716, 1994.

[LSW96]   G. Larcher, W.Ch. Schmid, and R. Wolf. Quasi-Monte Carlo methods for the numerical integration of multivariate Walsh series. *Math. Comput. Modelling*, **23**:55–67, 1996.

[LT94]   G. Larcher and C. Traunfellner. On the numerical integration of Walsh-series by number-theoretical methods. *Math. Comp.*, **63**:277–291, 1994.

[LW97]   H. Leeb and S. Wegenkittl. Inversive and linear congruential pseudorandom number generators in selected empirical tests. *ACM Trans. Modeling and Computer Simulation*, **7**:272–286, 1997.

[Mac92]   N.M. MacLaren. A limit on the usable length of a pseudorandom sequence. *J. Statist. Comput. Simul.*, **42**:47–54, 1992.

[Mar85]   G. Marsaglia. A current view of random number generators. In L. Brillard, editor, *Computer Science and Statistics: The Interface*, pages 3–10, Amsterdam, 1985. Elsevier Science Publishers B.V. (North Holland).

[MK92]   M. Matsumoto and Y. Kurita. Twisted GFSR generators. *ACM Trans. Modeling and Computer Simulation*, **2**:179–194, 1992.

[MK94]   M. Matsumoto and Y. Kurita. Twisted GFSR generators II. *ACM Trans. Modeling and Computer Simulation*, **4**:254–266, 1994.

[MN98]   M. Matsumoto and T. Nishimura. Mersenne Twister: A 623-dimensionally equidistributed uniform pseudorandom number generator. *ACM Trans. Modeling and Computer Simulation*, **8**:3–30, 1998.

[NHLZ97]   H. Niederreiter, P. Hellekalek, G. Larcher, and P. Zinterhof, editors. *Monte Carlo and Quasi-Monte Carlo Methods 1996*, volume 127 of *Springer Lecture Notes in Statistics*. Springer-Verlag, New York, 1997.

[Nie77]   H. Niederreiter. Pseudo-random numbers and optimal coefficients. *Adv. in Math.*, **26**:99–181, 1977.

[Nie78]   H. Niederreiter. Quasi-Monte Carlo methods and pseudorandom numbers. *Bull. Amer. Math. Soc.*, **84**:957–1041, 1978.

[Nie87]   H. Niederreiter. Point sets and sequences with small discrepancy. *Monatsh. Math.*, **104**:273–337, 1987.

[Nie88]   H. Niederreiter. Low-discrepancy and low-dispersion sequences. *J. Number Theory*, **30**:51–70, 1988.

[Nie92a]   H. Niederreiter. New methods for pseudorandom number and pseudorandom vector generation. In J.J. Swain et al., editor, *Proc. 1992 Winter Simulation Conference (Arlington, Va., 1992)*, pages 264–269. IEEE Press, Piscataway, N.J., 1992.

[Nie92b]   H. Niederreiter. *Random Number Generation and Quasi-Monte Carlo Methods*. SIAM, Philadelphia, 1992.

[Nie94a]   H. Niederreiter. On a new class of pseudorandom numbers for simulation methods. *J. Comp. Appl. Math.*, **56**:159–167, 1994.

[Nie94b]   H. Niederreiter. Pseudorandom vector generation by the inversive method. *ACM Trans. Modeling and Computer Simulation*, **4**:191–212, 1994.

[Nie95]   H. Niederreiter. New developments in uniform pseudorandom number and vector generation. In H. Niederreiter and P.J.-S. Shiue, editors, *Monte Carlo and Quasi-Monte Carlo Methods in Scientific Computing*, volume 106 of *Lecture Notes in Statistics*, pages 87–120. Springer-Verlag, New York, 1995.

[Rip87]   B.D. Ripley. *Stochastic Simulation*. John Wiley, New York, 1987.

[Sch71]   K. Schmidt. Eine Diskrepanz für Maßfolgen auf lokalkompakten Gruppen. *Z. Wahrscheinlichkeitstheorie verw. Geb.*, **17**:48–52, 1971.

[Sch96]   B. Schneier. *Applied Cryptography*. Wiley, New York, second edition, 1996.

[SJ94]   I.H. Sloan and S. Joe. *Lattice Methods for Multiple Integration*. Clarendon Press, Oxford, 1994.

[Sob67]   I.M. Sobol'. The distribution of points in a cube and the approximate evaluation of integrals. *Zh. Vychisl. Mat. i Mat. Fiz.*, **7**:784–802, 1967. (In Russian).

[Sob69]   I.M. Sobol'. *Multidimensional Quadrature Formulas and Haar Functions*. Izdat. "Nauka", Moscow, 1969. (In Russian).

[Str94]   O. Strauch. L^2 discrepancy. *Math. Slovaca*, **44**:601–632, 1994.

[SWS90] F. Schipp, W.R. Wade, and P. Simon. *Walsh Series. An Introduction to Dyadic Harmonic Analysis. With the collaboration of J. Pál.* Adam Hilger, Bristol and New York, 1990.

[Tez87] S. Tezuka. Walsh-spectral test for GFSR pseudorandom numbers. *Comm. ACM.*, 30:731–735, 1987.

[Tez95] S. Tezuka. *Uniform Random Numbers: Theory and Practice.* Kluwer Academic Publ., Norwell, Mass., 1995.

[vHKH97] A. van Hameren, R. Kleiss, and J. Hoogland. Gaussian limits for discrepancies. I: Asymptotic results. Preprint, 1997.

[Weg95] S. Wegenkittl. Empirical testing of pseudorandom number generators. Master's thesis, Institut für Mathematik, Universität Salzburg, Austria, 1995. Available from `http://random.mat.sbg.ac.at/`.

[WF95] P. Winker and K.-T. Fang. Application of threshold accepting to the evaluation of the discrepancy of a set of points. Research report, Universität Konstanz, 1995.

[Yue77] C.-K. Yuen. Testing random number generators by Walsh transform. *IEEE Trans. Comput.*, 26:329–333, 1977.

[Zin76] P. Zinterhof. Über einige Abschätzungen bei der Approximation von Funktionen mit Gleichverteilungsmethoden. *Sitzungsber. Österr. Akad. Wiss. Math.-Natur. Kl. II*, 185:121–132, 1976.

[ZS78] P. Zinterhof and H. Stegbuchner. Trigonometrische Approximation mit Gleichverteilungsmethoden. *Studia Sci. Math. Hungar.*, 13:273–289, 1978.

Author's address:
Peter Hellekalek, Institut für Mathematik, Universität Salzburg, Hellbrunner Straße 34, A-5020 Salzburg, Austria

e-mail address : `peter.hellekalek@sbg.ac.at`
WWW address: `http://random.mat.sbg.ac.at/`

Lattice Rules: How Well Do They Measure Up?

Fred J. Hickernell [1]

1 Introduction

A simple, but often effective, way to approximate an integral over the s-dimensional unit cube is to take the average of the integrand over some set P of N points. Monte Carlo methods choose P randomly and typically obtain an error of $O(N^{-1/2})$. Quasi-Monte Carlo methods attempt to decrease the error by choosing P in a deterministic (or quasi-random) way so that the points are more uniformly spread over the integration domain.

One popular family of such quasi-random points is the family of **integration lattices**, and the resulting quadrature formulae are called **lattice rules**. These lattice rules may be thought of as s-dimensional generalizations of the rectangle rule for one-dimensional quadrature. Lattice rules go back at least as far as Korobov [Kor59], but their general formulation and analysis began with the work of Sloan and his collaborators [Slo85, SK87]. Several excellent monographs review the large body of research on lattice rules [HW81, Nie92, SJ94].

The purpose of this chapter is to highlight some recent developments in the study of lattice rules. As the title suggests, the focus is on how well integration lattices perform for quadrature. The answer depends on how one measures performance.

The next section reviews some basic properties of lattice rules, including a popular quadrature error bound involving $\mathcal{P}_\alpha(L)$, a measure of the quality of an integration lattice. Section 3 shows this error bound to be a special case of quite general worst-case and average-case error analyses. Several other examples included in this general framework are given in Section 4, such as the Koksma-Hlawka inequality involving the star discrepancy. In Section 5 the notion of a shift-invariant discrepancy is introduced, and it is seen that a weighted form of $\mathcal{P}_\alpha(L)$ is closely connected to many interesting discrepancies.

Upper and lower bounds on $\mathcal{P}_\alpha(L)$ and related discrepancies are discussed in Section 6, and the discrepancies of integration lattices and nets

---

[1] Department of Mathematics, Hong Kong Baptist University, Kowloon Tong, Hong Kong, fred@hkbu.edu.hk, http://www.math.hkbu.edu.hk/~fred. This research was supported by an HKBU FRG grant 96-97/II-67.

are compared in Section 7. In Section 8 we look again at how the choice of weights in the definition of $\mathcal{P}_\alpha(L)$ can affect its size, in particular for high dimensions. The choice of weights both reflects and determines whether one takes an optimistic or pessimistic view of lattice rules for large s. Concluding remarks are given in the final section.

At several places in this chapter we quote results in somewhat more general form than they appear in the literature. Rather than proving the generalizations here, we refer the reader to the original proofs, which can be suitably modified without much difficulty.

Although this chapter touches on several recent developments in the study of lattice rules, the choice of topics is somewhat subjective and definitely not comprehensive. Some important topics not covered in this chapter include recent work on finding lattice rules with low Zaremba merit or trigonometric degree [CS96, Lan96, LS97], and canonical forms of lattice rules [Lan95, LJ96].

2   Some Basic Properties of Lattice Rules

The s-dimensional integral that we wish to evaluate has the form:

$$I(f) \equiv \int_{C^s} f(x)\,dx,$$

where $C^s = [0,1)^s$ is the unit cube. A general quasi-Monte Carlo quadrature rule to approximate this integral takes the average of the integrand values over some set $P \subset C^s$ with N points:

$$(2.1) \qquad Q(f) \equiv Q(f; P) \equiv \frac{1}{N} \sum_{z \in P} f(z).$$

Throughout this chapter P is allowed to have multiple copies of a point (see [Nie92, p. 14]). In this rule the weight given each integrand value is $1/N$. So, the weight is the same for integrand value and the sum of all the weights is one, which implies that the rule is exact for constant integrands. Quadrature rules with unequal weights and which may not be exact for constant integrands are considered in Section 6.3.

In the one-dimensional case the choice

$$P = \left\{ 0, \frac{1}{N}, \ldots, \frac{N-1}{N} \right\},$$

yields the left rectangle rule. There is more than one way to generalize this to a two-dimensional rule. The most straightforward way might be a grid,

$$P = \left\{ \left(\frac{i_1}{\sqrt{N}}, \frac{i_2}{\sqrt{N}} \right) : i_1, i_2 = 0, \ldots, \sqrt{N} - 1 \right\},$$

but another possibility is

$$P = \left\{ \left(\frac{i}{N}, \left\{ \frac{ih}{N} \right\} \right) : i = 0, \ldots, N - 1 \right\},$$

where h is some integer, and $\{x\} = x - \lfloor x \rfloor$ denotes the fractional part of x. Figures 2.1 and 2.2 show both of these sets P for the case $N = 64$ and $h = 51$. These P are the node sets of integration lattices as defined below:

Definition 2.1.

a. An s-dimensional **integration lattice** L is a discrete subset of $\mathbf{R}^s$ that is closed under addition and subtraction and which contains the integer vectors $\mathbf{Z}^s$ as a subset.

b. A **shifted lattice** with shift $\Delta \in \mathbf{R}^s$ is the set $L + \Delta \equiv \{z + \Delta : z \in L\}$ for some lattice L.

c. The **node set** for a shifted integration lattice, $L + \Delta$, is the set of points in the lattice that fall inside the unit cube, that is, $P = (L + \Delta) \cap C^s$.

d. The **dual lattice** of a lattice L is denoted $L^{\perp}$ and is defined as $\{k \in \mathbf{R}^s : k \cdot z \in \mathbf{Z} \text{ for all } z \in L\}$. (Here and below $\cdot$ denotes the dot product of two vectors.)

e. A **rank-1 lattice** is a lattice whose node set may be expressed as

(2.2) $$P = \{\{ih/N\} : i = 0, \ldots, N - 1\},$$

for some generating vector $h \in \mathbf{Z}^s$.

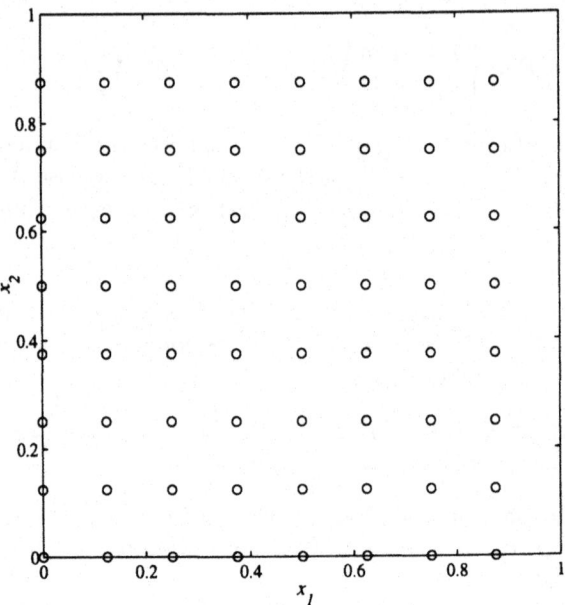

FIGURE 2.1. A grid of 64 points.

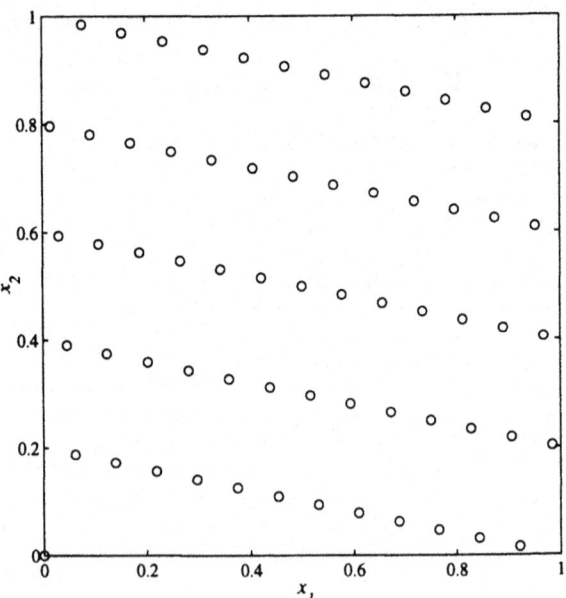

FIGURE 2.2. A rank-1 lattice with 64 points.

Although an integration lattice contains an infinite number of points, its node set, P, which defines the lattice quadrature rule through (2.1), is always finite. Below the notation $Q(f; L+\Delta)$ is sometimes used to denote a quadrature rule based on a (shifted) lattice, that is, $Q(f; L+\Delta) \equiv Q(f; P)$, where P is the node set for $L + \Delta$. The one-dimensional set

$$P = \left\{ \frac{1}{2N}, \frac{3}{2N}, \ldots, \frac{2N-1}{2N} \right\}$$

is the node set of the shifted lattice (with shift $\Delta = \frac{1}{2N}$) and gives the midpoint rule.

The left rectangle rule gives low accuracy in general, but it gives high accuracy if the integrand is periodic and has a sufficient degree of smoothness. The same general principle is true for lattice rules. Therefore, one often uses Fourier analysis to investigate the error of lattice rules. Suppose that the integrand, f, has an absolutely convergent Fourier series,

(2.3a)
$$f(x) = \sum_{k \in \mathbf{Z}^s} \hat{f}(k) e^{2\pi i k \cdot x},$$

where the Fourier coefficients, $\hat{f}(k)$ are defined by

(2.3b)
$$\hat{f}(k) \equiv \mathcal{F}\{f\}(k) \equiv \int_{C^s} f(x) e^{-2\pi i k \cdot x} \, dx.$$

The quadrature error for an absolutely convergent Fourier series is the sum of the error in each term. Since constants are integrated exactly, this implies that

(2.4)
$$I(f) - Q(f; P) = -{\sum_{k \in \mathbf{Z}^s}}' \hat{f}(k) Q(e^{2\pi i k \cdot x}; P)$$

$$= -{\sum_{k \in \mathbf{Z}^s}}' \left\{ \hat{f}(k) \left[\frac{1}{N} \sum_{z \in P} e^{2\pi i k \cdot z} \right] \right\},$$

where the notation $\sum'$ means that the zero term is to be omitted from the sum.

An intriguing feature of lattice rules is that they integrate the term $e^{2\pi i k \cdot x}$ with either no error or 100% error, depending on whether k is in the dual lattice:

Lemma 2.2. *[SJ94, Lemma 2.7] If L is an integration lattice, then*

$$Q\left(e^{2\pi i k \cdot x}; L\right) = 1_{\{k \in L^\perp\}} = \begin{cases} 1, & k \in L^\perp, \\ 0, & k \notin L^\perp. \end{cases}$$

Here and below, $1_{\{\cdot\}}$ denotes the indicator function. This lemma allows one to write the error of a shifted lattice rule in terms of the Fourier coefficients of the integrand evaluated on the dual lattice:

Theorem 2.3. *[SJ94, Theorem 2.10] The error of a shifted lattice rule is:*

$$I(f) - Q(f; L + \Delta) = -\sum_{k \in L^\perp}{}' \hat{f}(k) e^{2\pi i k \cdot \Delta}.$$

From this theorem one can easily derive a lattice rule error bound. First, we introduce the notation

(2.5)
$$\bar{k}_j = \begin{cases} |k_j|, & k_j \neq 0, \\ 1, & k_j = 0. \end{cases}$$

Note that his definition is somewhat different than the usual one, $\bar{k}_j = \max(|k_j|, 1)$, however, the two definitions are equivalent if k_j is an integer. An error bound for lattice rules follows by applying Hölder's inequality to Theorem 2.3:

Theorem 2.4. *[SJ94, Section 5.2] Suppose that the Fourier coefficients of the integrand decay fast enough so that:*

$$\left\| (\bar{k}_1 \cdots \bar{k}_s)^\alpha \hat{f}(k) \right\|_\infty < \infty \ \text{for some } \alpha > 1,$$

where $\|\cdot\|_p$ denotes the ℓ_p-norm. Then the error of a shifted lattice rule is bounded as follows:

(2.6)
$$|I(f) - Q(f; L + \Delta)| \leq \mathcal{P}_\alpha(L) \left\| 1_{\{k \neq 0\}} (\bar{k}_1 \cdots \bar{k}_s)^\alpha \hat{f}(k) \right\|_\infty,$$

where $\mathcal{P}_\alpha(L)$ is defined as:

(2.7)
$$\mathcal{P}_\alpha(L) \equiv \left\| \frac{1_{\{k \neq 0, k \in L^\perp\}}}{(\bar{k}_1 \cdots \bar{k}_s)^\alpha} \right\|_1 = \sum_{k \in L^\perp}{}' \frac{1}{(\bar{k}_1 \cdots \bar{k}_s)^\alpha}.$$

This theorem and the quantity $\mathcal{P}_\alpha(L)$ are the springboard for much of the analysis and discussion in the remainder of this chapter. To motivate what follows a couple of remarks are in order.

Remark 2.5. The error bound above is a product of *two* parts. One part measures the quality of the lattice. The other part measures the roughness of the integrand. The following sections investigate other error bounds that have this same general form.

Remark 2.6. Error bound (2.6) and $\mathcal{P}_\alpha(L)$ are specific to *lattice* rules. It is both desirable and possible to generalize these so that they apply to all quasi-Monte Carlo quadrature rules of the form (2.1). This will be done in Sections 3 and 4.

The quantity $\mathcal{P}_\alpha(L)$ may be written as the lattice rule error for the particular integrand $\xi_\alpha(x)$ defined as follows:

(2.8)
$$\xi_\alpha(x) = -\sum_{k \in Z^s} \frac{e^{2\pi i k \cdot x}}{(\bar{k}_1 \cdots \bar{k}_s)^\alpha}.$$

By Theorem 2.3 it follows that $\mathcal{P}_\alpha(L)$ is the quadrature error obtained when integrating ξ_α using an unshifted lattice L:

$$\mathcal{P}_\alpha(L) = I(\xi_\alpha) - Q(\xi_\alpha; L).$$

Remark 2.7. The error bound in Theorem 2.4 is tight, equality holding when the integrand is a constant multiple of ξ_α.

When α is a positive even integer one may write ξ_α and $\mathcal{P}_\alpha(L)$ in terms of Bernoulli polynomials. The definitions and some important properties of these polynomials are given in [AS64, Chapter 23]. The first few Bernoulli polynomials are:

$$B_0(x) = 1, \qquad B_1(x) = x - \frac{1}{2}, \qquad B_2(x) = x^2 - x + \frac{1}{6},$$

$$B_3(x) = x^3 - \frac{3}{2}x^2 + \frac{1}{2}x, \qquad B_4(x) = x^4 - 2x^2 + x - \frac{1}{30}.$$

The Fourier expansions of the Bernoulli polynomials of even degree are:

$$(2.9) \qquad B_\alpha(x) = \frac{-(-1)^{\alpha/2}\alpha!}{(2\pi)^\alpha} \sum_{k \neq 0} \frac{e^{2\pi i k x}}{|k|^\alpha}, \qquad 0 \leq x \leq 1$$

which implies that for even integer $\alpha \geq 2$:

$$\xi_\alpha(x) = -\prod_{j=1}^{s} \left[1 - \frac{(-1)^{\alpha/2}(2\pi)^\alpha}{\alpha!} B_\alpha(x_j) \right], \qquad x \in C^s,$$

$$\mathcal{P}_\alpha(L) = -1 + \frac{1}{N} \sum_{z \in P} \prod_{j=1}^{s} \left[1 - \frac{(-1)^{\alpha/2}(2\pi)^\alpha}{\alpha!} B_\alpha(z_j) \right],$$

where P is the node set of the lattice L.

Remark 2.8. The advantage of this formula for $\mathcal{P}_\alpha(L)$ is that it involves only a sum over the N points in the node set as opposed to the general formula (2.7) that requires a sum over an infinite number of points in the dual lattice.

3 A General Approach to Worst-Case and Average-Case Error Analysis

The previous section provides a formula for the error of lattice rules (Theorem 2.3) and an error bound for lattice rules (Theorem 2.4). In this section a general framework is presented that extends these results to general quasi-Monte Carlo quadrature rules and other spaces of integrands. Worst-case error analysis is discussed first, as this more closely parallels the development of the previous section. An average-case error analysis is also presented. For related presentations of worst-case and average-case error analyses see [Rit95, FH96, Hic98].

3.1 Worst-Case Quadrature Error for Reproducing Kernel Hilbert Spaces

In the worst-case error analysis the integrands are assumed to lie in a Hilbert space, W, of real-valued functions on C^s. This Hilbert space is assumed to have a reproducing kernel $K(x, y)$ and an inner product $\langle \cdot, \cdot \rangle_K$. The definition of a reproducing kernel is a function on $C^s \times C^s$ such that $K(\cdot, y) \in W$ for all $y \in C^s$, and

$$f(y) = \langle K(\cdot, y), f \rangle_K \text{ for all } f \in W \text{ and } y \in C^s.$$

For a more thorough discussion of reproducing kernels see [Sai88, Wah90]. It can be shown that any reproducing kernel is symmetric in its arguments and positive definite:

$$(3.1a) \qquad K(x, y) = K(y, x) \quad \text{for all } x, y \in C^s$$

$$(3.1b) \qquad \sum_{i,k=1}^{N} a^{(i)} a^{(k)} K(x^{(i)}, x^{(k)}) \geq 0 \quad \text{for all } a^{(i)} \in \mathbf{R}, \ x^{(i)} \in C^s.$$

Conversely, any function $K(x, y)$ satisfying these two conditions uniquely defines a reproducing kernel Hilbert space.

The advantage of assuming the integrands to be in a reproducing kernel Hilbert space is that the quadrature error is then a bounded linear functional on that space. Furthermore, its representer, ξ_K, may be easily found in terms of the reproducing kernel. By the definition of the reproducing kernel, $\xi_K(x) = \langle K(\cdot, x), \xi_K \rangle_K$, and by the definition of representer, $\langle K(\cdot, x), \xi_K \rangle_K = I(K(\cdot, x)) - Q(K(\cdot, x); P)$, so

$$(3.2) \qquad \xi_K(x) = I(K(\cdot, x)) - Q(K(\cdot, x); P)$$

(The dependence of ξ_K on the point set P defining the quadrature rule is suppressed for ease of notation). The Cauchy-Schwarz inequality then yields a worst-case error bound:

$$|I(f) - Q(f; P)| = |\langle \xi_K, f \rangle_K| \leq \|\xi_K\|_K \|f\|_K.$$

This error bound is tight since equality holds when f is a multiple of ξ_K, the worst-case integrand. The discrepancy is defined as the norm of the worst-case integrand, which is equivalent to the square root of the quadrature error for the worst-case integrand. The discrepancy may be written in terms of integrals and sums involving the reproducing kernel:

$$(3.3) \qquad D(P; K) \equiv \|\xi_K\|_K = \langle \xi_K, \xi_K \rangle_K^{1/2} = [I(\xi_K) - Q(\xi_K; P)]^{1/2}$$

$$= \left\{ \int_{C^s \times C^s} K(x, y) \, dx \, dy - \frac{2}{N} \sum_{z \in P} \int_{C^s} K(z, y) \, dy \right.$$

$$+\frac{1}{N^2} \sum_{z,z' \in P} K(z,z') \Bigg\}^{1/2}.$$

Since quadrature rule (2.1) is exact for constants, one may replace the norm of the integrand by the variation of the integrand, which is defined as the norm of the nonconstant part, that is, the part which is orthogonal to the function 1.

$$(3.4) \qquad V(f;K) \equiv \|f_\perp\|_K, \quad \text{where } f_\perp \equiv f - \frac{1 \langle f,1 \rangle_K}{\langle 1,1 \rangle_K}.$$

Note that the discrepancy is the variation of the worst-case integrand. The following theorem gives the quadrature error bound as derived above.

Theorem 3.1. *Suppose that W is a reproducing kernel Hilbert space of real-valued functions on C^s with reproducing kernel $K(x,y)$, inner product $\langle \cdot, \cdot \rangle_K$, and norm $\|\cdot\|_K$. Then the error for the quadrature rule defined in (2.1) has the following upper bound:*

$$|I(f) - Q(f;P)| \le D(P;K)V(f;K),$$

in terms of the discrepancy of P and the variation of f as defined above. This bound is attained for the worst-case integrand, ξ_K, defined above.

Remark 3.2. In this theorem the set P is an arbitrary subset of the unit cube and need not come from a lattice. Therefore the error bound applies to any quasi-Monte Carlo quadrature rule. However, the special case where P is the node set of a lattice will be discussed below.

Remark 3.3. The discrepancy is uniquely determined by any $K(x,y)$ satisfying (3.1). Moreover, the choice of K also uniquely determines the space of integrands W and the accompanying inner product (see [Wah90]).

Remark 3.4. The formula for the discrepancy in (3.3) requires only $O(N^2)$ operations to evaluate. In fact, for the case of the $\mathcal{L}_2$-star discrepancy, which is a special case of (3.3) (see Section 4.5), the number of operations required can be reduced to $O(N[\log N]^s)$ [Hei96].

Remark 3.5. The worst-case integrand is the quadrature error obtained upon integrating the reproducing kernel.

This theorem provides many worst-case error bounds, each arising from a different choice of the reproducing kernel $K(x,y)$. To illustrate this theorem consider an inner product motivated by Theorem 2.3 with the associated norm:

$$(3.5) \qquad \langle f, g \rangle_{\mathcal{F},\alpha} \equiv \sum_{k \in \mathbf{Z}^s} (\bar{k}_1 \cdots \bar{k}_s)^\alpha \hat{f}(k) \hat{g}^*(k),$$

$$\|f\|_{\mathcal{F},\alpha} = \left\{ \sum_{k \in \mathbf{Z}^s} (\bar{k}_1 \cdots \bar{k}_s)^\alpha |\hat{f}(k)|^2 \right\}^{1/2},$$

where * denotes the complex conjugate. The subscript $\mathcal{F}$ denotes the fact that this inner product is defined in terms of the Fourier coefficients of a function. The space of integrands is a Hilbert space of real-valued, periodic functions, whose Fourier coefficients decay to zero sufficiently fast:

$$\mathcal{W}_{\mathcal{F},\alpha} \equiv \{ f : \|f\|_{\mathcal{F},\alpha} < \infty \}, \quad \alpha > 1.$$

The reproducing kernel for this space is:

$$(3.6) \qquad K_{\mathcal{F},\alpha}(x, y) = \sum_{k \in \mathbf{Z}^s} \frac{e^{2\pi i k \cdot (x-y)}}{(\bar{k}_1 \cdots \bar{k}_s)^\alpha}.$$

This can be verified by noting that $K_{\mathcal{F},\alpha}(\cdot, y) \in \mathcal{W}_{\mathcal{F},\alpha}$ for all y and that

$$\langle K(\cdot, y), f \rangle_{\mathcal{F},\alpha} = \sum_{k \in \mathbf{Z}^s} (\bar{k}_1 \cdots \bar{k}_s)^\alpha \frac{e^{-2\pi i k \cdot y}}{(\bar{k}_1 \cdots \bar{k}_s)^\alpha} \hat{f}^*(k)$$

$$= \sum_{k \in \mathbf{Z}^s} e^{2\pi i k \cdot y} \hat{f}(k) = f(y).$$

According to Theorem 3.1 the quadrature error bound is

$$(3.7) \qquad |I(f) - Q(f; P)| \le D_{\mathcal{F},\alpha}(P) V_{\mathcal{F},\alpha}(f),$$

where the worst-case integrand, (3.2), the discrepancy, (3.3), and the variation of the integrand, (3.4) are:

$$(3.8) \qquad \xi_{\mathcal{F},\alpha}(x) = I(K(\cdot, x)) - Q(K(\cdot, x); P)$$

$$= -\frac{1}{N} \sum_{z \in P} {\sum_{k \in \mathbf{Z}^s}}' \frac{e^{2\pi i k \cdot (z-x)}}{(\bar{k}_1 \cdots \bar{k}_s)^\alpha},$$

$$(3.9) \qquad D_{\mathcal{F},\alpha}(P) = \left\{ \frac{1}{N^2} \sum_{z,z' \in P} {\sum_{k \in \mathbf{Z}^s}}' \frac{e^{2\pi i k \cdot (z-z')}}{(\bar{k}_1 \cdots \bar{k}_s)^\alpha} \right\}^{1/2},$$

$$(3.10) \qquad V_{\mathcal{F},\alpha}(f) \equiv \|f_\perp\|_{\mathcal{F},\alpha} = \left\{ {\sum_{k \in \mathbf{Z}^s}}' (\bar{k}_1 \cdots \bar{k}_s)^\alpha |\hat{f}(k)|^2 \right\}^{1/2}.$$

Since $\xi_{\mathcal{F},\alpha}$ is also the representer of the quadrature error functional, the quadrature error for the integrand f is $\langle f, \xi_{\mathcal{F},\alpha} \rangle_{\mathcal{F},\alpha}$, which is just the right-hand side of (2.4).

For positive, even integers α the formulas for the reproducing kernel, the worst-case integrand, and the discrepancy may be written in terms of Bernoulli functions. Using the Fourier expansion (2.9), one obtains for all even integers $\alpha \geq 2$:

$$K_{\mathcal{F},\alpha}(x,y) = \prod_{j=1}^{s} \left[1 - \frac{(-1)^{\alpha/2}(2\pi)^{\alpha}}{\alpha!} B_{\alpha}(\{x_j - y_j\}) \right],$$

$$\xi_{\mathcal{F},\alpha}(x) = 1 - \frac{1}{N} \sum_{z \in P} \prod_{j=1}^{s} \left[1 - \frac{(-1)^{\alpha/2}(2\pi)^{\alpha}}{\alpha!} B_{\alpha}(\{x_j - z_j\}) \right],$$

(3.11) $\quad D_{\mathcal{F},\alpha}(P)$

$$= \left\{ -1 + \frac{1}{N^2} \sum_{z,z' \in P} \prod_{j=1}^{s} \left[1 - \frac{(-1)^{\alpha/2}(2\pi)^{\alpha}}{\alpha!} B_{\alpha}(\{z_j - z'_j\}) \right] \right\}^{1/2}$$

Remark 3.6. When computing $D_{\mathcal{F},\alpha}(P)$ in practice it is advisable to choose α to be an even integer. Formula (3.11) requires $O(N^2)$ operations, whereas the number of operations required by formula (3.9) is $O(N^2)$ times the work required to sum an infinite series.

If P is the node set of a shifted lattice, $L + \Delta$, then the worst-case integrand and the discrepancy may be written in terms of the dual lattice, $L^{\perp}$, by using Lemma 2.2 as follows:

$$\xi_{\mathcal{F},\alpha}(x) = -\sum_{k \in L^{\perp}}' \frac{e^{2\pi i k \cdot (x - \Delta)}}{(\bar{k}_1 \cdots \bar{k}_s)^{\alpha}},$$

$$D_{\mathcal{F},\alpha}(P) = \left\{ \sum_{k \in L^{\perp}}' \frac{1}{(\bar{k}_1 \cdots \bar{k}_s)^{\alpha}} \right\}^{1/2} = \sqrt{P_{\alpha}(L)},$$

and when α is a positive even integer:

$$\xi_{\mathcal{F},\alpha}(x) = -\prod_{j=1}^{s} \left[1 - \frac{(-1)^{\alpha/2}(2\pi)^{\alpha}}{\alpha!} B_{\alpha}(\{x_j - \Delta_j\}) \right],$$

$$D_{\mathcal{F},\alpha}(P) = \left\{ -1 + \frac{1}{N} \sum_{z \in P} \prod_{j=1}^{s} \left[1 - \frac{(-1)^{\alpha/2}(2\pi)^{\alpha}}{\alpha!} B_{\alpha}(\{z_j - \Delta_j\}) \right] \right\}^{1/2}$$

Remark 3.7. The number of operations required to compute $D_{\mathcal{F},\alpha}(P)$ is only $O(N)$ when α is an even integer and P is the node set of a lattice. This is substantially faster than the case when P is arbitrary (see Remark 3.6).

3.2   A More General Worst-Case Quadrature Error Analysis

The error bound in (3.7) that came from Theorem 3.1 may be obtained by applying the Cauchy-Schwarz inequality to Theorem 2.3. This *does not* correspond to the lattice rule error bound in Theorem 2.4, which is derived by applying Hölder's inequality. However, one can extend the idea behind Theorem 3.1 to cover Theorem 2.4.

Suppose that the space of integrands, $\mathcal{W}$, is a Banach space of real-valued functions on the unit cube C^s with norm $\|\cdot\|_{\mathcal{W}}$. Instead of an inner product, $\langle \cdot, \cdot \rangle_K$, suppose, that one has a bilinear, non-negative definite function, $\langle \cdot, \cdot \rangle$ from $\mathcal{W} \times \mathcal{W}'$ into $\mathbf{R}$, where $\mathcal{W}'$ is another Banach space of real-valued functions on the unit cube C^s. The norm for $\mathcal{W}'$ is assumed to be related to $\|\cdot\|_{\mathcal{W}}$ and $\langle \cdot, \cdot \rangle$ in the following way:

$$(3.12) \qquad \|f\|_{\mathcal{W}} \|g\|_{\mathcal{W}'} = \sup_{f \in \mathcal{W}} |\langle f, g \rangle| = \sup_{g \in \mathcal{W}'} |\langle f, g \rangle|.$$

Thus, for each $g \in \mathcal{W}'$, the function $\langle \cdot, g \rangle$ is a bounded linear functional on $\mathcal{W}$, and $\mathcal{W}'$ is isomorphic to the dual space of $\mathcal{W}$. Conversely, for each $f \in \mathcal{W}$, the function $\langle f, \cdot \rangle$ is a bounded linear functional on $\mathcal{W}'$. Suppose that the evaluation functional is bounded on $\mathcal{W}'$, so that there exists a real-valued function K on $C^s \times C^s$ such that

$$(3.13a) \qquad K(\cdot, x) \in \mathcal{W} \quad \text{for all } x \in C^s,$$
$$(3.13b) \qquad g(x) = \langle K(\cdot, x), g \rangle \quad \text{for all } x \in C^s, g \in \mathcal{W}'.$$

Then the quadrature error may be written in terms of the representer for the quadrature error, ξ_K, which in turn is found from the kernel as in (3.2) above:

$$I(f) - Q(f; P) = \langle f, \xi_K \rangle,$$
$$(3.14) \qquad \xi_K(x) = \langle K(\cdot, x), \xi_K \rangle = I(K(\cdot, x)) - Q(K(\cdot, x); P).$$

Applying condition (3.12) then yields the desired error bound.

Theorem 3.8. *Suppose that $\mathcal{W}$ and $\mathcal{W}'$ are two Banach spaces of real-valued functions on the unit cube C^s, as described above, and that $\langle \cdot, \cdot \rangle$ is a bilinear, non-negative definite function from $\mathcal{W} \times \mathcal{W}'$ into $\mathbf{R}$ satisfying (3.12). Suppose that both $\mathcal{W}$ and $\mathcal{W}'$ contain the function 1, and that a kernel $K(\cdot, \cdot)$ exists that satisfies conditions (3.13). Then the error for the quadrature rule defined in (2.1) has the following upper bound:*

$$|I(f) - Q(f; P)| = |\langle f, \xi_K \rangle| \leq D(P; \mathcal{W}') V(f; \mathcal{W}).$$

The discrepancy of P is defined as the norm of the function ξ_K defined in (3.14), and the variation of f is defined as the norm of its non-constant part:

$$D(P; \mathcal{W}') \equiv \|\xi_K\|_{\mathcal{W}'},$$

$$V(f; \mathcal{W}) \equiv \|f_\perp\|_{\mathcal{W}}, \qquad f_\perp \equiv f - \frac{1\langle f, 1\rangle}{\langle 1, 1\rangle}.$$

Remark 3.9. Theorem 3.8 contains Theorem 3.1 as a special case when $\mathcal{W} = \mathcal{W}'$ is a reproducing kernel Hilbert space with reproducing kernel $K(x, y)$, and the bilinear function $\langle \cdot, \cdot \rangle$ is $\langle \cdot, \cdot \rangle_K$.

Remark 3.10. For Theorem 3.1 one only needs to know the reproducing kernel to compute the discrepancy. In contrast Theorem 3.8 seems more difficult to apply because one must specify two Banach spaces with their norms and $\langle \cdot, \cdot \rangle$. However, if one starts with a reproducing kernel Hilbert space satisfying the hypotheses of Theorem 3.1, one is often able to use $\langle \cdot, \cdot \rangle_K$ as the $\langle \cdot, \cdot \rangle$ in Theorem 3.8 along with the same $K(\cdot, \cdot)$. The definitions of the $\mathcal{W}$ and $\mathcal{W}'$ often suggest themselves in the same way as an $\mathcal{L}_p$ space generalizes an $\mathcal{L}_2$ space.

Theorem 2.4 can now be derived as a special case of Theorem 3.8. The appropriate bilinear function $\langle \cdot, \cdot \rangle$ is just $\langle \cdot, \cdot \rangle_{\mathcal{F}, \gamma}$ as defined in (3.5). The relevant Banach spaces and norms are defined in terms of the ℓ_p-norm. For any non-negative number α, define the norm

$$\|f\|_{\mathcal{F}, \alpha, p} \equiv \left\| (\bar{k}_1 \cdots \bar{k}_s)^\alpha \hat{f}(k) \right\|_p, \qquad 1 \leq p \leq \infty,$$

and the Banach space $\mathcal{W}_{\mathcal{F}, \alpha, p} = \{f : \|f\|_{\mathcal{F}, \alpha, p} < \infty\}$. If the space of integrands $\mathcal{W}$ in Theorem 3.8 is $\mathcal{W}_{\mathcal{F}, \alpha, q}$, then the corresponding $\mathcal{W}'$ is $\mathcal{W}_{\mathcal{F}, \gamma-\alpha, p}$, where $\frac{1}{p} + \frac{1}{q} = 1$. Condition (3.12) is satisfied by applying Hölder's inequality to $\langle \cdot, \cdot \rangle_{\mathcal{F}, \gamma}$. The appropriate kernel in Theorem 3.8 is $K_{\mathcal{F}, \gamma}$ as defined above in (3.6), so ξ_K in Theorem 3.8 is just $\xi_{\mathcal{F}, \gamma}$ as defined in (3.8). The quadrature error bound is

$$(3.15) \qquad |I(f) - Q(f; P)| = |\langle f, \xi_{\mathcal{F}, \gamma} \rangle_{\mathcal{F}, \gamma}|$$

$$= \left| \sum_{k \in \mathbf{Z}^s} \left\{ \hat{f}(k) \left[\frac{1}{N} \sum_{z \in P} e^{2\pi i k \cdot z} \right] \right\} \right|$$

$$\leq D_{\mathcal{F}, \alpha, p}(P) V_{\mathcal{F}, \alpha, q}(f) \quad \forall f \in \mathcal{W}_{\mathcal{F}, \alpha, q},$$

where the corresponding discrepancy and variation are:

$$(3.16) \quad D_{\mathcal{F}, \alpha, p}(P) \equiv \|\xi_{\mathcal{F}, \gamma}\|_{\mathcal{F}, \gamma-\alpha, p}$$

$$= \left\| 1_{\{k \neq 0\}} (\bar{k}_1 \cdots \bar{k}_s)^{-\alpha} \frac{1}{N} \sum_{z \in P} e^{2\pi i k \cdot z} \right\|_p,$$

$$(3.17) \quad V_{\mathcal{F},\alpha,q}(f) \equiv \left\| 1_{\{k \neq 0\}} (\bar{k}_1 \cdots \bar{k}_s)^\alpha \hat{f}(k) \right\|_q, \quad 1 \leq p \leq \infty, \quad \frac{1}{p} + \frac{1}{q} = 1.$$

For $p < \infty$ there is a worst-case integrand, $\xi_{\mathcal{F},\alpha,p}$, that is, an integrand for which the above error bound is attained. Its Fourier coefficients satisfy the condition:

$$(3.18) \quad \hat{\xi}_{\mathcal{F},\alpha,p}(k) \left[\frac{1}{N} \sum_{z \in P} e^{2\pi i k \cdot z} \right]$$

$$= 1_{\{k \neq 0\}} (\bar{k}_1 \cdots \bar{k}_s)^{-\alpha p} \left| \frac{1}{N} \sum_{z \in P} e^{2\pi i k \cdot z} \right|^p,$$

Remark 3.11. Error bound (3.15) may be obtained in a more straightforward manner by applying Hölder's inequality directly to (2.4). However, Theorem 3.8 also has many other interesting cases, some of which are explored in the next section.

Remark 3.12. The computational complexity of calculating the discrepancy, (3.16), is high unless one chooses $p = 2$. In this case one obtains $D_{\mathcal{F},\alpha,2} = D_{\mathcal{F},2\alpha}(P)$ (see (3.9)).

In the error bound above there should be some condition relating α and p to insure that the discrepancy is well-defined and that the space of integrands is a subset of absolutely convergent Fourier series. To derive this condition note that for $1 \leq r \leq p$, and $\alpha, \gamma \geq 0$ it follows by Hölder's inequality that

$$\|f\|_{\mathcal{F},\gamma,r} = \left\| (\bar{k}_1 \cdots \bar{k}_s)^\gamma \hat{f}(k) \right\|_r$$

$$\leq \left\| (\bar{k}_1 \cdots \bar{k}_s)^{-\alpha} \right\|_{pr/(p-r)} \left\| (\bar{k}_1 \cdots \bar{k}_s)^{\alpha+\gamma} \hat{f}(k) \right\|_p$$

$$= \left[1 + 2\zeta \left(\frac{\alpha p r}{p-r} \right) \right]^{s(p-r)/(rp)} \|f\|_{\mathcal{F},\alpha+\gamma,p},$$

provided that $\alpha p r/(p-r) > 1$, where ζ is the Riemann zeta function. Therefore,

$$\mathcal{W}_{\mathcal{F},\alpha+\gamma,p} \subseteq \mathcal{W}_{\mathcal{F},\gamma,r} \text{ for } \alpha > \frac{1}{r} - \frac{1}{p}.$$

By choosing $r = 1$ and $\gamma = 0$, it follows that $\mathcal{W}_{\mathcal{F},\alpha,p}$ is a subset of absolutely convergent Fourier series when $\alpha > 1 - \frac{1}{p}$. To insure that the two spaces $\mathcal{W}_{\mathcal{F},\alpha,q}$ and $\mathcal{W}_{\mathcal{F},\gamma-\alpha,p}$ contain only absolutely convergent Fourier series, the following constraint is made on α:

$$(3.19) \quad \frac{1}{p} = 1 - \frac{1}{q} < \alpha < \gamma - 1 + \frac{1}{p}.$$

Since error bound (3.15) does not depend explicitly on γ, one may always choose γ large enough so that the upper bound on α is satisfied.

If P is the node set of a shifted lattice $L + \Delta$, then the discrepancy above can be written in terms of $\mathcal{P}_\alpha(L)$. By Lemma 2.2 it follows that:

$$D_{\mathcal{F},\alpha,p}(P) = \left\| 1_{\{k \neq 0, k \in L^\perp\}} (\bar{k}_1 \cdots \bar{k}_s)^{-\alpha} \right\|_p$$

$$= \left\{ {\sum_{k \in L^\perp}}' (\bar{k}_1 \cdots \bar{k}_s)^{-p\alpha} \right\}^{1/p} = [\mathcal{P}_{p\alpha}(L)]^{1/p}, \quad 1 \leq p < \infty,$$

$$D_{\mathcal{F},\alpha,\infty}(P) = \left\| 1_{\{k \neq 0, k \in L^\perp\}} (\bar{k}_1 \cdots \bar{k}_s)^{-\alpha} \right\|_\infty = [\rho(L)]^{-\alpha},$$

where $\rho(L)$ is the Zaremba figure of merit for lattice rules [Nie92, Definition 5.31] defined as

$$\rho(L) = \min_{\substack{k \in L^\perp \\ k \neq 0}} \bar{k}_1 \cdots \bar{k}_s.$$

In fact one may consider $\rho(L)$ to be the limit of $[\mathcal{P}_p(L)]^{-1/p}$ as p tends to infinity. Error bound (3.15) when applied to lattice rules may be written as:

$$|I(f) - Q(f; L + \Delta)| \leq \begin{cases} [\mathcal{P}_{p\alpha}(L)]^{1/p} \left\| 1_{\{k \neq 0\}} (\bar{k}_1 \cdots \bar{k}_s)^\alpha \hat{f}(k) \right\|_q, & 1 \leq p < \infty, \\ & \frac{1}{p} + \frac{1}{q} = 1, \\ [\rho(L)]^{-\alpha} \left\| 1_{\{k \neq 0\}} (\bar{k}_1 \cdots \bar{k}_s)^\alpha \hat{f}(k) \right\|_1, & p = \infty. \end{cases}$$

The case $p = 1$, $q = \infty$ corresponds to Theorem 2.4.

Remark 3.13. For lattice rules the worst-case integrand $\xi_{\mathcal{F},\alpha,p}$, defined in (3.18), is just $\xi_{\mathcal{F},p\alpha}$, as defined in (3.8). Furthermore, $\xi_{\mathcal{F},\alpha,p}$ is similar to $\xi_{p\alpha}$ defined in (2.8).

Remark 3.14. According to restriction (3.19), α must be positive. However, $p = \infty$ and $\alpha = 0$ is allowable since $\mathcal{W}_{\mathcal{F},0,1}$ and $\mathcal{W}_{\mathcal{F},\gamma,\infty}$, $(\gamma > 1)$ contain only absolutely convergent Fourier series. In this case the discrepancy of the node set of a shifted lattice is one, no matter what the lattice is. So, for any lattice L, one may construct an integrand $f \in \mathcal{W}_{\mathcal{F},0,1}$ such that the quadrature error is the size of variation of the integrand, $V_{\mathcal{F},0,1}(f) = {\sum_{k \in \mathbf{Z}^s}}' |\hat{f}(k)|$. The construction is straightforward. Since lattice rules integrate Fourier coefficients with $k \notin L^\perp$ exactly, but give 100% error for Fourier coefficients with $k \in L^\perp$, one chooses f to have nonzero Fourier coefficients only for $k \in L^\perp$.

3.3 Average-Case Quadrature Error Analysis

The error bounds given in Theorems 2.4, 3.1, and 3.8 give an upper bound on the quadrature error. Another approach is to study the **average or**

mean error over some space of random integrands. In this average-case analysis the same discrepancy as defined above serves as a measure of the quality of the quadrature rule.

Theorem 3.15. *Suppose that $\mathcal{A}$ is a space of random functions such that*

$$E_{f \in \mathcal{A}}[f(x)] = 0, \quad E_{f \in \mathcal{A}}[f(x)f(y)] = K(x,y) \quad \forall x, y \in C^s,$$

for some given covariance kernel, K. Then the mean square quadrature error for quadrature rule (2.1) is the discrepancy as defined in (3.3):

$$E_{f \in \mathcal{A}}\{[I(f) - Q(f; P)]^2\} = [D(P; K)]^2.$$

Proof. Proofs of this theorem or special cases are given in [Woź91, MC94, Rit95, HH98] and elsewhere. The conclusion follows by interchanging the order of the expectation and the integration or summation:

$$E_{f \in \mathcal{A}}\{[I(f) - Q(f; P)]^2\}$$

$$= E_{f \in \mathcal{A}}\left[\iint_{C^s \times C^s} f(x)f(y) \, dx \, dy \right.$$

$$\left. - \frac{2}{N} \sum_{z \in P} \int_{C^s} f(z)f(y) \, dy + \frac{1}{N^2} \sum_{z, z' \in P} f(z)f(z') \right]$$

$$= \int_{C^s \times C^s} K(x,y) \, dx \, dy - \frac{2}{N} \sum_{z \in P} \int_{C^s} K(z,y) \, dy + \frac{1}{N^2} \sum_{z, z' \in P} K(z, z')$$

$$= [D(P)]^2.$$

$\square$

Remark 3.16. It may seem odd that the same discrepancy acts as a measure of both average-case and worst-case quadrature error. The reason why this is possible is that the space of integrands for the average-case analysis, $\mathcal{A}$, is larger than the space of integrands for the worst-case analysis, $\mathcal{W}$.

Remark 3.17. The double integral over the reproducing/covariance kernel that appears in the definition of the discrepancy in (3.3) has interesting interpretations in both the worst-case and average-case error analyses. In the worst-case analysis the discrepancy is the norm of the linear functional corresponding to the quadrature error, i.e. $\|I - Q\|_K = D(P; K)$. The norm of the integration functional itself is:

$$\|I\|_K^2 = \int_{C^s \times C^s} K(x,y) \, dx \, dy.$$

This corresponds to the expected value of the square of the integral in the average-case analysis:

$$E_{f \in \mathcal{A}}\{[I(f)]^2\} = \int_{C^s \times C^s} K(x,y) \, dx \, dy.$$

The expected value of the variance of the function can also be written in terms of the covariance kernel:

$$(3.20) \quad E_{f \in A}\{\text{Var}(f)\} = E_{f \in A}\{I(f^2) - [I(f)]^2\}$$

$$= \int_{C^s} K(x, x) \, dx - \int_{C^s \times C^s} K(x, y) \, dx \, dy.$$

Because quasi-Monte Carlo methods, such as lattice rules, are intended to be more accurate than simple Monte Carlo, one might wonder how large the discrepancy is for a simple uniform random sample, P. Elementary calculations give the following result.

Theorem 3.18. *The mean square of the discrepancy defined in* (3.3) *for simple, uniform random samples, P is*

$$E_P[D(P)]^2 = \frac{1}{N} \left\{ \int_{C^s} K(x, x) \, dx - \int_{C^s \times C^s} K(x, y) \, dx \, dy \right\}$$

$$= \frac{1}{N} E_{f \in A}\{\text{Var}(f)\}.$$

This theorem implies that the discrepancy of a simple random sample is typically of $O(N^{-1/2})$. For quasi-random points, such as lattices, it is expected to be much smaller.

4 Examples of Other Discrepancies

The previous section provided only one example of the error bounds and discrepancies that may be obtained from Theorems 3.1, 3.8, and 3.15. This section gives some more examples of error bounds and discrepancies that can be derived using these theorems. These examples include the popular star discrepancy and the related unanchored discrepancy.

4.1 The ANOVA Decomposition

The first example, however, is a generalization of the one given in the previous section. One observation made by several researchers is that the nominal dimension of an integration domain, s, may not be as important as some effective dimension of the integrand. Consider a possible integrand on C^4:

$$(4.1) \qquad f(x_1, x_2, x_3, x_4) = 5x_1 \cos(3\pi x_2) - 4e^{x_3}.$$

What is the effective dimension of this integrand? Only x_1, x_2 and x_3 appear, so one could argue that the effective dimension is three. On the other hand f is a sum of two terms, each of which depends on only two variables, so one could also argue that the dimension is two. The analysis of variance (ANOVA) decomposition provides a way to look more closely at which variables and combinations of variables are important.

First, some notation must be introduced. Let $S = \{1, 2, \ldots, s\}$. For any set of coordinate indices $u \subseteq S$, let $|u|$ denote the number of elements in u. Let x_u denote the $|u|$-vector containing the components of x indexed by u, and let C^u denote the $|u|$-dimensional cube which is the projection of C^s into the coordinates indexed by u. Any function $f \in \mathcal{L}_2(C^s)$ may be written as the sum of its ANOVA effects, f_u, defined recursively as follows [ES81, Owe92, Hic96b]:

$$(4.2) \quad f(x) = \sum_{u \subseteq S} f_u(x), \text{ where } f_u(x) = \int_{C^{S-u}} f(x) \, dx_{S-u} - \sum_{v \subset u} f_v(x).$$

(By convention $\int_{C^0} f \, dx_\emptyset = f(x)$.) Every function has a total of 2^s ANOVA effects. The ANOVA effect f_u is the part of the function depending only on the x_j with $j \in u$. The constant $f_\emptyset$ is the average value or integral of the function:

$$f_\emptyset = \int_{C^s} f(x) \, dx = I(f).$$

The main effect along the jth axis is $f_{\{j\}}$, while $f_{\{j,m\}}$ is the interaction along the axes j and m. From the definition it can be shown that

$$(4.3) \qquad \int_0^1 f_u(x_u) \, dx_j = 0 \text{ for any } j \in u.$$

The order of an effect f_u is $|u|$. The function f defined in (4.1) has ANOVA effects:

$$f_\emptyset = 4(1 - e), \quad f_{\{2\}} = \frac{5}{2} \cos(3\pi x_2),$$

$$f_{\{3\}} = 4 \left(e - 1 - e^{x_3} \right), \quad f_{\{1,2\}} = 5 \left(x_1 - \frac{1}{2} \right) \cos(3\pi x_2),$$

$$f_{\{1\}} = f_{\{4\}} = f_{\{1,3\}} = f_{\{1,4\}} = f_{\{2,3\}} = f_{\{2,4\}} = f_{\{3,4\}} = 0,$$
$$f_{\{1,2,3\}} = f_{\{1,2,4\}} = f_{\{1,3,4\}} = f_{\{2,3,4\}} = f_{\{1,2,3,4\}} = 0.$$

Remark 4.1. One would expect that it is easier to approximate the integrals of low order ANOVA effects because they are essentially functions on low-dimensional cubes. This point is developed further in Section 8.

The notation $\| \cdot \|_p$ may correspond to either the $\mathcal{L}_p$-norm of a function or the ℓ_p-norm of a sequence, depending on the context. This notation is extended to the case of a vector of ANOVA effects (f_u) or their Fourier coefficients $(\hat{f}_u)$, where u is an index running over some or all of the subsets of S, each f_u is a function on C^u, and each $\hat{f}_u$ is a function on $\mathbf{Z}^u$:

$$\|(f_u)\|_p = \left[\sum_u \|f_u\|_p^p\right]^{1/p}, \quad \|(\hat{f}_u)\|_p = \left[\sum_u \|\hat{f}_u\|_p^p\right]^{1/p}, \quad 1 \le p < \infty,$$

$$\|(f_u)\|_\infty = \max_u \|f_u\|_\infty, \quad \|(\hat{f}_u)\|_\infty = \max_u \|\hat{f}_u\|_\infty.$$

In the case where (f_u) is a vector of constants $\|(f_u)\|_p$ corresponds to the ℓ_p-norm, and when (f_u) is a single function $\|(f_u)\|_p$ corresponds to the $\mathcal{L}_p$-norm.

Now we take another look at the quadrature error, as given in (3.15), by writing it in terms of the ANOVA effects of the integrand and of $\xi_{\mathcal{F},\alpha}$. Let $\hat{f}_u(k_u)$ denote the Fourier coefficients of the ANOVA effect $f_u(x_u)$. From (4.2) and (4.3) one can show that

$$\hat{f}_u(k_u) = \begin{cases} \hat{f}(k_u) & \text{if } k_j \ne 0 \text{ for all } j \in u, \\ 0 & \text{otherwise.} \end{cases}$$

This implies that $\langle f_u, g_v \rangle_{\mathcal{F},\alpha} = 0$ for any ANOVA effects f_u and g_v where $u \ne v$. Thus, error bound (3.15) may be written as:

$$(4.4) \quad |I(f) - Q(f; P)| = \left| \sum_{\emptyset \subset u \subseteq S} [I(f_u) - Q(f_u; P)] \right|$$

$$= \left| \sum_{\emptyset \subset u \subseteq S} \langle f_u, \xi_{\mathcal{F},\alpha,u} \rangle_{\mathcal{F},\alpha} \right|$$

$$\le \sum_{\emptyset \subset u \subseteq S} D_{\mathcal{F},\alpha,p,u}(P) V_{\mathcal{F},\alpha,q,u}(f), \quad \frac{1}{p} + \frac{1}{q} = 1,$$

where

$$(4.5) \quad D_{\mathcal{F},\alpha,p,u}(P) \equiv D_{\mathcal{F},\alpha,p,u}(P_u) \equiv \|\xi_{\mathcal{F},\alpha,u}\|_{\mathcal{F},\gamma-\alpha,p}$$

$$= \left\| \left[\prod_{j \in u} 1_{\{k_j \ne 0\}} \bar{k}_j^{-\alpha} \right] \frac{1}{N} \sum_{z \in P} e^{2\pi i k_u \cdot z_u} \right\|_p,$$

$$(4.6) \quad V_{\mathcal{F},\alpha,q,u}(f) \equiv V_{\mathcal{F},\alpha,q,u}(f_u) \equiv \|f_u\|_{\mathcal{F},\alpha,q}$$

$$= \left\| \left[\prod_{j \in u} \bar{k}_j^{\alpha} \right] \hat{f}_u(k_u) \right\|_q,$$

and P_u denotes the projection of P into C^u.

Remark 4.2. Consider the space of integrands:

$$W_{\mathcal{F},\alpha,q,u} \equiv \left\{ f \in W_{\mathcal{F},\alpha,q} : \int_0^1 f \, dx_j = 0 \text{ for all } j \in u \right\}.$$

This is a subspace of $W_{\mathcal{F},\alpha,q}$, and in fact, is the space consisting of all ANOVA effects, f_u, of all $f \in W_{\mathcal{F},\alpha,q}$. The quadrature error bound for integrands in $W_{\mathcal{F},\alpha,q,u}$ involves the quantities defined in (4.5) and (4.6), specifically:

$$|I(f) - Q(f;P)| \leq D_{\mathcal{F},\alpha,p,u}(P)V_{\mathcal{F},\alpha,q,u}(f) \quad \text{for all } f \in W_{\mathcal{F},\alpha,q,u}.$$

Remark 4.3. Another viewpoint is that the $D_{\mathcal{F},\alpha,p,u}$ are pieces of the discrepancy $D_{\mathcal{F},\alpha,p}$, and the $V_{\mathcal{F},\alpha,q,u}$ are pieces of the variation $V_{\mathcal{F},\alpha,q}$, because

$$(4.7a) \qquad D_{\mathcal{F},\alpha,p}(P) = \left\| (D_{\mathcal{F},\alpha,p,u}(P_u))_{u \neq \emptyset} \right\|_p,$$

$$(4.7b) \qquad V_{\mathcal{F},\alpha,q}(f) = \left\| (V_{\mathcal{F},\alpha,q,u}(f_u))_{u \neq \emptyset} \right\|_q.$$

By applying Hölder's inequality to the last line of (4.4) one then arrives back at (3.15)

When the quadrature error bound is written as (4.4), one can identify how each ANOVA effect contributes to the quadrature error. For the integrand f in (4.1), the only nonzero ANOVA effects, f_u, are for $u = \emptyset, \{2\}, \{3\}$ and $\{1,2\}$. Therefore, the variation terms $V_{\mathcal{F},\alpha,q,u}(f_u)$ are zero for all other $u \subseteq \{1,2,3,4\}$, and the only necessary discrepancy terms in the quadrature error bound are $D_{\mathcal{F},\alpha,p,u}(P_u)$ for $u = \{2\}, \{3\}$ and $\{1,2\}$. If the set P is evenly distributed along the unit interval in the second and third coordinates and in the unit square in the first and second coordinates, then the quadrature rule based on P will be accurate for this particular integrand f.

4.2 A Generalization of $\mathcal{P}_\alpha(L)$ with Weights

One may generalize the discrepancy and variation as written in (4.7) by introducing positive weights, β_u, $\emptyset \subset u \subseteq S$, specifically,

$$(4.8a) \quad D_{\mathcal{F},\alpha,p}(P) = \left\| (\beta_u D_{\mathcal{F},\alpha,p,u}(P_u))_{u \neq \emptyset} \right\|_p$$

$$= \left\| \left(\beta_u \left[\prod_{j \in u} 1_{\{k_j \neq 0\}} |k_j|^{-\alpha} \right] \frac{1}{N} \sum_{z \in P} e^{2\pi i k_u \cdot z_u} \right)_{u \neq \emptyset} \right\|_p,$$

$$(4.8b) \quad V_{\mathcal{F},\alpha,q}(f) = \left\| (\beta_u^{-1} V_{\mathcal{F},\alpha,q,u}(f_u))_{u \neq \emptyset} \right\|_q.$$

The dependence of the discrepancy and variation on the weights is suppressed for ease of notation. Under these modified definitions one still has the quadrature error bound:

$$(4.8c) \qquad |I(f) - Q(f; P)| \le D_{\mathcal{F},\alpha,p}(P) V_{\mathcal{F},\alpha,q}(f).$$

Weighted discrepancies were proposed by [Hic96b, Hic98], with $\beta_u = \beta^{|u|}$, and by [SW97] for the star discrepancy, with $\beta_u = \prod_{j \in u} \beta_j$.

It is natural to ask how best to choose the weights. To make the error bound as tight as possible, that is,

$$\sum_{\emptyset \subset u \subseteq S} D_{\mathcal{F},\alpha,p,u}(P_u) V_{\mathcal{F},\alpha,q,u}(f_u) = D_{\mathcal{F},\alpha,p}(P) V_{\mathcal{F},\alpha,q}(f),$$

one should choose the β_u such that

$$\beta_u [D_{\mathcal{F},\alpha,p,u}(P_u)]^{1/q} [V_{\mathcal{F},\alpha,q,u}(f_u)]^{-1/p} = \text{constant} \quad \forall \emptyset \subset u \subseteq S.$$

In practice this condition is difficult to satisfy exactly since it requires detailed knowledge about the integrand. However, this condition suggests a useful qualitative principle.

Remark 4.4. A set of points with low discrepancy $D_{\mathcal{F},\alpha,p}(P)$ must have low values for its components $D_{\mathcal{F},\alpha,p,u}(P_u)$ (see (4.8)), but the relative importance of each component depends on the weight β_u assigned to it. If one expects the variation of the ANOVA effect f_u to be relatively large, then one should make β_u relatively large according to the condition above. This gives heavier weight to the component of the discrepancy $D_{\mathcal{F},\alpha,p,u}(P_u)$, which implies that a low-discrepancy set must do better in integrating the ANOVA effect f_u.

For the weighted form of the discrepancy $D_{\mathcal{F},\alpha,p}(P)$ a convenient special case of β_u is

$$(4.9) \qquad \beta_u = \beta_0 \prod_{j \in u} \beta_j^\alpha,$$

for some positive $\beta_0, \beta_1, \ldots, \beta_s$. Weights of this form are called **product-type weights**. For this type of weights $D_{\mathcal{F},\alpha,p}(P)$ becomes

$$(4.10) \qquad D_{\mathcal{F},\alpha,p}(P) = \beta_0 \left\| \left[\overline{(\beta_1^{-1} k_1)} \cdots \overline{(\beta_s^{-1} k_s)} \right]^{-\alpha} \frac{1}{N} \sum_{z \in P} e^{2\pi i k \cdot z} \right\|_p,$$

where the definition of the notation $\bar{k}$ is given in (2.5).

The discrepancy, $D_{\mathcal{F},\alpha}(P)$ defined in (3.9) with product-type weights now takes the form:

$$(4.11) \quad D_{\mathcal{F},\alpha}(P) = \beta_0 \left\{ {\sum_{k \in \mathbf{Z}^s}}' \left(\left[\overline{(\beta_1^{-1}k_1)} \cdots \overline{(\beta_s^{-1}k_s)} \right]^{-\alpha} \right. \right.$$

$$\left. \left. \times \frac{1}{N^2} \sum_{z,z' \in P} e^{2\pi i k \cdot (z-z')} \right) \right\}^{1/2}$$

The corresponding reproducing/covariance kernel originally defined in (3.6) becomes:

$$(4.12) \quad K_{\mathcal{F},\alpha}(x,y) = \beta_0^2 \sum_{k \in \mathbf{Z}^s} \left(e^{2\pi i k \cdot (x-y)} \left[\overline{(\beta_1^{-1}k_1)} \cdots \overline{(\beta_s^{-1}k_s)} \right]^{-\alpha} \right).$$

The relationship $D_{\mathcal{F},\alpha,p}(P) = D_{\mathcal{F},2\alpha}(P)$ continues to hold for these weighted discrepancies.

If $p = 2$ and α is a positive integer, then the above discrepancies can be written in terms of Bernoulli polynomials as was done in (3.11):

$$(4.13) \quad D_{\mathcal{F},\alpha,2}(P) = D_{\mathcal{F},2\alpha}(P)$$

$$= \tilde{\beta}_0 \left\{ -1 + \frac{1}{N^2} \sum_{z,z' \in P} \prod_{j=1}^{s} \left[1 - \frac{(-\tilde{\beta}_j^2)^\alpha}{(2\alpha)!} B_{2\alpha}(\{z_j - z_j'\}) \right] \right\}^{1/2},$$

where

$$(4.14) \qquad \beta_0 = \tilde{\beta}_0, \quad \beta_j = \frac{\tilde{\beta}_j}{2\pi}, \; j = 1, \ldots, s.$$

The weighted form of $\mathcal{P}_\alpha(L)$ for the above product-type weights is defined as

$$\mathcal{P}_\alpha(L) = \beta_0 {\sum_{k \in L^\perp}}' \left[\overline{(\beta_1^{-1}k_1)} \cdots \overline{(\beta_s^{-1}k_s)} \right]^{-\alpha}.$$

This may also be written in terms of Bernoulli polynomials as

$$(4.15) \quad \mathcal{P}_{2\alpha}(L) = \tilde{\beta}_0 \left\{ -1 + \frac{1}{N} \sum_{z \in P} \prod_{j=1}^{s} \left[1 - \frac{(-\tilde{\beta}_j^2)^\alpha}{(2\alpha)!} B_{2\alpha}(z_j) \right] \right\}^{1/2},$$

for integers α, where P is the node set of the lattice L. The relationship between the weighted form of $\mathcal{P}_\alpha(L)$ and the weighted forms of the discrepancies $D_{\mathcal{F},\alpha,p}(P)$ and $D_{\mathcal{F},\alpha}(P)$ for lattices is

$$(4.16) \qquad \mathcal{P}_\alpha(L) = \beta_0^{-1}[D_{\mathcal{F},\alpha}(P)]^2 = \beta_0^{1-p}[D_{\mathcal{F},\alpha/p,p}(P)]^p,$$

for any $\alpha > 1$ and any $1 \leq p < +\infty$.

Remark 4.5. There is always an implicit, if not explicit, choice of weights for the discrepancies $D_{\mathcal{F},\alpha,p}(P)$, $D_{\mathcal{F},\alpha}(P)$ and $\mathcal{P}_\alpha(L)$ and the other weighted discrepancies introduced below. If one prefers the simplicity of the original definition of $\mathcal{P}_\alpha(L)$ in (2.7), then one would choose $\beta_u = 1$. However, if one prefers the formulas for discrepancy in terms of Bernoulli polynomials (see (3.11) and (4.13)) to be elegant, then one would choose the product-type weights $\beta_u = (2\pi)^{-\alpha|u|}$. It is difficult to argue for one "natural" choice of weights. It is better to realize that regardless which choice is made, the relative values of the weights β_u reflect one's assumptions about the relative sizes of the variations of the ANOVA effects f_u of the integrands one expects to encounter as discussed in Remark 4.4.

The space $\mathcal{W}_{\mathcal{F},\alpha,p}$ consists of functions, f, with finite variation, $V_{\mathcal{F},\alpha,p}(f)$. If one writes the weights in (4.8) as $\beta_u = (2\pi)^{-\alpha|u|}\tilde{\beta}_u$, then the definition of the variation may be written as:

$$V_{\mathcal{F},\alpha,p}(f) = \left\| \left(\tilde{\beta}_u^{-1} \left\{ \prod_{j\in u} [2\pi\bar{k}_j]^\alpha \right\} \hat{f}_u(k) \right)_{\emptyset \subset u \subseteq S} \right\|_p$$

$$= \left\| \left(\tilde{\beta}_u^{-1} \mathcal{F}\left\{ \frac{\partial^{\alpha|u|} f_u}{\partial x_u^\alpha} \right\}(k_u) \right)_{\emptyset \subset u \subseteq S} \right\|_p ,$$

for integer α. Here $\mathcal{F}$ denotes the Fourier transform as defined in (2.3). If $p = 2$ this norm may also be written without referring to the Fourier coefficients:

$$V_{\mathcal{F},\alpha,2}(f) = \left\{ \sum_{\emptyset \subset u \subseteq S} \tilde{\beta}_u^{-2} \left\| \frac{\partial^{\alpha|u|} f_u}{\partial x_u^\alpha} \right\|_2^2 \right\}^{1/2} .$$

Therefore, $\mathcal{W}_{\mathcal{F},\alpha,2}$ consists of all periodic integrands where the α-order mixed partial derivatives of the ANOVA effects are square integrable.

4.3 The Periodic Bernoulli Discrepancy — Another Generalization of $\mathcal{P}_\alpha(L)$

A family of error bounds and discrepancies that are related to $\mathcal{P}_\alpha$ is given in [Hic96b, Hic98]. These may be derived from Theorem 3.8 by using the ANOVA decomposition. Let $\mathcal{W}_{\mathrm{per},\alpha,p}$ be a space of periodic functions whose first α mixed partial derivatives are in $\mathcal{L}_p(C^s)$, that is, let

$$(4.17a) \quad \mathcal{W}_{\mathrm{per},\alpha,p} \equiv \left\{ f : \frac{\partial^{\alpha|u|} f}{\partial x_u^\alpha} \in \mathcal{L}_p(C^u) \text{ and } \int_0^1 \frac{\partial^{\gamma|u|} f}{\partial x_u^\gamma} dx_j = 0 \right.$$
$$\left. \text{for all } \gamma \le \alpha, \ j \in u, \ \emptyset \subset u \subseteq S \right\}.$$

Then one has the quadrature error bound:

$$|I(f) - Q(f; P)| \le \sum_{\emptyset \subset u \subseteq S} D_{\text{per},\alpha,p,u}(P) V_{\text{per},\alpha,q,u}(f)$$

$$\le D_{\text{per},\alpha,p}(P) V_{\text{per},\alpha,q}(f), \quad \forall f \in \mathcal{W}_{\text{per},\alpha,p}, \quad \frac{1}{p} + \frac{1}{q} = 1,$$

where the discrepancy and variation are defined as:

$$(4.17b) \qquad V_{\text{per},\alpha,p,u}(f) \equiv \left\| \frac{\partial^{\alpha|u|} f_u}{\partial x_u^\alpha} \right\|_p,$$

$$(4.17c) \qquad V_{\text{per},\alpha,p}(f) \equiv \left\| \left(\tilde{\beta}_u^{-1} V_{\text{per},\alpha,p,u}(f) \right)_{\emptyset \subset u \subseteq S} \right\|_p,$$

$$(4.17d) \qquad D_{\text{per},\alpha,p,u}(P) \equiv \left\| \frac{1}{N} \sum_{z \in P} \prod_{j \in u} \left[\frac{(-1)^{\alpha+1}}{\alpha!} B_\alpha(\{x_j - z_j\}) \right] \right\|_p,$$

$$(4.17e) \qquad D_{\text{per},\alpha,p}(P) \equiv \left\| \left(\tilde{\beta}_u D_{\text{per},\alpha,p,u}(P) \right)_{\emptyset \subset u \subseteq S} \right\|_p.$$

Because the definition of the discrepancy involves Bernoulli polynomials, it can be called a periodic Bernoulli discrepancy. The error bound, discrepancy and variation are equivalent to those in (4.8) for $p = q = 2$, but they differ for other values of p and q. For $p = q = 2$ and for product-type weights the discrepancy is the same as (4.13).

4.4   The Non-Periodic Bernoulli Discrepancy

The discrepancies derived so far appear in error bounds for *periodic* integrands. Although lattice rules are particularly suited for such integrands, they can perform satisfactorily for non-periodic integrands. Extensions of the periodic Bernoulli discrepancy and the corresponding error bounds to spaces of non-periodic integrands are derived by the author [Hic96b, Hic98, HH98] on the basis of Theorems 3.1, 3.8, and 3.15. These make use of the ANOVA decomposition introduced above. They are introduced briefly below.

As a generalization to (4.17) for $\alpha = 1$, let

$$\mathcal{W}_{\text{non},1,p} \equiv \left\{ f : \frac{\partial^{|u|} f}{\partial x_u} \in \mathcal{L}_p(C^u) \; \forall u \subseteq S \right\},$$

$$V_{\text{non},1,p,u}(f) \equiv \left\| \frac{\partial^{|u|} f_u}{\partial x_u} \right\|_p,$$

$$V_{\text{non},1,p}(f) \equiv \left\| \left(\hat{\beta}_u^{-1} V_{\text{non},1,p,u}(f) \right)_{\emptyset \subset u \subseteq S} \right\|_p,$$

$$D_{\text{non},1,p,u}(P) \equiv \left\| \frac{1}{N} \sum_{z \in P} \prod_{j \in u} [x_j - 1_{\{x_j > z_j\}}] \right\|_p,$$

$$D_{\text{non},1,p}(P) \equiv \left\| \left(\hat{\beta}_u D_{\text{non},1,p,u}(P) \right)_{\emptyset \subset u \subseteq S} \right\|_p.$$

Then one has the quadrature error bound:

$$|I(f) - Q(f; P)| \leq \sum_{\emptyset \subset u \subseteq S} D_{\text{non},1,p,u}(P) V_{\text{non},1,q,u}(f)$$

$$\leq D_{\text{non},1,p}(P) V_{\text{non},1,q}(f) \quad \forall f \in \mathcal{W}_{\text{non},1,q}, \quad \frac{1}{p} + \frac{1}{q} = 1.$$

For integers $\alpha > 1$ the generalization for all p is tedious, but the generalization for $p = 2$ with product-type weights is manageable. The reproducing/covariance kernel and corresponding $\mathcal{L}_2$-non-periodic Bernoulli discrepancy are:

$$K_{\text{non},\alpha}(x,y) = \hat{\beta}_0^2 \prod_{j=1}^s \left[-\frac{(-\hat{\beta}_j^2)^\alpha}{(2\alpha)!} B_{2\alpha}(\{x_j - y_j\}) \right.$$

$$\left. + \sum_{i=0}^\alpha \frac{\hat{\beta}_j^{2i}}{(i!)^2} B_i(x_j) B_i(y_j) \right],$$

$$D_{\text{non},\alpha}(P) \equiv \hat{\beta}_0 \left\{ -1 + \frac{1}{N^2} \sum_{z,z' \in P} \prod_{j=1}^s \left[-\frac{(-\hat{\beta}_j^2)^\alpha}{(2\alpha)!} B_{2\alpha}(\{z_j - z_j'\}) \right. \right.$$

$$\left. \left. + \sum_{i=0}^\alpha \frac{\hat{\beta}_j^{2i}}{(i!)^2} B_i(z_j) B_i(z_j') \right] \right\}^{1/2}.$$

This reproducing/covariance kernel has been used by Wahba [Wah90, Section 10.2] and others.

4.5 The Star Discrepancy

The Koksma-Hlawka inequality [Hla61], [Nie92, Theorem 2.11] is perhaps the most well-known and popular multidimensional quadrature error bound. The original inequality is the case $p = \infty$ below, but it has been extended by [Zar68] (for $p = 2$), [Sob69, Chapter 8] (for $1 \leq p \leq \infty$) and [SW97] (by introducing weights). The following error bound may be derived by using

Theorem 3.8 (see [Hic98]):

$$|I(f) - Q(f; P)| \le \sum_{\emptyset \subset u \subseteq S} D_{p,u}^*(P) V_{q,u}^*(f)$$

$$\le D_p^*(P) V_q^*(f) \quad \forall f \in \mathcal{W}_{\mathrm{non},1,q}, \quad \frac{1}{p} + \frac{1}{q} = 1,$$

where,

$$V_{p,u}^*(f) \equiv \left\| \left. \frac{\partial^{|u|} f}{\partial x_u} \right|_{x_{S-u} = (1,\ldots,1)} \right\|_p,$$

$$V_p^*(f) \equiv \left\| \left(\hat{\beta}_u^{-1} V_{p,u}^*(f) \right)_{u \ne 0} \right\|_p,$$

$$D_{p,u}^*(P) \equiv \left\| \prod_{j \in u} x_j - \frac{1}{N} \sum_{z \in P} 1_{\{x_u > z_u\}} \right\|_p,$$

$$D_p^*(P) \equiv \left\| \left(\hat{\beta}_u D_{p,u}^*(P) \right)_{u \ne 0} \right\|_p.$$

Here $x_u > z_u$ means that $x_j > z_j$ for all $j \in u$. The variation $V_1(f)$ with unit weights corresponds to the variation in the sense of Hardy and Krause.

The piece of the discrepancy $D_{p,u}^*(P)$ is the $\mathcal{L}_p$-norm of the volume of the rectangular solid $[0, x_u)$ minus the proportion of points in P that are also in $[0, x_u)$. (See Figure 4.1 for a graphical representation of the star discrepancy.) In fact, $D_{p,S}^*(P)$ is often defined as the $\mathcal{L}_p$-star discrepancy. However, such a definition has the disadvantage that the corresponding error bound,

$$|I(f) - Q(f; P)| \le D_{p,S}^*(P) V_{q,S}^*(f), \quad \frac{1}{p} + \frac{1}{q} = 1,$$

applies only to integrands which vanish for any $x_j = 1$. For unit weights $D_{p,S}^*(P) < D_p^*(P)$ in general, but $D_{\infty,S}^*(P) = D_\infty^*(P)$.

The $\mathcal{L}_2$-star discrepancy is a special case of Theorems 3.1 and 3.15. The discrepancy and the corresponding kernel on which it is based are given below for the case of product-type weights:

$$K^*(x, y) = \hat{\beta}_0^2 \prod_{j=1}^{s} \left\{ 1 + \hat{\beta}_j^2 [1 - \max(x_j, y_j)] \right\},$$

$$D_2^*(P) = \hat{\beta}_0 \left\{ \prod_{j=1}^{s} \left(1 + \frac{\hat{\beta}_j^2}{3} \right) - \frac{2}{N} \sum_{z \in P} \prod_{j=1}^{s} \left[1 + \frac{\hat{\beta}_j^2 (1 - z_j^2)}{2} \right] \right.$$

$$\left. + \frac{1}{N^2} \sum_{z, z' \in P} \prod_{j=1}^{s} \left(1 + \hat{\beta}_j^2 [1 - \max(z_j, z_j')] \right) \right\}^{1/2}.$$

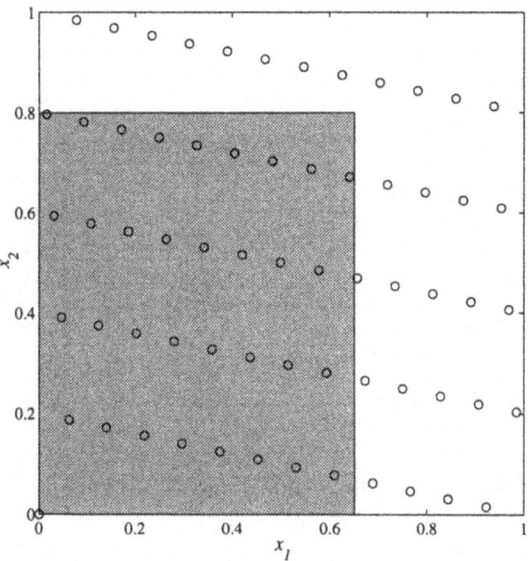

FIGURE 4.1. For $x_{\{1,2\}} = (0.65, 0.8)$ the volume of the box is 0.52, and the proportion of points in the box is $34/64 = 0.53125$. The difference between these two is $x_1 x_2 - \frac{1}{64} \sum_{z \in P} 1_{\{x_{\{1,2\}} > z_{\{1,2\}}\}} = -0.01125$.

4.6   The Unanchored Discrepancy

Although the star discrepancy enjoys widespread popularity as a figure of merit for quasi-random points, it lacks the following property:

Definition 4.6. A discrepancy, $D(P)$, is said to be **reflection invariant** if and only if $D(P) = D(P')$, where P' is obtained by reflecting the points in P about any plane $x_j = 1/2$ passing through the center of the cube, that is

$$P' = \{(z_1, \dots, z_{j-1}, 1 - z_j, z_{j+1}, \dots, z_s) : (z_1, \dots, z_s) \in P\}.$$

The star discrepancy is not reflection invariant because it counts points in a box *anchored at the origin*. One modification of the star discrepancy that is reflection invariant is the unanchored discrepancy. This discrepancy, defined in [Nie92, Definition 2.2] and [MC94], counts points in an unanchored box $[x'_u, x_u] \subseteq C^u$ with $x'_u < x_u$ (see Figure 4.2). In the formula below $|A|$ is the number of points in a set A. The corresponding error bound for the $\mathcal{L}_2$-unanchored discrepancy can be found using Theorem 3.1, and is given as follows:

$$|I(f) - Q(f; P)| \leq \sum_{\emptyset \subset u \subseteq S} D_{\text{un},u}(P) V_{\text{un},u}(f)$$

$$\leq D_{\text{un}}(P) V_{\text{un}}(f) \quad \forall f \in \mathcal{W}_{\text{per},1,2},$$

where,

$$V_{\mathrm{un},u}(f) \equiv \left\| \frac{\partial^{|u|} f}{\partial x_u} \right|_{x_{S-u}=(1,\dots,1)} \right\|_2,$$

$$V_{\mathrm{un}}(f) \equiv \left\| \left(\hat{\beta}_u^{-1} V_{\mathrm{un},u}(f) \right)_{u \neq \emptyset} \right\|_2,$$

$$D_{\mathrm{un},u}(P) \equiv \left\{ \int_{C^{2s}, x' < x} \left[\prod_{j \in u} (x_j - x_j') - \frac{|P_u \cap [x_u', x_u)|}{N} \right]^2 dx' dx \right\}^{1/2},$$

$$D_{\mathrm{un}}(P) \equiv \left\| \left(\hat{\beta}_u D_{\mathrm{un},u}(P) \right)_{u \neq \emptyset} \right\|_2.$$

For product-type weights the discrepancy and its kernel are:

$$K_{\mathrm{un}}(x,y) = \hat{\beta}_0^2 \prod_{j=1}^{s} \left\{ 1 + \hat{\beta}_j^2 [\min(x_j, y_j) - x_j y_j] \right\},$$

$$D_{\mathrm{un}}(P) = \hat{\beta}_0 \left\{ \prod_{j=1}^{s} \left(1 + \frac{\hat{\beta}_j^2}{12} \right) - \frac{2}{N} \sum_{z \in P} \prod_{j=1}^{s} \left[1 + \frac{\hat{\beta}_j^2 z_j (1 - z_j)}{2} \right] \right.$$

$$\left. + \frac{1}{N^2} \sum_{z, z' \in P} \prod_{j=1}^{s} \left(1 + \hat{\beta}_j^2 [\min(z_j, z_j') - z_j z_j'] \right) \right\}^{1/2}.$$

Remark 4.7. As is the case for the star discrepancy, the unanchored discrepancy is and ought to be defined by looking at boxes not only in C^s, but in all possible lower dimensional unit cubes $C^u, \emptyset \subset u \subseteq S$. Otherwise, the corresponding quadrature error analyses apply only to integrands that vanish on all faces of the cube C^s.

Remark 4.8. The worst-case and average-case error analyses for the unanchored discrepancy apply to *periodic* integrands. This follows from the fact that the reproducing/covariance kernel satisfies $K_{\mathrm{un}}(x,y)|_{x_j=0} = K_{\mathrm{un}}(x,y)|_{x_j=1}$ for all $j \in S$.

4.7   The Wrap-Around Discrepancy

A variation on the unanchored discrepancy that is also reflection-invariant arises by counting points in a "box" $[x_u', x_u)$, allowing for wrap-around

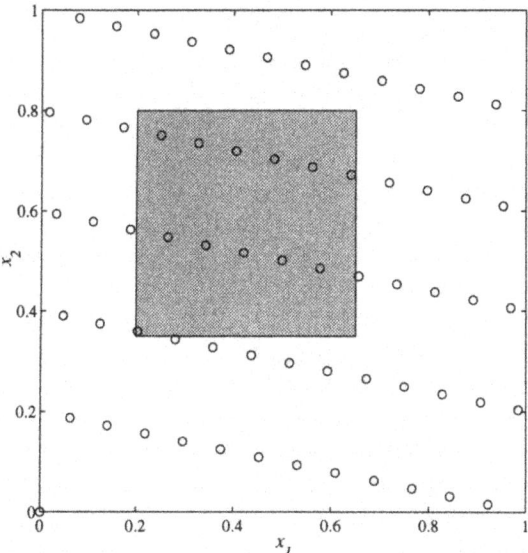

FIGURE 4.2. For $x'_{\{1,2\}} = (0.2, 0.35)$ and $x_{\{1,2\}} = (0.65, 0.8)$ the volume of the box is 0.2025, and the proportion of points in the box is $12/64 = 0.1875$. The difference between these two is $(x_1 - x'_1)(x_2 - x'_2) - \frac{1}{64}|P_{\{1,2\}} \cap [x'_{\{1,2\}}, x_{\{1,2\}})| = 0.015$.

when $x'_j > x_j$ (see Figure 4.3). Specifically, let

$$J_{\text{wrap}}(x'_j, x_j) = \begin{cases} [x'_j, x_j) & x'_j < x_j \\ [0, x_j) \cup [x'_j, 1) & x_j \le x'_j, \end{cases}$$

$$J_{\text{wrap}}(x'_u, x_u) = \bigotimes_{j \in u} J_{\text{wrap}}(x'_j, x_j).$$

The volume of $J_{\text{wrap}}(x'_u, x_u)$ is $\prod_{j \in u} \{x_j - x'_j\}$. The $\mathcal{L}_2$-wrap-around discrepancy is then defined analogously to the unanchored discrepancy. Assuming product-type weights the resulting discrepancy is:

$$D_{\text{wrap},u}(P) \equiv \left\{ \int_{C^{2s}} \left[\prod_{j \in u} \{x_j - x'_j\} - \frac{|P_u \cap J_{\text{wrap}}(x'_u, x_u)|}{N} \right]^2 dx' dx \right\}^{1/2},$$

$$D_{\text{wrap}}(P) \equiv \hat{\beta}_0 \left\| \left(\left[\prod_{j \in u} \hat{\beta}_j \right] D_{\text{wrap},u}(P) \right)_{u \ne \emptyset} \right\|_2$$

$$= \hat{\beta}_0 \left\{ -\prod_{j=1}^{s} \left(1 + \frac{\hat{\beta}_j^2}{3} \right) \right.$$

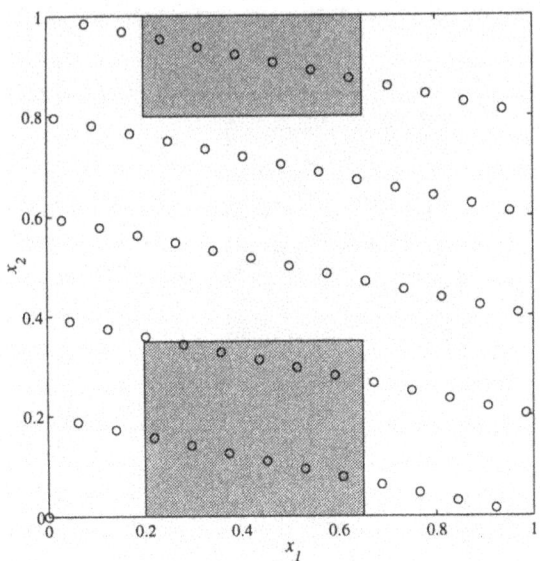

FIGURE 4.3. For $x'_{\{1,2\}} = (0.2, 0.8)$ and $x_{\{1,2\}} = (0.65, 0.35)$ the volume of the wrapped-around box is 0.2475, and the proportion of points in the box is $17/64 = 0.265625$. The difference between these two is $\{x_1 - x'_1\}\{x_2 - x'_2\} - \frac{1}{64}|P_{\{1,2\}} \cap J_{\mathrm{wrap}}(x'_{\{1,2\}}, x_{\{1,2\}})| = -0.018125$.

$$+ \frac{1}{N^2} \sum_{z, z' \in P} \prod_{j=1}^{s} \left(1 + \hat{\beta}_j^2 \left[\frac{1}{3} + B_2(\{z_j - z'_j\}) \right] \right) \Bigg\}^{1/2},$$

$$K_{\mathrm{wrap}}(x, y) = \hat{\beta}_0^2 \prod_{j=1}^{s} \left\{ 1 + \hat{\beta}_j^2 \left[\frac{1}{3} + B_2(\{x_j - y_j\}) \right] \right\}.$$

Remark 4.9. This discrepancy is the same as the weighted definition of $D_{\mathcal{F},1,2}(P) = D_{\mathcal{F},2}(P)$ in (4.13) if one chooses

$$\tilde{\beta}_0^2 = \hat{\beta}_0^2 \prod_{j=1^s} \left(1 + \frac{\hat{\beta}_j^2}{3} \right), \qquad \tilde{\beta}_j^2 = \frac{2\hat{\beta}_j^2}{1 + \hat{\beta}_j^2/3}.$$

Recall, too, from (4.16) that $D_{\mathcal{F},2}(P)$ is equivalent to $\sqrt{\mathcal{P}_2(L)}$ if P is the node set of a lattice. Thus, the wrap-around discrepancy provides a geometric interpretation for a weighted version of $\mathcal{P}_2(L)$.

4.8   The Symmetric Discrepancy

Another reflection invariant discrepancy is the symmetric discrepancy [Hic98, Section 5.3]. Let $\{0, 1\}^s$ denote the vertices of the unit cube C^s. Any point

x_u, divides the cube C^u into $2^{|u|}$ rectangular solids, each being defined as the region between a vertex $a_u \in \{0,1\}^u$ and x_u (see Figure 4.4). Let $J_e(x_u)$ be the union of the "even" rectangular solids, that is those touching a vertex a_u where $\sum_{j \in u} a_j$ is an even number. The symmetric discrepancy is defined by comparing the proportion of points in $P_u \cap J_e(x_u)$, to the volume of $J_e(x_u)$. (An equivalent definition would be obtained by looking at odd rather than even solids.) The associated quadrature error bound is:

$$|I(f) - Q(f;P)| \leq \sum_{\emptyset \subset u \subseteq S} D_{\text{sym},p,u}(P) V_{\text{sym},q,u}(f)$$

$$\leq D_{\text{sym},p}(P) V_{\text{sym},q}(f) \quad \forall f \in W_{\text{non},1,q}, \quad \frac{1}{p} + \frac{1}{q} = 1,$$

where,

$$V_{\text{sym},p,u}(f) \equiv \left\| \frac{1}{2^s} \sum_{a_{S-u} \in \{0,1\}^{S-u}} \left. \frac{\partial^{|u|} f}{\partial x_u} \right|_{x_{S-u} = a_{S-u}} \right\|_p ,$$

$$V_{\text{sym},p}(f) \equiv \left\| \left(\hat{\beta}_u^{-1} V_{\text{sym},p,u}(f) \right)_{u \neq \emptyset} \right\|_p ,$$

$$D_{\text{sym},p,u}(P) \equiv 2 \left\| \text{Vol}\,(J_e(x_u)) - \frac{|P_u \cap J_e(x_u)|}{N} \right\|_p ,$$

$$D_{\text{sym},p}(P) \equiv \left\| \left(\hat{\beta}_u D_{\text{sym},p,u}(P) \right)_{u \neq \emptyset} \right\|_p .$$

For unit weights $\hat{\beta}_u$ it was shown in [Hic98] that $D_{\text{sym},\infty}(P) = D_{\text{sym},\infty,S}(P)$, as is the case for the star discrepancy.

The $\mathcal{L}_2$-symmetric discrepancy and its defining kernel are given below for the case of product-type weights:

$$K_{\text{sym}}(x,y) = \hat{\beta}_0^2 \prod_{j=1}^s [1 + \hat{\beta}_j^2(1 - |x_j - y_j|)],$$

$$D_{\text{sym},2}(P) = \hat{\beta}_0 \left\{ \prod_{j=1}^s \left(1 + \frac{\hat{\beta}_j^2}{3} \right) - \frac{2}{N} \sum_{z \in P} \prod_{j=1}^s [1 + \hat{\beta}_j^2(2z_j - 2z_j^2)] \right.$$

$$\left. + \frac{1}{N^2} \sum_{z,z' \in P} \prod_{j=1}^s [1 + \hat{\beta}_j^2(1 - |z_j - z_j'|)] \right\}^{1/2} .$$

5 Shift-Invariant Kernels and Discrepancies

The previous section provides several examples of discrepancies for measuring the quality of quasi-random point sets and the associated quadrature

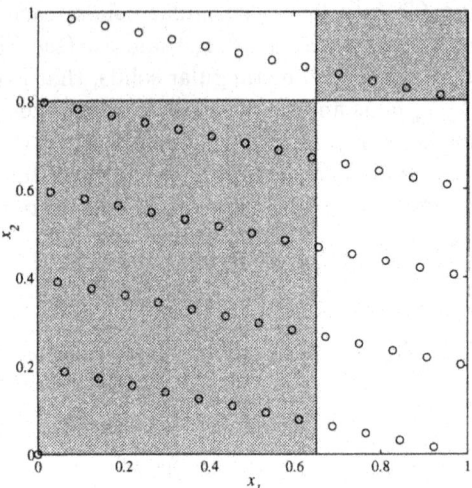

FIGURE 4.4. For $x_{\{1,2\}} = (0.65, 0.8)$ the total volume of the even rectangles is 0.59, and the proportion of points in these rectangles is $38/64 = 0.59375$. The difference between these two is $\text{Vol}\left(J_e(x_{\{1,2\}})\right) - \frac{1}{64}|P_{\{1,2\}} \cap J_e(x_{\{1,2\}})| = -0.00375$.

rules. This presents the difficulty of knowing which discrepancy to use in practice. This section shows how nearly all of the discrepancies described in the previous section are in some way related to $D_{\mathcal{F},\alpha}(P)$, which corresponds to $\mathcal{P}_\alpha(L)$ for lattices by (4.16).

The discrepancy $D_{\mathcal{F},\alpha}(P)$, defined in (3.9) and (4.11), is based on a kernel whose value remains the same when an arbitrary shift (modulo 1) is added to both of the arguments simultaneously, that is,

$$K(\{x + \Delta\}, \{y + \Delta\}) = K(x, y) \quad \text{for all } x, y, \Delta \in C^s.$$

This **shift-invariance** condition may be written equivalently as

(5.1) $$K(x, y) = K(\{x - y\}, 0) \quad \text{for all } x, y \in C^s.$$

A discrepancy based on a shift-invariant kernel is called a **shift-invariant discrepancy**. The formula for such a discrepancy is simpler than the original definition (3.3), and further simplifications can be made for the case of lattices.

Theorem 5.1. *If a kernel, K, satisfies the shift-invariance condition (5.1), then the discrepancy as defined in (3.3) may be written as:*

$$D(P; K) = \left\{ -\int_{C^s} K(x, 0) \, dx + \frac{1}{N^2} \sum_{z,z' \in P} K(\{z - z'\}, 0) \right\}^{1/2}.$$

Furthermore, if P is a node set of a shifted lattice $L + \Delta$, then this formula for the discrepancy may be further simplified as:

$$(5.2) \qquad D(P; K) = \left\{ -\int_{C^s} K(x, 0) \; dx + \frac{1}{N} \sum_{z \in P} K(\{z - \Delta\}, 0) \right\}^{1/2}.$$

Proof. The first part of the theorem follows directly by substituting condition (5.1) into (3.3) and performing a change of integration variable. The second part of the theorem follows from the first part by noting that the difference of any two node points of the shifted lattice, $\{z - z'\}$, must be some node point of the unshifted lattice. □

Remark 5.2. This theorem shows that the shift-invariant discrepancy of a set P is unchanged when one adds a shift Δ (modulo 1) to every point.

Remark 5.3. Because (5.2) contains only a single sum, it requires only $O(N)$ operations to evaluate, as opposed to the general discrepancy as defined by (3.3), which requires $O(N^2)$ operations to evaluate.

It might seem that a shift-invariant discrepancy is a rather special case. However, suppose one considers random shifts of a set P. Then according to the theorem below the mean square discrepancy defined by an arbitrary kernel, is the square discrepancy of P defined by a suitable shift-invariant kernel. The proof of this theorem follows from the definition of discrepancy and the shift-invariance condition.

Theorem 5.4. *Given an arbitrary kernel, K, define an associated shift-invariant kernel, K_{shift}, as follows:*

$$(5.3) \qquad K_{shift}(x, y) = \int_{C^s} K(\{x + \Delta\}, \{y + \Delta\}) \; d\Delta.$$

For a given finite set of points $P \subset C^s$, and any $\Delta \in C^s$, let $P + \Delta \equiv \{\{z + \Delta\} : z \in P\}$ denote a shifted copy of P. If Δ is a random variable uniformly distributed on C^s, then

$$E_\Delta \{ [D(P + \Delta; K)]^2 \} = [D(P; K_{shift})]^2.$$

The discrepancy based on K_{shift} is called the associated shift-invariant discrepancy.

This theorem implies that $D(P; K_{shift})$ gives an upper bound on the value of $D(P + \Delta; K)$ that is possible by a good choice of Δ. Many different reproducing kernels have the same associated shift-invariant kernel, as is demonstrated below. Thus, knowing about the shift-invariant discrepancy can provide useful information about many different discrepancies. When one is interested in lattice rules, there is the added benefit that the computational complexity of evaluating a shift-invariant discrepancy reduces from $O(N^2)$ to $O(N)$.

Theorem 5.5. *The kernels defining the $\mathcal{L}_2$-non-periodic Bernoulli ($\alpha = 1$), $\mathcal{L}_2$-star, $\mathcal{L}_2$-unanchored, and $\mathcal{L}_2$-symmetric discrepancies with product-type weights all have $K_{\mathcal{F},\alpha}$ as their associated shift-invariant kernel, assuming the following choices of weights:*

$\mathcal{L}_2$-non-periodic Bernoulli ($\alpha = 1$): $\quad \tilde{\beta}_0^2 = \hat{\beta}_0^2, \quad \tilde{\beta}_j^2 = 2\hat{\beta}_j^2,$

$$\mathcal{L}_2\text{-star:} \quad \tilde{\beta}_0^2 = \hat{\beta}_0^2 \prod_{j=1}^{s}\left(1 + \frac{\hat{\beta}_j^2}{3}\right), \quad \tilde{\beta}_j^2 = \frac{2\hat{\beta}_j^2}{1 + \hat{\beta}_j^2/3},$$

$$\mathcal{L}_2\text{-unanchored:} \quad \tilde{\beta}_0^2 = \hat{\beta}_0^2 \prod_{j=1}^{s}\left(1 + \frac{\hat{\beta}_j^2}{12}\right), \quad \tilde{\beta}_j^2 = \frac{\hat{\beta}_j^2}{1 + \hat{\beta}_j^2/12},$$

$$\mathcal{L}_2\text{-symmetric:} \quad \tilde{\beta}_0^2 = \hat{\beta}_0^2 \prod_{j=1}^{s}\left(1 + \frac{2\hat{\beta}_j^2}{3}\right), \quad \tilde{\beta}_j^2 = \frac{4\hat{\beta}_j^2}{1 + 2\hat{\beta}_j^2/3}.$$

Recall that β_j is related to $\tilde{\beta}_j$ by (4.14).

Remark 5.6. The quantities in Remark 3.17 can be written in terms of the associated shift-invariant kernel defined in (5.3) as follows:

$$\|I\|_K^2 = E_{f \in \mathcal{A}}\{[I(f)]^2\} = \int_{C^s} K_{\text{shift}}(x, 0) \, dx,$$

$$E_{f \in \mathcal{A}}\{\text{Var}(f)\} = K_{\text{shift}}(0, 0) - \int_{C^s} K_{\text{shift}}(x, 0) \, dx.$$

The implication is that different reproducing/covariance kernels that share the same associated shift-invariant kernel have the same $\|I\|_K$, etc.

6   Discrepancy Bounds

Sections 3 and 4 have presented several worst-case error bounds for quasi-Monte Carlo quadrature. Each bound is a product of a suitably defined discrepancy, $D(P)$, which depends only on the set P defining the quadrature rule, and the variation of the integrand, $V(f)$ which is a measure of the integrand's roughness. The average-case quadrature error is also a suitably defined discrepancy, $D(P)$. Since a smaller discrepancy tends to imply greater accuracy, one would like to know how small the discrepancy can be. This chapter presents lower and upper bounds.

Remark 6.1. An easy way to make the discrepancy smaller is to redefine it by multiplying it by some positive constant $a < 1$. In the worst-case analysis one must then redefine the variation by multiplying it by a^{-1}. Thus the error bound, which is the product of the discrepancy and the variation, remains the same for any particular integrand f. In the average-case analysis redefining the discrepancy in this way is equivalent to multiplying the covariance kernel by a. Thus, one is assuming that the integrands are not as rough as before. One way to avoid the ambiguities raised by an arbitrary constant multiple is to define the discrepancy with the following normalization:

$$\|I\|_K^2 = E_{f \in \mathcal{A}}\{[I(f)]^2\} = \int_{C^s \times C^s} K(x, y) \, dx \, dy = 1.$$

For the discrepancy $D_{\mathcal{F},\alpha}(P)$ this is equivalent to choosing $\beta_0 = 1$.

Our focus in this and the following sections is on the weighted version of the discrepancy $D_{\mathcal{F},\alpha,p}(P)$. The reasons are two-fold. For lattice rules this discrepancy can be written in terms of the familiar $\mathcal{P}_\alpha(L)$ (see (4.16)). Also, $D_{\mathcal{F},\alpha,2}(P) = D_{\mathcal{F},2\alpha}(P)$ is related to several of the other discrepancies introduced in Section 4 via Theorem 5.5.

Before looking at lattices we consider the discrepancies of simple random samples used in Monte Carlo quadrature. For product-type weights Theorem 3.18 can be applied to obtain a benchmark value for such samples.

Theorem 6.2. *The mean square of the discrepancy $D_{\mathcal{F},\alpha}(P)$ with product-type weights for simple uniform random samples, P, is:*

$$E_P\{[D_{\mathcal{F},\alpha}(P)]^2\} = \frac{\beta_0^2}{N} \left\{ -1 + \prod_{j=1}^{s} [1 + 2\beta_j^\alpha \zeta(\alpha)] \right\}.$$

The mean square discrepancy, which is comparable to $\mathcal{P}_\alpha(L)$, is plotted in Figures 6.1–6.4 for different values of the dimension, s, and the weights, β_j. Although the magnitude of this discrepancy is smaller when β_j and/or s are smaller, the asymptotic order in N is the same for all cases, as is clear from the formula above.

6.1 Upper Bounds for $\mathcal{P}_\alpha(L)$

Upper bounds on $\mathcal{P}_\alpha(L)$ for rank-1 lattice rules are given by several authors (see [Nie92, SJ94]). Here we consider the bounds given by Disney and Sloan [DS91] for prime numbers of points, N, because of the relative simplicity of their proofs. Extending their results to the case of product-type weights we have two lemmas. The first is proved by computing the average $\mathcal{P}_\alpha(L)$ over a family of rank-1 lattices. The second lemma is proved by invoking Jensen's inequality.

Lemma 6.3. *[SJ94, Proposition 4.6 and Theorem 4.8] If N is a prime number, then there exists a generating vector $h \in \mathbf{Z}^s$ such that the rank-1* *lattice with node set $P = \{\{ih/N\} : i = 0, \ldots, N-1\}$ has discrepancy:*

$$\mathcal{P}_\alpha(L) = \beta_0^{-1}[D_{\mathcal{F},\alpha}(P)]^2 = \beta_0^{1-p}[D_{\mathcal{F},\alpha/p,p}(P)]^p$$

$$\leq \beta_0 \left\{ -1 + \frac{1}{N} \prod_{j=1}^s [1 + 2\beta_j^\alpha \zeta(\alpha)] \right.$$

$$\left. + \frac{N-1}{N} \prod_{j=1}^s \left[1 - \frac{(1-N^{1-\alpha})2\beta_j^\alpha \zeta(\alpha)}{N-1} \right] \right\}$$

$$\leq \frac{\beta_0}{N} \prod_{j=1}^s [1 + 2\beta_j^\alpha \zeta(\alpha)] \quad \text{for } N > 1 + 2\zeta(\alpha) \max_j \beta_j.$$

Lemma 6.4. *[SJ94, Proposition 4.7] If $N \geq 2$, then for any lattice:*

$$\mathcal{P}_\alpha(L) \leq \beta_0^{1-\alpha/\gamma}[\mathcal{P}_\gamma(L)]^{\alpha/\gamma}.$$

As noted by Disney and Sloan the first result by itself is not very strong because it indicates that asymptotic order of the discrepancy $D_{\mathcal{F},\alpha}(P)$ for node sets of lattices is the same as that for simple random samples, that is, $O(N^{-1/2})$. However, one may combine the previous two lemmas as follows:

Theorem 6.5. *If N is a prime number, then there exists a generating vector $h \in \mathbf{Z}^s$ such that the rank-1 lattice with node set $P = \{\{ih/N\} : i = 0, \ldots, N-1\}$ has discrepancy:*

$$\mathcal{P}_\alpha(L) = \beta_0^{-1}[D_{\mathcal{F},\alpha}(P)]^2 = \beta_0^{1-p}[D_{\mathcal{F},\alpha/p,p}(P)]^p$$

$$\leq \beta_0 \min_{1 < \gamma \leq \alpha} \left\{ -1 + \frac{1}{N} \prod_{j=1}^s [1 + 2\beta_j^\gamma \zeta(\gamma)] \right.$$

$$\left. + \frac{N-1}{N} \prod_{j=1}^s \left[1 - \frac{(1-N^{1-\gamma})2\beta_j^\gamma \zeta(\gamma)}{N-1} \right] \right\}^{\alpha/\gamma}.$$

Disney and Sloan used this theorem to show that there exist rank-1 lattices with $\mathcal{P}_\alpha(L) = O(N^{-\alpha}[\log N]^{\alpha s})$ for prime N. However, the value of N at which one sees the nearly $N^{-\alpha}$ decay may be quite large, even for moderate dimensions.

Figures 6.1–6.4 show the upper bound on $\mathcal{P}_2(L)$ from Theorem 6.5 for dimensions 2 and 13 and two cases of product-type weights. One case is $\beta_0 = \beta_1 = \cdots = \beta_s = 1$, which corresponds to the original (or unweighted) $\mathcal{P}_2(L)$. The second case is $\beta_0 = 1, \beta_1 = \cdots = \beta_s = \sqrt{3/(8\pi^2)}$. These weights correspond to setting $\hat{\beta}_1 = \cdots = \hat{\beta}_s = 1$ for the $\mathcal{L}_2$-star discrepancy, and then computing the corresponding β_j according to Theorem 5.5. Although

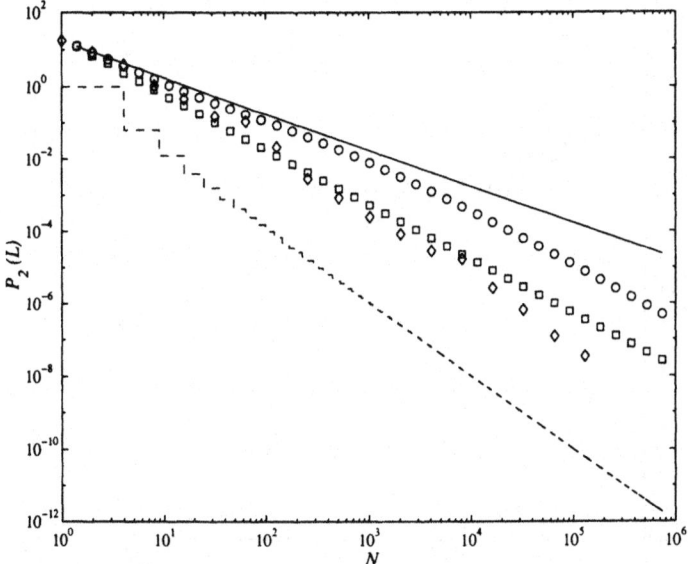

FIGURE 6.1. The mean square of $D_{\mathcal{F},2}(P)$ for simple random samples by Theorem 6.2 (solid); the upper bound on $\mathcal{P}_2(L)$ for rank-1 lattices from Theorem 6.5 ($\circ$); the upper bound on $\mathcal{P}_2(L)$ for copy rules using Theorem 6.7 ($\square$) ; the lower bound on $D_{\mathcal{F},2,1}(P)$ from Theorem 6.6 (dashed); $\mathcal{P}_2(L)$ for the lattice sequence described in Section 7.2 ($\diamond$); for $s = 2$, $\beta_0 = \beta_1 = \cdots = \beta_s = 1$.

Theorem 6.5 applies strictly to prime numbers N, this restriction is ignored for convenience in these figures

The upper bounds on $\mathcal{P}_2(L)$ vary substantially with β_j and s. Although the upper bounds are smaller than the corresponding discrepancy of a simple random sample, the difference is small for higher dimensions for the range of N plotted here. Also, the eventual asymptotic decay rate of nearly $O(N^{-2})$ is not evident for moderate N.

6.2   A Lower Bound on $D_{\mathcal{F},\alpha,1}(P)$

In contrast to the upper bounds on $\mathcal{P}_\alpha(L)$ above, Sloan and Woźniakowski [SW97] have given some lower bounds on the discrepancy $D_{\mathcal{F},\alpha,1}(P)$ for any P and for $N < 2^s$. Here we generalize their argument for the case of a weighted discrepancy and for other values of N.

Theorem 6.6. *For any set $P \subset C^s$ with N points there is the following lower bound on the weighted discrepancy:*

$$D_{\mathcal{F},\alpha,1}(P) \geq \min_{\emptyset \subset u \subseteq S}(\beta_u \lfloor N^{1/s} \rfloor^{-\alpha|u|}).$$

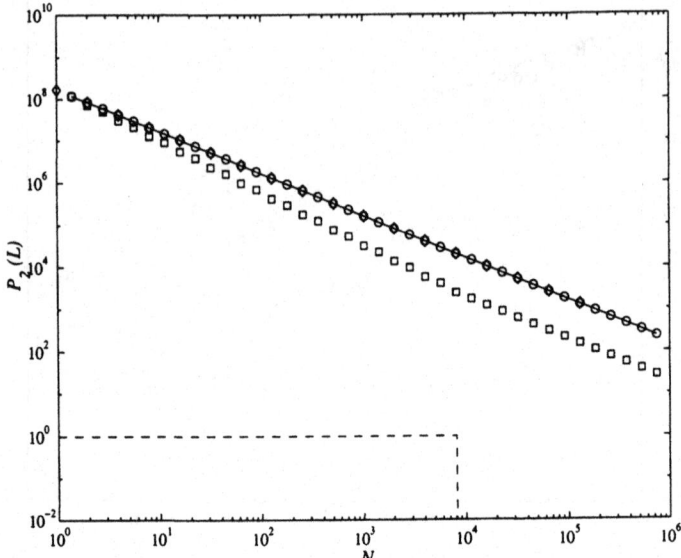

FIGURE 6.2. The same as Figure 6.1, except that $s = 13$.

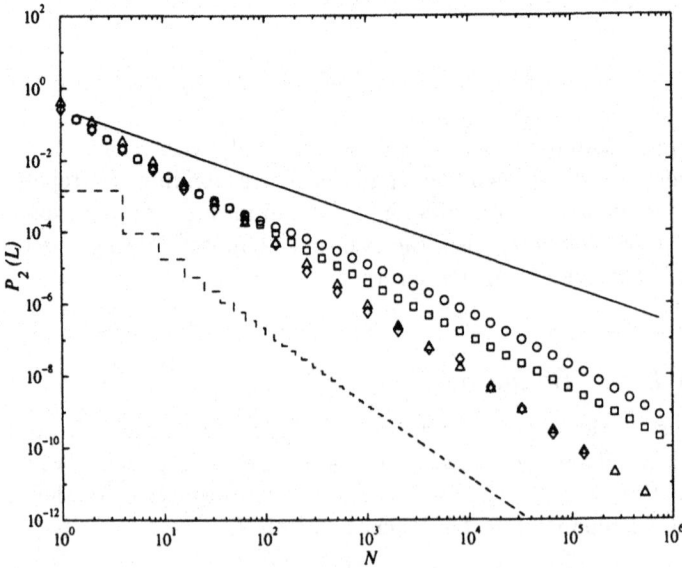

FIGURE 6.3. The same as Figure 6.1, except that $\beta_1 = \cdots = \beta_s = \sqrt{3/(8\pi^2)}$. Also, includes the mean square $\mathcal{L}_2$-star discrepancy of a randomized $(0, m, s)$-net in base s according to (7.1) ($\triangle$).

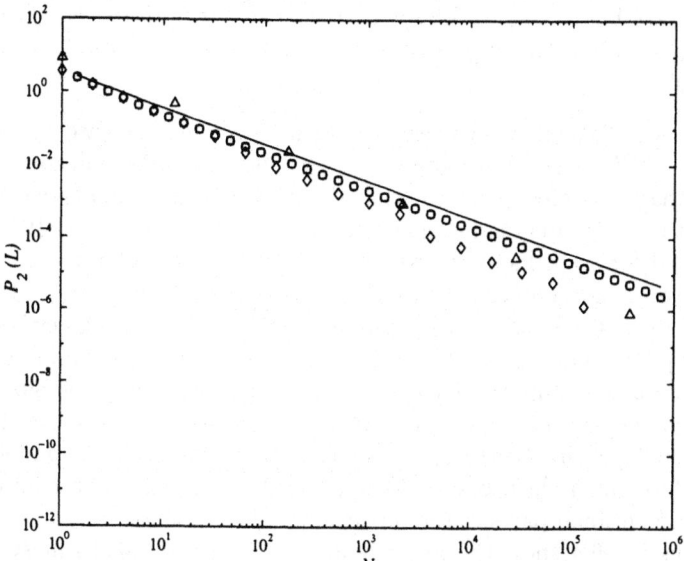

FIGURE 6.4. The same as Figure 6.3, except that $s = 13$.

Proof. Suppose that $r = \lfloor N^{1/s} \rfloor$, which implies that $N < (r+1)^s$. For any set $P \subset C^s$ with N points, the system of N linear equations,

$$\sum_{k \in \{0,1,\dots,r\}^s} a_k e^{2\pi i k \cdot z} = 0 \quad \text{for all } z \in P,$$

has a nontrivial solution for the $(r+1)^s$ coefficients a_k. This nontrivial solution may be scaled so that $|a_k| \le a_{k^*} = 1$, $\forall k \in \{0,1,\dots,r\}^s$ for some particular k^*. Now form the integrand

$$f(x) = \sum_{k \in \{0,1,\dots,r\}^s} a_k e^{2\pi i (k-k^*) \cdot x}.$$

Because of the above conditions on the a_k the quadrature rule applied to f gives 0, but the integral of f is 1. So,

$$1 = |I(f) - Q(f;P)| \le D_{\mathcal{F},\alpha,1}(P) V_{\mathcal{F},\alpha,\infty}(f).$$

A lower bound on the variation of f implies an upper bound on the discrepancy:

$$D_{\mathcal{F},\alpha,1}(P) \ge [V_{\mathcal{F},\alpha,\infty}(f)]^{-1} = \left\| (\beta_u^{-1} V_{\mathcal{F},\alpha,\infty,u}(f_u))_{u \neq \emptyset} \right\|_\infty^{-1}$$

$$= \left\| \left(\beta_u^{-1} \left[\prod_{j \in u} (\overline{k_j - k_j^*})^\alpha \right] \hat{f}_u(k_u) \right)_{\emptyset \subset u \subseteq S} \right\|_\infty^{-1}.$$

For $k, k^* \in \{0, 1, \ldots, r\}^s$ it follows that $\overline{k_j - k_j^*} \leq r^\alpha$. Moreover, $|\hat{f}_u(k_u)| \leq \max_k |a_k| = 1$. Substituting these bounds into the above formula completes the proof. $\square$

Sloan and Woźniakowski's version of Theorem 6.6 is the case $\beta_u = 1$ and $N < 2^s$, which implies that $D_{\mathcal{F},\alpha,1}(P) \geq 1$. Sloan and Woźniakowski thus argued that it is safer to assume that the integral is zero unless one cannot afford at least 2^s integrand evaluations. This means that at least 10^3 points are needed for $s = 10$ and at least 10^6 points are needed for $s = 20$.

However, if one chooses product-type weights of the form (4.9), with $\beta_0 = 1$, $\beta_1 = \cdots = \beta_s = \beta < 1$, then the lower bound on the discrepancy becomes $(\beta \lfloor N^{1/s} \rfloor)^{-\alpha s}$, which is $\beta^{-\alpha s}$ for $N < 2^s$. This lower bound vanishes as α or s approach infinity. Therefore, the pessimistic result of Sloan and Woźniakowski [SW97] is strongly dependent on the choice of weights in the definition of the discrepancy. As mentioned in Remarks 4.4 and 4.5 the choice of weights reflects one's assumptions about the sizes of the lower and higher dimensional parts of the integrand.

Figures 6.1–6.4 show the lower bound in Theorem 6.6. In most of the cases considered the lower bound is not close to the known upper bound, except for small N and s. In some of the figures the lower bound is too small to appear within the plotting window. This suggests the need for improved upper and/or lower bounds.

6.3   Quadrature Rules with Different Weights

Even if one interprets the lower bound on the discrepancy in Theorem 6.6 pessimistically, there is a better alternative to choosing the quadrature rule $Q(f) = 0$ for $N < 2^s$. Consider a quadrature rule that is the *weighted average* of the values of the integrand, that is,

$$Q(f; P, \{w^{(i)}\}) = \sum_{i=1}^{N} w^{(i)} f(z^{(i)}).$$

(The choice $w^{(i)} = N^{-1}$ is the original quadrature rule.) The worst-case error and average-case analyses in Theorems 3.1 and 3.15 can be modified to accommodate this new quadrature rule. If the weights do not sum up to 1, that is, the new rule is not exact for constants, then the variation of the integrand must be defined as its norm, that is, $V(f; K) = \|f\|_K$. The definition of the discrepancy is modified as follows:

$$D(P, \{w^{(i)}\}; K) = \left\{ \int_{C^s \times C^s} K(x, y) \, dx \, dy - 2 \sum_{i=1}^{N} w^{(i)} \int_{C^s} K(z^{(i)}, y) \, dy \right.$$
$$\left. + \sum_{i,i'=1}^{N} w^{(i)} w^{(i')} K(z^{(i)}, z^{(i')}) \right\}^{1/2}.$$

This formula is a quadratic function of the weights. Therefore, the weights which minimize the discrepancy satisfy the following system of N linear equations for the N unknown $w^{(i)}$:

$$-\int_{C^s} K(z^{(i)}, y)\, dy + \sum_{i'=1}^{N} w^{(i')} K(z^{(i)}, z^{(i')}) = 0, \quad i = 1, \ldots, N.$$

This system of equations may be too large to solve in practice. However, in the case where P is the node set of a lattice, and the kernel is shift-invariant, it can be shown that the optimal weights are uniform, i.e. $w^{(1)} = \cdots = w^{(N)}$. Even in cases where the optimal weights are not uniform, one may easily solve for the best uniform weights:

$$w^{(1)} = \cdots = w^{(N)} = \frac{w_{\mathrm{opt}}}{N} = \left[\sum_{z \in P} \int_{C^s} K(z, y)\, dy\right] \left[\sum_{z, z' \in P} K(z, z')\right]^{-1}.$$

In general, this weight satisfies $0 < w_{\mathrm{opt}} < 1$. For small N the weight w_{opt} is close to 0, which corresponds to a cautious approach in estimating the integral with too few function evaluations. As N tends to infinity, w_{opt} tends to 1, which corresponds to the original quadrature rule, (2.1). The discrepancy for the best uniform weights is:

$$D(P, w_{\mathrm{opt}} N^{-1}; K) = \left\{ \int_{C^s \times C^s} K(x, y)\, dx\, dy \right.$$

$$\left. - \left[\sum_{z \in P} \int_{C^s} K(z, y)\, dy\right]^2 \left[\sum_{z, z' \in P} K(z, z')\right]^{-1} \right\}^{1/2}.$$

For shift-invariant kernels, the discrepancy for the best uniform weights can be written in terms of the original discrepancy with each integrand value weighted by N^{-1}:

$$D(P, w_{\mathrm{opt}} N^{-1}; K) = \left\{ [D(P; K)]^{-2} + \|I\|_K^{-2} \right\}^{-1/2}.$$

Therefore, by choosing optimal uniform weights one obtains a smaller discrepancy than either $D(P; K) = D(P, N^{-1}; K)$ or $\|I\|_K = D(P, 0; K)$. The former case corresponds to the original quadrature rule, and the latter case corresponds to pessimistically choosing $Q(f) = 0$.

6.4 Copy Rules

The monograph of Sloan and Joe [SJ94] and the papers cited therein recommend copy rules for two reasons: i) they provide a way to estimate quadrature error, and ii) they tend to give smaller $\mathcal{P}_\alpha(L)$ for the same number of points than rank-1 lattice rules. Here we generalize some of the results in [SJ94] on copy rules and comment on the assertion that they give smaller $\mathcal{P}_\alpha(L)$ values.

Define a rectangular grid with spacing n_j^{-1} in the jth coordinate as follows:

$$G = \left\{ \left(\frac{m_1}{n_1}, \dots , \frac{m_s}{n_s} \right) : m_j = 0, \dots , n_j - 1 \right\},$$

for some integers $n_1, \dots , n_s$. For any set $P \in C^s$ (P is not necessarily the node set of a lattice) let $P + G \in C^s$ denote $n_1 \cdots n_s$ shifted copies of P, specifically,

$$P + G = \{\{z + z'\} : z \in P, z' \in G\}.$$

Let $D_{\mathcal{F},\alpha,p}(P + G; \beta_0, \dots , \beta_s)$ denote the weighted version of the discrepancy of this set as defined in (4.10), where the dependence on β_i is given explicitly. This discrepancy can be computed in terms of the discrepancy of a set related to P as follows:

Theorem 6.7. *[Hic96b, Lemma 4.1] For any set $P \in C^s$ and a grid G as defined above,*

$$(6.1) \quad D_{\mathcal{F},\alpha,p}(P + G; \beta_0, \dots , \beta_s) = D_{\mathcal{F},\alpha,p}(\tilde{P}; \beta_0, n^{-1}\beta_1, \dots , n^{-s}\beta_s),$$

where $\tilde{P} = \{\{(n_1 z_1, \dots , n_s z_s)\} : z \in P\}$.

Proof. According to (4.10) the discrepancy of $P + G$ involves the following term, which can be rewritten using some straightforward algebraic manipulations:

$$\frac{1}{n_1 \cdots n_s N} \sum_{z \in P+G} e^{2\pi i k \cdot z} = \frac{1}{n_1 \cdots n_s N} \sum_{z \in P} e^{2\pi i k \cdot z} \sum_{z' \in G} e^{2\pi i k \cdot z'}$$

$$= \frac{1}{n_1 \cdots n_s N} \sum_{z \in P} e^{2\pi i k \cdot z} \prod_{j=1}^{s} \sum_{m_j=0}^{n_j} e^{2\pi i k_j m_j / n_j}$$

$$= \frac{1}{n_1 \cdots n_s N} \sum_{z \in P} e^{2\pi i k \cdot z} \prod_{j=1}^{s} n_j 1_{\{k_j \in n_j \mathbf{Z}\}}$$

$$= \prod_{j=1}^{s} 1_{\{k_j \in n_j \mathbf{Z}\}} \frac{1}{N} \sum_{z \in P} e^{2\pi i k \cdot z}.$$

Since one need only consider those k_j which are integer multiples of n_j, one can replace the vector k by $(\tilde{k}_1 n_1, \dots , \tilde{k}_s n_s)$. Making this substitution into the formula for the discrepancy of $P + G$ completes the proof. $\square$

The essence of this theorem is that copy rules correspond to reducing the sizes of the weights β_j. Since the discrepancy $D_{\mathcal{F},\alpha,p}(P; \beta_0, \dots , \beta_s)$ is

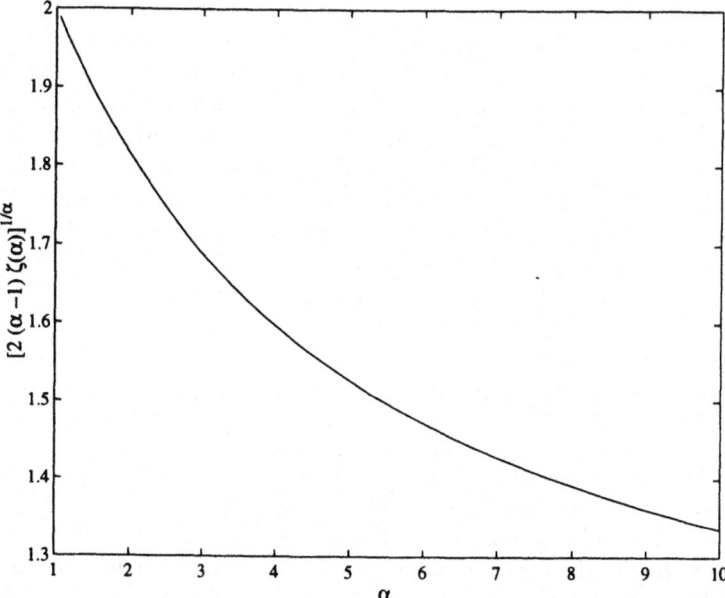

FIGURE 6.5. The optimal n_j for $\beta_j = 1$.

monotonically increasing in each β_j for any set P, this result would appear to favor copy rules. However, one must realize that the cost of reducing the β_j is that $P + G$ in the left-hand side of (6.1) contains $n_1 \cdots n_s$ times as many points as the $\tilde{P}$ in the right-hand side. It is not immediately obvious whether or not some other set, P', with the same number of points as $P+G$ might have an even lower discrepancy value.

For lattice rules Sloan and Joe [SJ94] look at the simpler, though not as tight, upper bound $\mathcal{P}_\alpha(L)$ based on Lemma 6.3 and Theorem 6.7 for a copy lattice rule with N points:

$$\frac{\beta_0}{N} \prod_{j=1}^{s} [n_j + 2n_j^{1-\alpha} \beta_j^\alpha \zeta(\alpha)].$$

The choice of n_j that minimizes this bound can be found using simple calculus, and corresponds to:

$$n_j = \beta_j [2(\alpha - 1)\zeta(\alpha)]^{1/\alpha}.$$

The quantity multiplying β_j is plotted in Figure 6.5. Recognizing that in practice n_j must be an integer it is easy to see from the Figure 6.5 that $n_j = 2$ is a good choice for $\beta_j = 1$. In fact, Sloan and Joe showed that if $\beta_j = 1$, then $n_j = 2$ is better than $n_j = 1$ (no copies) for all $\alpha > 1$. However, for small enough values of β_j the optimum is $n_j = 1$ (no copies).

Figures 6.1–6.4 show the upper bound on $\mathcal{P}_2(L)$ found by combining Theorems 6.5 and 6.7 and searching for copy lattice rules where $n_1 = \cdots = n_j = n_{j+1} + 1 = \cdots = n_s + 1$. Typical values of the total number of points, N, are chosen, where $N > n_1 \cdots n_s$, but N is not constrained to be an integer multiple of $n_1 \cdots n_s$, for convenience. As N increases, so do the n_j. While this might support the assertion that copy rules are preferable, the author believes that it is due to the fact that the upper bounds being minimized are not tight. An obvious disadvantage of copy rules is that while increasing the number of points by a factor $n_1 \cdots n_s$, you only get n_j times as many new values in the jth coordinate direction.

7   Discrepancies of Integration Lattices and Nets

Rather than relying on upper bounds one may compute the discrepancies of specific integration lattices and other quasi-random sets and compare their values. This has been done by [MC94, Hic95] and others. This section considers the discrepancies of two kinds of quasi-random points. For a look at the discrepancies corresponding to Smolyak quadrature rules see [FH96].

7.1   The Expected Discrepancy of Randomized $(0, m, s)$-Nets

Besides integration lattices, (t, m, s)-nets and (t, s)-sequences in base b are popular choices of quasi-random points. This family includes the sequences of Sobol' [Sob67], Faure [Fau82] and Niederreiter [BFN92]. For a review of nets and sequences see [Nie92] and elsewhere in this volume.

Although computing a discrepancy of the form (3.3) for a particular (t, m, s)-net requires $O(sN^2)$ operations (or somewhat less when the algorithm in [Hei96] can be used), the root mean square discrepancy over a randomized $(0, m, s)$-nets has a computationally simple formula (see [Hic96a]). Here the nets are assumed to be randomized according to Owen's procedure [Owe95, Owe97a, Owe97b]. For the $\mathcal{L}_2$-star discrepancy defined in Section 4.5 with weights $\hat{\beta}_1 = \cdots = \hat{\beta}_s = \hat{\beta}$ the mean square discrepancy is:

$$(7.1)\quad E_P\{[D_2^*(P)]^2\} = \hat{\beta}_0^2 \left(1 + \frac{\hat{\beta}^2}{2}\right)^s b^{-2m}$$

$$\times \sum_{l=0}^{s-1} \sum_{k=0}^{l} \sum_{j=0}^{l-k} \left[\binom{s}{l+1} \binom{m}{k} \binom{l}{j} \left(\frac{\hat{\beta}^2}{6+3\hat{\beta}^2}\right)^{l+1} b^{-l}(b^2-1)^k(-b-1)^j \right].$$

Figures 6.3–6.4 show plots of the mean square $\mathcal{L}_2$-star discrepancy for $(0, m, s)$-nets with $\hat{\beta} = 1$. The corresponding shift-invariant kernel according to Theorem 5.5 is $K_{\mathcal{F},\alpha}$ with weights $\beta_1 = \cdots = \beta_s = \sqrt{3/(8\pi^2)}$, so the mean square $\mathcal{L}_2$-star discrepancy is comparable to $\mathcal{P}_\alpha(L)$ with these weights. The $(0, m, s)$-nets have significantly smaller discrepancy than the upper bounds for lattice rules, especially as N increases.

Remark 7.1. Formula (7.1) requires only $O(s \max(s, \log N))$ operations to evaluate.

7.2 Infinite Sequences of Embedded Lattices

One advantage of (t, s)-sequences over integration lattices is that the number of points required, N, need not be chosen in advance. If according to some quadrature error estimate the original N is too small, one may take the next $N' - N$ points of the (t, s)-sequence to get a total of N' points without discarding the original N points. However, this is typically not the case for integration lattices. Since the generating vector is a function of the number of points, an N' point lattice cannot normally be obtained by adding $N' - N$ points to an N point lattice.

The author and his collaborator have proposed a method for overcoming this disadvantage of lattice rules by defining an infinite lattice sequence that is analogous to a (t, s)-sequence [HH97]. Let $\phi_b(i)$ denote the Van der Corput sequence, that is, if $\cdots i_3 i_2 i_1$ is the base-b representation of the non-negative integer i with digits $i_1, i_2, \ldots$, then

$$\phi_b(i) = 0.i_1 i_2 i_3 \ldots \quad \text{(base } b).$$

For example, the first several terms of the base-2 Van der Corput sequence are:

$$0, \frac{1}{2}, \frac{1}{4}, \frac{3}{4}, \frac{1}{8}, \frac{5}{8}, \ldots$$

A rank-1 infinite lattice sequence is defined by replacing i/N in (2.2) by $\phi_b(i)$, that is,

$$P_\infty = \{\{\phi_b(i)h\} : i = 0, 1, \ldots\}.$$

Each component of the generating vector is now considered to be an infinite string of digits,

$$h = (\ldots h_{12} h_{11}, \ldots h_{22} h_{21}, \ldots, \ldots h_{s2} h_{s1}) \quad \text{(base } b).$$

The first b^m points of P_∞ are the node set of a lattice for any non-negative integer m. Moreover, the succeeding runs of b^m points are shifted copies of the first b^m points. Furthermore, only the first m digits of each component of h are needed to generate the first $N = b^m$ points of P_∞.

Remark 7.2. Although one may in theory use the first N points from P_∞ for quadrature for any value of N, the lattice property holds only if N is an integer power of the base, b. An analogous property holds for (t, s)-sequences, that is, the (t, m, s)-net is the first b^m points of the (t, s)-sequence. For both lattices and nets one expects the discrepancy to be smaller when choosing $N = b^m$ points.

Finding good generating vectors for lattices of the above form is an area of ongoing research. Some promising results have been obtained by using optimization methods to find h which minimize $\mathcal{P}_\alpha(L)$. To make the optimization problem tractable one may restrict the class of possible generating vectors, for example, requiring h to be of Korobov-type, that is,

$$h = (1, h_2, h_2^2, h_2^3, \ldots, h_2^{s-1}).$$

The values of $P_\alpha(L)$ for the first $N = 1, 2, 4, \ldots, 2^{17}$ points of an infinite lattice sequence of this type are shown in Figures 6.1–6.4. The generator is

$$h_2 = 89715 = 10101111001110011 \text{ (base 2)}$$

The infinite lattice has similar discrepancy values to the $(0, m, s)$-nets.

8   Tractability of High Dimensional Quadrature

The discussion so far has assumed the dimension s to be some fixed, finite quantity. It is also interesting to explore what happens if s becomes arbitrarily large.

8.1   Quadrature in Arbitrarily High Dimensions

Quasi-Monte Carlo methods have proven to be effective in evaluating some integrals arising in the valuation of financial securities (see [PT95, NT96, PT96, CMO97] and elsewhere in this volume). In such problems the dimension, s, may be in the hundreds or thousands. The asymptotic order of the discrepancy for a fixed s and letting N tend to infinity typically contains $\log N$ to some power that increases with s. Such expressions for the asymptotic order of the quadrature error are quite pessimistic for arbitrarily large s. This has motivated some scholars to try other approaches to understand quadrature error in arbitrarily high dimensions.

The theorems in Section 6 can be used to determine the behavior of the discrepancy when s tends to infinity. For Monte Carlo methods Theorem 6.2 says that the mean square discrepancy is uniformly bounded in s, if and only if

$$\lim_{s \to +\infty} \prod_{j=1}^{s} [1 + 2\beta_j^\alpha \zeta(\alpha)] < \infty.$$

By taking logarithms of the product, this condition can be shown to be equivalent to

$$(8.1) \qquad \qquad \sum_{j=1}^{+\infty} \beta_j^\alpha < \infty.$$

The upper bound on $\mathcal{P}_\alpha(L)$ is also uniformly bounded in s if and only if condition (8.1) holds. A comparable condition is obtained by [SW98] for the $\mathcal{L}_2$-star discrepancy. The condition that the β_j tend to zero fast enough means that the integrand varies less and less with respect to x_j as j tends to infinity.

8.2 The Effective Dimension of an Integrand

Figures 6.1–6.4 show the dependence of the discrepancy or $\mathcal{P}_\alpha(L)$ on the dimension, s, and the weights, β_j. One can imagine integrands, such as that given in (4.1), whose effective dimension is smaller than the nominal dimension s. The ANOVA decomposition introduced in Section 4.1 provides a way of defining the effective dimension of an integrand.

The effective dimension of a particular integrand has been defined by Caflisch, Morokoff and Owen [CMO97]. It can be shown from the definition in (4.2) that the variance of a function is the sum of the variances of its ANOVA effects:

$$\mathrm{Var}(f) = \sum_{\emptyset \subset u \subseteq S} \mathrm{Var}(f_u).$$

(Note that $\mathrm{Var}(f_\emptyset) = 0$ because $f_\emptyset$ is a constant.) The dimension of an integrand may be defined in terms of the dimension of the ANOVA effects that contain some proportion of the variance of the integrand.

Definition 8.1. An integrand f is said to be proportion p_t of **truncated dimension** s_t if and only if

$$\sum_{u \subseteq \{1,\ldots,s_t\}} \mathrm{Var}(f_u) = p_t \, \mathrm{Var}(f),$$

where $0 \le p_t \le 1$. An integrand f is said to be proportion p_s of **superposition dimension** s_s if and only if

$$\sum_{|u| \le s_s} \mathrm{Var}(f_u) = p_s \, \mathrm{Var}(f),$$

where $0 \le p_s \le 1$.

The example in (4.1) is a relatively simple function, so its dimension can be computed easily. This function has 100% truncation dimension 3 and 100% superposition dimension 2.

An alternative to computing the dimension of a single integrand is to compute the average dimension of a family of integrands. For the spaces of random functions, $\mathcal{A}$, introduced in Section 3.3, recall formula (3.20) for the mean value of the variance of a function. An analogous definition to the one above is:

Definition 8.2. The integrands in a space of random functions, $\mathcal{A}$, are said to be proportion p_t of **truncated dimension** s_t if and only if

$$\sum_{u \subseteq \{1,\ldots,s_t\}} E_{f \in \mathcal{A}}[\mathrm{Var}(f_u)] = p_t E_{f \in \mathcal{A}}[\mathrm{Var}(f)].$$

The integrands in a space of random functions, $\mathcal{A}$, are said to be proportion p_s of **superposition dimension** s_s if and only if

$$\sum_{|u| \leq s_s} E_{f \in \mathcal{A}}[\mathrm{Var}(f_u)] = p_s E_{f \in \mathcal{A}}[\mathrm{Var}(f)].$$

Remark 8.3. In the previous two definitions one may either fix the proportions, p_s and p_t, and then solve for s_s and s_t, or vice versa. In [CMO97] the choice $p_s = p_t = 0.95$ is recommended.

Remark 8.4. If $p_s = p_t$, then $s_s \leq s_t$, and if $s_s = s_t$, then $p_s \geq p_t$.

Computing the truncated or superposition dimension of integrands in a space $\mathcal{A}$ requires formulae for $E_{f \in \mathcal{A}}[\mathrm{Var}(f_u)]$ in terms of the covariance kernel. From the definition of the ANOVA decomposition in (4.2) it follows that:

$$\sum_{(u',v') \subseteq (u,v)} E_{f \in \mathcal{A}}[f_{u'}(x_{u'}) f_{v'}(y_{v'})] = E_{f \in \mathcal{A}}\left[\sum_{u' \subseteq u} f_{u'}(x_{u'}) \sum_{v' \subseteq v} f_{v'}(y_{v'})\right]$$

$$= E_{f \in \mathcal{A}}\left[\int_{C^{S-u}} f_u(x_u)\, dx_{S-u} \int_{C^{S-v}} f_v(y_v)\, dy_{S-v}\right]$$

$$= \int_{C^{S-u} \times C^{S-v}} K(x,y)\, dx_{S-u}\, dy_{S-v}.$$

The mean over $\mathcal{A}$ of $\mathrm{Var}(f_u)$ is obtained by setting $v = u$ and $y_v = x_u$ in the above equation and integrating with respect to x_u over C^u. Furthermore, one may use the associated shift-invariant kernel, as defined in (5.3), to simplify the resulting formula:

$$(8.2) \quad E_{f \in \mathcal{A}}[f_\emptyset^2] + \sum_{\emptyset \subset u' \subseteq u} E_{f \in \mathcal{A}}[\mathrm{Var}(f_{u'})]$$

$$= \int_{C^s \times C^{S-u}} K(x,y)\big|_{y_u = x_u}\, dx\, dy_{S-u}$$

$$= \int_{C^{S-u}} K_{\mathrm{shift}}(x,0)\big|_{x_u = 0}\, dx_{S-u}.$$

Further manipulation of this equation leads to the conclusion that:

$$(8.3) \quad E_{f \in \mathcal{A}}[\mathrm{Var}(f_u)] = \sum_{v \subseteq u}(-1)^{|u-v|} \int_{C^{S-v}} K_{\mathrm{shift}}(x,0)\big|_{x_v = 0}\, dx_{S-v}.$$

Using this formula and some counting arguments one may arrive at the following formulae for the truncated and superposition dimensions of functions in $\mathcal{A}$.

Theorem 8.5. *Consider a space of random functions, $\mathcal{A}$, with covariance kernel, K, and associated shift-invariant kernel, K_{shift}. Let $S_t = \{1, \ldots, s_t\}$. The truncated dimension, s_t, and the superposition dimension, s_s, with proportions p_t and p_s for the space $\mathcal{A}$ satisfy the following equations:*

$$\int_{C^{S-S_t}} K_{shift}(x, 0)|_{x_{S_t}=0} \; dx_{S-S_t}$$
$$= p_t K_{shift}(0, 0) + (1 - p_t) \int_{C^s} K_{shift}(x, 0) \; dx,$$

$$\sum_{|u| \leq s_s} \left[(-1)^{s_s - |u|} \binom{s - 1 - |u|}{s_s - |u|} \int_{C^{S-u}} K_{shift}(x, 0)|_{x_u=0} \; dx_{S-u} \right]$$
$$= p_s K_{shift}(0, 0) + (1 - p_s) \int_{C^s} K_{shift}(x, 0) \; dx.$$

Remark 8.6. A function space $\mathcal{A}$ may be said to contain purely s-dimensional integrands if the proportions p_t and p_s are zero for s_t and s_s less than s. This occurs if and only if $\int_0^1 K_{shift}(x_j, 0) \; dx_j = 0$, for all $j = 1, \ldots, s$. An example of such a kernel would be the following weighted version of $K_{\mathcal{F},\alpha}$:

$$K_{\mathcal{F},\alpha}(x, y) = \beta_S \sum_{k_j \neq 0} \frac{e^{2\pi i k \cdot (x-y)}}{(\bar{k}_1 \cdots \bar{k}_s)^\alpha}.$$

Many interesting covariance kernels are of product form, that is,

$$K(x, y) = \prod_{j=1}^s \tilde{K}_j(x_j, y_j).$$

In this case the associated shift-invariant kernel is also of product form:

$$K_{shift}(x, 0) = \prod_{j=1}^s \tilde{K}_{shift,j}(x_j, 0).$$

Let $a_j = \int_0^1 \tilde{K}_{shift,j}(x_j, 0) \; dx_j$, and $b_j = \tilde{K}_{shift,j}(0, 0)$. Then some of the relevant formulae above may be written in terms of these a_j and b_j. Equation (8.2) becomes:

$$E_{f \in \mathcal{A}}[f_0^2] + \sum_{\emptyset \subset u' \subseteq u} E_{f \in \mathcal{A}}[\text{Var}(f_{u'})] = \prod_{j \in u} b_j \prod_{j' \in S-u} a_{j'}.$$

Thus, the formula for the truncated dimension is:

$$(b_1 \cdots b_{s_t})(a_{s_t+1} \cdots a_s) = p_t(b_1 \cdots b_s) + (1 - p_t)(a_1 \cdots a_s).$$

The formula for the superposition dimension is best computed using a recursion relation. Let

$$A(s, s_s) = E_{f \in \mathcal{A}}[f_0^2] + \sum_{0 < |u| \le s_s} E_{f \in \mathcal{A}}[\mathrm{Var}(f_u)],$$

so that the superposition dimension satisfies the formula:

$$A(s, s_s) = p_s(b_1 \cdots b_s) - (1 - p_s)(a_1 \cdots a_s).$$

Because (8.3) can be written as

$$E_{f \in \mathcal{A}}[\mathrm{Var}(f_u)] = \prod_{j \in u}(b_j - a_j) \prod_{j' \in S-u} a_{j'},$$

it follows that $A(s, s_s)$ satisfies the following recursion relationship:

$$A(t, 0) = a_1 \cdots a_t, \quad t = 1, 2, \ldots,$$
$$A(1, 1) = A(1, 2) = \cdots = b_1,$$
$$A(t, s_s) = A(t - 1, s_s)a_s + A(t - 1, s_s - 1)(b_s - a_s).$$

For the case of product-type weights, (4.9), the covariance kernel $K_{\mathcal{F},\alpha}$ is a shift-invariant, product-type kernel with

$$a_j = 1, \quad b_j = 1 + 2\beta_j^\alpha \zeta(\alpha), \quad j = 1, \ldots, s.$$

The values of p_t and p_s are plotted in Figures 8.1–8.6 for a range of s_t, s_s and s, and for three different types of weights. For $\beta_1 = \cdots = \beta_s = 1$ the integrands are only about 25% of truncation dimension $s - 1$. The superposition dimension, s_s, is also close to the nominal dimension, s, for p_s in the range 10-90%. For the other two cases where the weights are smaller, the truncation dimension takes on most of the range of 1 through s for p_t in the range 10-90%, and the superposition dimension is only 1 through 3 for p_s in the range 10-90%. Therefore, the effective truncation and superposition dimensions of the integrands depend strongly on the choices of the weights, β_j.

9   Discussion and Conclusion

How well do lattice rules measure up? The answer depends on the measuring stick. In this chapter several discrepancies have been introduced, but

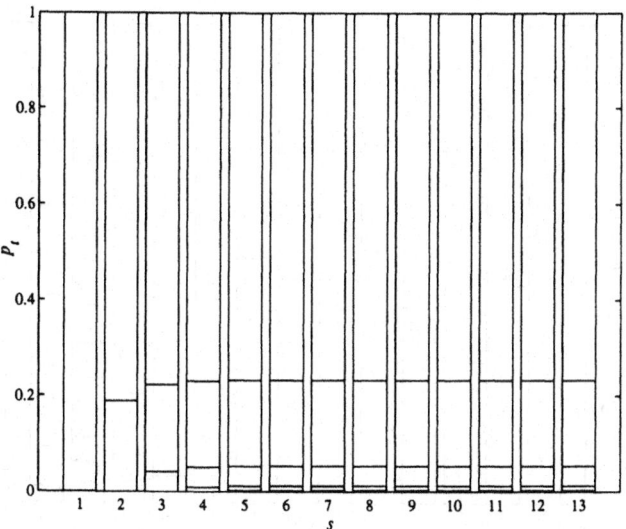

FIGURE 8.1. The proportions p_t for truncation dimensions $s_t = 1, \ldots, s$ and $s = 1, \ldots, 13$ for the space of random functions with covariance kernel $K_{\mathcal{F},\alpha}$ and $\beta_1 = \cdots = \beta_s = 1$.

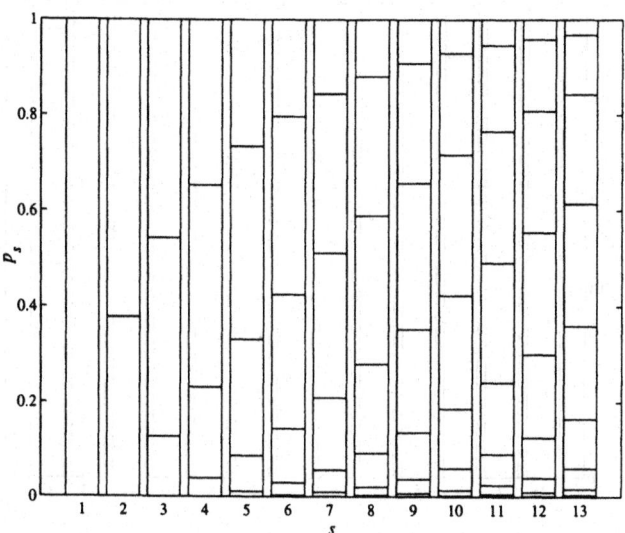

FIGURE 8.2. The same as Figure 8.1 but for the superposition dimension.

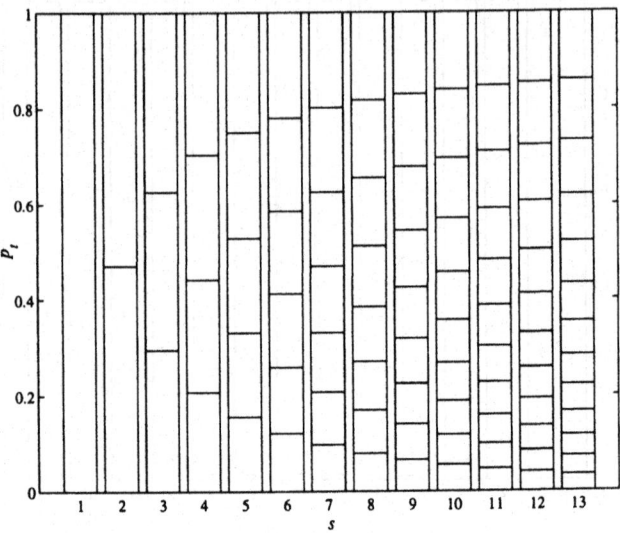

FIGURE 8.3. The same as Figure 8.1 but for $\beta_1 = \cdots = \beta_s = \sqrt{3/(8\pi^2)}$.

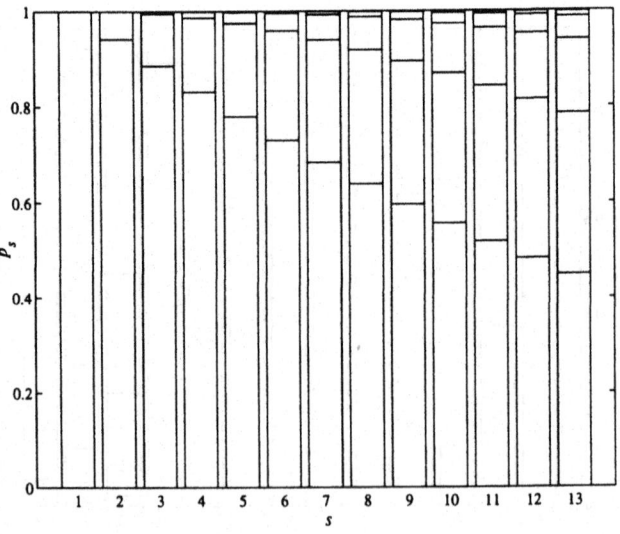

FIGURE 8.4. The same as Figure 8.3 but for the superposition dimension.

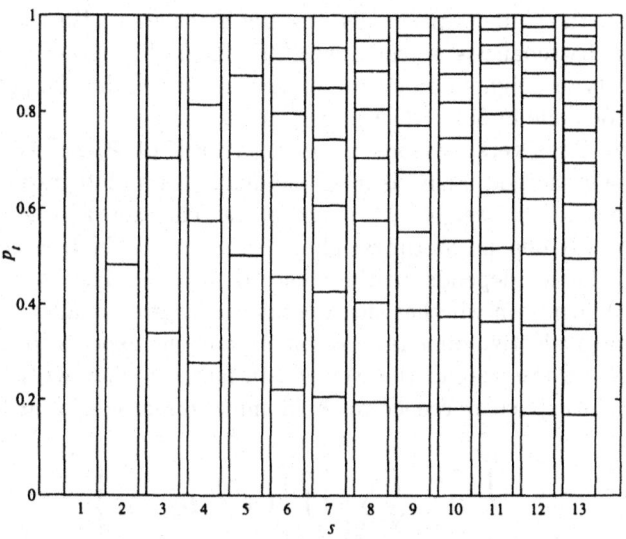

FIGURE 8.5. The same as Figure 8.1 but for $\beta_j = j^{-1}$, $j = 1, \ldots, s$.

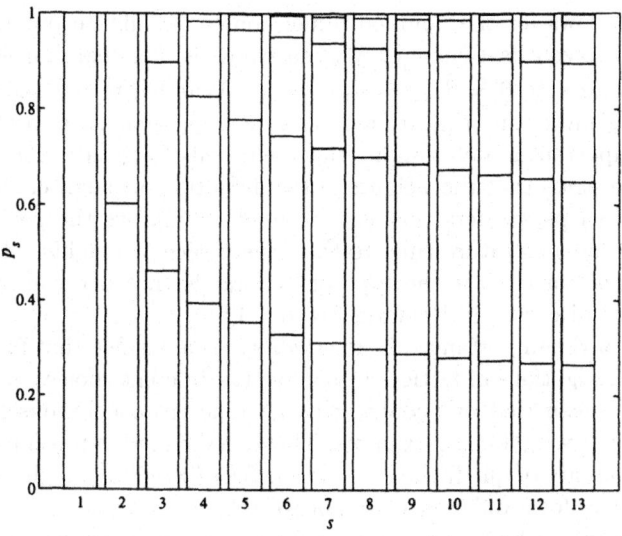

FIGURE 8.6. The same as Figure 8.5 but for the superposition dimension.

most of them are related to a weighted form of $\mathcal{P}_\alpha(L)$, either as a direct generalization or through the associated shift-invariant discrepancy (Theorem 5.5). In fact, the shift-invariant discrepancy associated with the star and unanchored discrepancies is just a weighted form of $\mathcal{P}_2(L)$. Furthermore, for product-type weights and even, positive integer α, the computation of $\mathcal{P}_\alpha(L)$ can be done in only $O(sN)$ operations.

Therefore, for several reasons a weighted form of $\mathcal{P}_2(L)$ seems to be the best single method for measuring the quality of an integration lattice. However, the choice of weights, β_j, is important. The value and rate of decay of $\mathcal{P}_2(L)$ with N depend on the weights. The choice of the best generating vector for a lattice depends on the choice of weights. The desirability of copy rules depends on whether the weights are large or small.

The average effective dimension of integrands one expects to encounter also depends on the sizes of the weights. In fact, one may write $\mathcal{P}_2(L)$ in (4.15) without explicit reference to the nominal dimension, s, as follows:

$$
\mathcal{P}_2(L) = \left\{ -1 + \frac{1}{N} \sum_{z \in P} \prod_{j=1}^{+\infty} \left[1 + \frac{\tilde{\beta}_j^2}{2} B_2(z_j) \right] \right\}^{1/2},
$$

with the condition on the weights, (8.1), that

$$
\sum_{j=1}^{+\infty} \tilde{\beta}_j^2 < \infty.
$$

This form allows for the nominal dimension to be infinite provided that the β_j tend to zero fast enough. Alternatively, the nominal dimension will be some finite s, if, $0 = \tilde{\beta}_{s+1} = \tilde{\beta}_{s+2} = \cdots$. Whether or not the nominal dimension is finite, the truncation or superposition dimensions are likely to be more important in determining the tractability of lattice rules.

Again we stress that the weights in the definition of the discrepancy are not additional parameters that can be used to improve the performance of a particular quadrature rule. Rather, the choice of weights reflects the user's assumptions about the type of integrands that one can expect to encounter as discussed in Remarks 4.4 and 4.5.

Because of the importance of the weights it is advisable that future theoretical investigations of lattice rules allow for different choices of weights. This might mean that one could arrive at different conclusions based on one's choice of weights. For example, Theorem 6.6 gives a pessimistic view of the tractability of quadrature for one choice of weights, but the situation is much better for a different choice of weights.

There seems to be much room for improvement in determining what values of $\mathcal{P}_\alpha(L)$ are possible. There is a sizable gap between the lower and upper bounds that begs to be closed. Moreover, the "average" $(0, m, s)$-net has a much smaller discrepancy than the "average" rank-1 lattice. However, a few specific lattices considered here have $\mathcal{P}_\alpha(L)$ substantially better than the average and have similar discrepancy values to $(0, m, s)$-nets. Perhaps, there is room to improve the upper bound on $\mathcal{P}_\alpha(L)$ by a more selective average that ignores bad lattices.

Although the discrepancies of lattices are similar to those of nets, the nets still have one practical advantage over lattices. Whereas there are several known explicit constructions of infinite (t, s)-sequences, whose first N points have low discrepancy, the situation is much less developed for lattices. Section 7.2 provides a way of constructing infinite lattice sequences. What is lacking is an explicit construction of a good generating vector.

Acknowledgments

I am grateful to many experts in quasi-Monte Carlo methods and lattice rules for their valuable comments and hearty encouragement, which have greatly benefitted my research. I would especially like to thank Professors Kai-Tai Fang and Yuan Wang for introducing me to integration lattices, and Professor Ian H. Sloan, who gave me helpful advice as I was writing my first paper on lattice rules.

10 REFERENCES

[AS64]   M. Abramowitz and I. A. Stegun (eds.), *Handbook of mathematical functions with formulas, graphs and mathematical tables*, U. S. Government Printing Office, Washington, DC, 1964.

[BFN92]   P. Bratley, B. L. Fox, and H. Niederreiter, *Implementation and tests of low-discrepancy sequences*, ACM Trans. Model. Comput. Simul. **2** (1992), 195–213.

[CMO97]   R. E. Caflisch, W. Morokoff, and A. Owen, *Valuation of mortgage backed securities using Brownian bridges to reduce effective dimension*, J. Comput. Finance **1** (1997), 27–46.

[CS96]   R. Cools and I. H. Sloan, *Minimal cubature formulae of trigonometric degree*, Math. Comp. **65** (1996), 1583–1600.

[DS91]   S. Disney and I. H. Sloan, *Error bounds for the method of good lattice points*, Math. Comp. **56** (1991), 257–266.

[ES81]   B. Efron and C. Stein, *The jackknife estimate of variance*, Ann. Stat. **9** (1981), 586–596.

[Fau82]   H. Faure, *Discrépance de suites associées à un système de numération (en dimension s)*, Acta Arith. **41** (1982), 337–351.

[FH96]   K. Frank and S. Heinrich, *Computing discrepancies of Smolyak quadrature rules*, J. Complexity **12** (1996), 287–314.

[Hei96]   S. Heinrich, *Efficient algorithms for computing the L_2-discrepancy*, Math. Comp. **65** (1996), 1621–1633.

[HH97]   F. J. Hickernell and H. S. Hong, *Computing multivariate normal probabilities using rank-1 lattice sequences*, Proceedings of the Workshop on Scientific Computing (Hong Kong) (G. H. Golub, S. H. Lui, F. T. Luk, and R. J. Plemmons, eds.), Springer-Verlag, Singapore, 1997, pp. 209–215.

[HH98]   F. J. Hickernell and H. S. Hong, *The asymptotic efficiency of randomized nets for quadrature*, Math. Comp. **67** (1998), to appear.

[Hic95]   F. J. Hickernell, *A comparison of random and quasirandom points for multidimensional quadrature*, Monte Carlo and Quasi-Monte Carlo Methods in Scientific Computing (H. Niederreiter and P. J.-S. Shiue, eds.), Lecture Notes in Statistics, vol. 106, Springer-Verlag, New York, 1995, pp. 213–227.

[Hic96a] F. J. Hickernell, *The mean square discrepancy of randomized nets*, ACM Trans. Model. Comput. Simul. **6** (1996), 274–296.

[Hic96b] F. J. Hickernell, *Quadrature error bounds with applications to lattice rules*, SIAM J. Numer. Anal. **33** (1996), 1995–2016, corrected printing of Sections 3-6 in ibid., **34** (1997), 853–866.

[Hic98]   F. J. Hickernell, *A generalized discrepancy and quadrature error bound*, Math. Comp. **67** (1998), 299–322.

[Hla61]   E. Hlawka, *Funktionen von beschränkter Variation in der Theorie der Gleichverteilung*, Ann. Mat. Pura Appl. **54** (1961), 325–333.

[HW81]   L. K. Hua and Y. Wang, *Applications of number theory to numerical analysis*, Springer-Verlag and Science Press, Berlin and Beijing, 1981.

[Kor59]   N. M. Korobov, *The approximate computation of multiple integrals*, Dokl. Adad. Nauk. SSR **124** (1959), 1207–1210, (Russian).

[Lan95]   T. N. Langtry, *The determination of canonical forms for lattice quadrature rules*, J. Comput. Appl. Math. **59** (1995), 129–143.

[Lan96]   T. N. Langtry, *An application of Diophantine approximation to the construction of rank-1 lattice quadrature rules*, Math. Comp. **65** (1996), 1635–1662.

[LJ96]   J. N. Lyness and S. Joe, *Triangular canonical forms for lattice rules of prime-power order*, Math. Comp. **65** (1996), 65–178.

[LS97]   J. N. Lyness and I. H. Sloan, *Cubature rules of prescribed merit*, SIAM J. Numer. Anal. **34** (1997), 586–602.

[MC94]   W. J. Morokoff and R. E. Caflisch, *Quasi-random sequences and their discrepancies*, SIAM J. Sci. Comput. **15** (1994), 1251–1279.

[Nie92]   H. Niederreiter, *Random number generation and quasi-Monte Carlo methods*, SIAM, Philadelphia, 1992.

[NT96]   S. Ninomiya and S. Tezuka, *Toward real-time pricing of complex financial derivatives*, Appl. Math. Finance **3** (1996), 1–20.

[Owe92]   A. B. Owen, *Orthogonal arrays for computer experiments, integration and visualization*, Statist. Sinica **2** (1992), 439–452.

[Owe95]   A. B. Owen, *Randomly permuted (t, m, s)-nets and (t, s)-sequences*, Monte Carlo and Quasi-Monte Carlo Methods in Scientific Computing (H. Niederreiter and P. J.-S. Shiue, eds.), Lecture Notes in Statistics, vol. 106, Springer-Verlag, New York, 1995, pp. 299–317.

[Owe97a] A. B. Owen, *Monte Carlo variance of scrambled equidistribution quadrature*, SIAM J. Numer. Anal. **34** (1997), 1884 – 1910.

[Owe97b] A. B. Owen, *Scrambled net variance for integrals of smooth functions*, Ann. Stat. **25** (1997), 1541–1562.

[PT95]   S. Paskov and J. Traub, *Faster valuation of financial derivatives*, J. Portfolio Management **22** (1995), 113–120.

[PT96]   A. Papageorgiou and J. F. Traub, *Beating monte carlo*, Risk **9** (1996), no. 6, 63–65.

[Rit95]   K. Ritter, *Average case analysis of numerical problems*, Ph.D. thesis, Universität Erlangen-Nürnberg, Erlangen, Germany, 1995.

[Sai88]   S. Saitoh, *Theory of reproducing kernels and its applications*, Longman Scientific & Technical, Essex, England, 1988.

[SJ94]   I. H. Sloan and S. Joe, *Lattice methods for multiple integration*, Oxford University Press, Oxford, 1994.

[SK87]   I. H. Sloan and P. J. Kachoyan, *Lattice methods for multiple integration: Theory, error analysis and examples*, SIAM J. Numer. Anal. **24** (1987), 116–128.

[Slo85]   I. H. Sloan, *Lattice methods for multiple integration*, J. Comput. Appl. Math. **12 & 13** (1985), 131–143.

[Sob67]   I. M. Sobol', *The distribution of points in a cube and the approximate evaluation of integrals*, U.S.S.R. Comput. Math. and Math. Phys. **7** (1967), 86–112.

[Sob69]   I. M. Sobol', *Multidimensional quadrature formulas and Haar functions (in Russian)*, Izdat. "Nauka", Moscow, 1969.

[SW97]   I. H. Sloan and H. Woźniakowski, *An intractability result for multiple integration*, Math. Comp. **66** (1997), 1119–1124.

[SW98]   I. H. Sloan and H. Woźniakowski, *When are quasi-Monte Carlo algorithms efficient for high dimensional integrals*, J. Complexity **14** (1998), 1–33.

[Wah90]   G. Wahba, *Spline models for observational data*, SIAM, Philadelphia, 1990.

[Woź91]   H. Woźniakowski, *Average case complexity of multivariate integration*, Bull. Amer. Math. Soc. **24** (1991), 185–194.

[Zar68]   S. K. Zaremba, *Some applications of multidimensional integration by parts*, Ann. Polon. Math. **21** (1968), 85–96.

Digital Point Sets: Analysis and Application

Gerhard Larcher, Salzburg

Acknowledgments: Work supported by the Austrian Research Foundation (FWF), Projects P 11009 MAT and P 12441 MAT

1 Introduction

In recent years the concept of (t, m, s)-nets and of (t, s)-sequences has turned out to be the – until now – by far most powerful basic concept for the construction of low-discrepancy point sets and point sequences in an s-dimensional unit cube.

Various concrete construction methods for (t, m, s)-nets and (t, s)-sequences of highest quality were given, analysed and used with great success in various kinds of application.

Essentially all these concrete constructions of relevance for applications are based on a general construction scheme, which is the concept of digital point sets. So for example Sobol sequences [Sob67], Faure sequences [Fau82], Niederreiter sequences [Nie92b], Niederreiter-Xing sequences [NX96], the sequences based on formal Laurent series over finite fields [LN93], the (t, m, s)-nets based on optimal polynomials [LLNS96], the shift nets [Sch97b], all of them are digital point sets.

The concept of digital point sets for the first time was explicitly given by Niederreiter in [Nie87a]. However, it was already implicitly used by Sobol and by Faure.

The special structure of digital point sets, besides their uniform distribution properties, allows, firstly, a fast implementation of the point sets, secondly the derivation of numerous of theoretical properties, and thirdly their use in various additional applications.

In our article we will give a survey of recent theoretical investigations on digital point sets and of applications of digital point sets, which are based on their special (group-) structure. However, we will not emphasize on concrete construction methods. The most powerful and fascinating concrete construction method for digital point sets, the method of Niederreiter-Xing is presented in the article of Niederreiter and Xing in this volume.

In Section 2 we will revise the concept of (t, m, s)-nets and of (t, s)-sequences, we will give the basic notions for digital point sets and we will state the basic properties. Further, some remarks on the implementation of digital point sets will be given.

The discrepancy of digital point sets can be estimated by the discrepancy bounds given for arbitrary (t, m, s)-nets and (t, s)-sequences by Niederreiter (see [Nie92b]). In Section 3 we give discrepancy bounds exclusively for digital nets and sequences, and we show how they can be applied to yield improved metrical results for the discrepancy of digital point sets.

In Section 4 we introduce some interesting special sub-classes of digital nets and digital sequences: digital sequences based on formal Laurent series over finite fields, digital nets based on optimal polynomials, and shift nets. We shortly refer to other important concrete construction methods, and we give known upper and lower estimates for the quality parameters of digital nets and of digital sequences.

The investigation of digital sequences based on formal Laurent series over finite fields leads to investigations in the theory of non-archimedean diophantine approximation. So these sequences are of considerable interest especially from a number-theoretical point of view. In Section 5 we will report on this topic.

In the following two sections we will give two mathematical applications of digital point sets.

In Section 6 we apply the results on digital nets to the analysis of the quality of certain pseudo-random-number generators.

In Section 7 we provide a survey on the digital lattice rule for the numerical integration of multivariate Walsh series, based on digital nets, which has been developed by the author of this article and several co-authors in recent years.

Finally, in Section 8, we point out several open problems in the field making also some suggestions for further theoretical work on digital point sets.

2   The concept and basic properties of digital point sets

We will give here the concept of digital nets and of digital sequences in its most general form, namely over a finite commutative ring R. Later on, in most cases we will restrict ourselves to the most important case where R is a finite field of prime order.

First we will give the definition of a digital net over a finite commutative ring R.

Definition 1 *Let $b \geq 2, s \geq 1$, and $m \geq 1$ be integers. We consider the following construction principle for point sets P consisting of b^m points in the s-dimensional unit cube $I^s := [0, 1)^s$. We choose*

(i) *a commutative ring R with identity and with cardinality b;*

(ii) *bijections*

$$\psi_r : Z_b := \{0, 1, \ldots, b-1\} \longrightarrow R$$

for $0 \leq r \leq m - 1$;

(iii) *bijections*

$$\eta_{ij} : R \longrightarrow Z_b$$

for $1 \leq i \leq s$ and $1 \leq j \leq m$;

(iv) *$m \times m$-matrices $C_1, \ldots, C_s$ over R.*

For $n = 0, 1, \ldots, b^m - 1$ let

$$n = \sum_{r=0}^{m-1} a_r b^r$$

with $a_r \in Z_b$ be the digit expansion of n in base b.

To construct the i-th coordinate $x_n(i)$ of the n-th point x_n of P, take the vector

$$\mathbf{n}_i := (\psi_0(a_0), \ldots, \psi_{m-1}(a_{m-1}))^T \in R^m,$$

calculate

$$C_i \cdot \mathbf{n}_i =: (b_1, b_2, \ldots, b_m)^T$$

and set

$$x_n^{(i)} := \frac{\eta_{i1}(b_1)}{b} + \cdots + \frac{\eta_{im}(b_m)}{b^m} \in [0, 1).$$

Any point set, constructed in this way, is called digital net over R.

In practice we will take one fixed bijection $\psi : Z_b \to R$ instead of the ψ_r and we will use $\eta_{ij} = \psi^{-1}$ for all i and j.

If R is the finite field $\mathbb{Z}_b$, with prime order b, then we usually apply the natural "identical" identification between the elements of $\mathbb{Z}_b$ and the digits of $\mathbb{Z}_b$. We then just have

$$(b_1, \ldots, b_m)^T = C_i \cdot (a_0, \ldots, a_{m-1})^T$$

and

$$x_n^{(i)} = \frac{b_1}{b} + \cdots + \frac{b_m}{b^{m-1}}.$$

With the help of the following definition it is possible to quantify the distribution quality of a digital net.

Definition 2 *Let $b, s, m, R, \psi_r, \eta_{ij}$ and $C_1, \ldots, C_s$ be like in Definition 1. Let P be a digital net over R based on $C_1, \ldots, C_s$.*

For arbitrary non-negative integers $d_1, \ldots, d_s \leq m$ let $C_{d_1, \ldots, d_s}$ be the $(d_1 + \cdots + d_s) \times m$-matrix consisting of the first d_1 rows of C_1, and the first d_2 rows of C_2, and so on, and finally of the first d_s rows of C_s.

Let t be a non-negative integer such that the following property is satisfied:

for any integers $d_1, \ldots, d_s \geq 0$ with $\sum_{i=1}^s d_i = m - t$, and any $f \in R^{m-t}$, the system

$$C_{d_1, \ldots, d_s} \cdot \mathbf{z} = f$$

has exactly b^t solutions $\mathbf{z} \in R^m$.

Then P is called a digital (t, m, s)-net over R.

If t is the minimal integer with the above property, then P is called a digital strict (t, m, s)-net over R.

If R is a finite field, then P is a strict digital (t, m, s)-net, if and only if t is the minimal integer such that $C_{d_1, d_2, \ldots, d_s}$ has rank $m - t$ for all $d_1, \ldots, d_s$ with $d_1 + \cdots + d_s = m - t$.

This definition of a digital (t, m, s)-net over a ring R fits the notion of an ordinary (t, m, s)-net in the sense of Niederreiter. We recall the definition:

Definition 3 *Let $b \geq 2, s \geq 1$, and $0 \leq t \leq m$ be integers. Then a point set P consisting of b^m points of I^s, is said to form a (t, m, s)-net in base b, if for every subinterval*

$$J = \prod_{i=1}^s [a_i b^{-d_i}, (a_i + 1)b^{-d_i})$$

of I^s with integers $d_i \geq 0$ and $0 \leq a_i < b^{d_i}$ for $1 \leq i \leq s$, and with volume b^{t-m}, the number of points of P lying in J is equal to b^t.

Furthermore, a (t, m, s)-net P in base b is a strict (t, m, s)-net in base b if there does not exist a u with $0 \leq u < t$ such that P is a (u, m, s)-net in base b.

Considering Lemma 4 in [LNS96] we have:

   a digital (strict) (t, m, s)-net over a ring R of order b is a (strict) (t, m, s)-net in base b.

   In a similar way the concept of (infinite) digital $(\mathbf{T}, s)$-sequences over a ring R is introduced.

Definition 4 *Let $b \geq 2$ and $s \geq 1$ be integers. We consider the following construction principle for an infinite point sequence P in I^s. We choose*

(i) *a commutative ring R with identity and cardinality b;*

(ii) *bijections*
$$\psi_r : Z_b := \{0, 1, \ldots, b - 1\} \longrightarrow R$$
for $r \geq 0$, and with $\psi_r(0) = 0$ for all sufficiently large r;

(iii) *bijections*
$$\eta_{ij} : R \longrightarrow Z_b$$
for $1 \leq i \leq s$ and $j \geq 1$;

(iv) $\mathbb{N} \times \mathbb{N}$*-matrices $C_1, \ldots, C_s$ over R.*

   For $n = 0, 1, \ldots$ let
$$n = \sum_{r=0}^{\infty} a_r b^r$$
with $a_r \in Z_b$ be the digit expansion of n in base b.

   To construct the i-th coordinate $x_n(i)$ of the n-th point x_n of P, take the vector
$$\mathbf{n}_i := (\psi_0(a_0), \psi_0(a_1), \ldots)^T \in R^{\mathbb{N}}$$
(note: almost all coordinates are zero),
calculate
$$C_i \cdot \mathbf{n}_i =: (b_1, b_2, \ldots)^T$$
and set
$$x_n^{(i)} := \frac{\eta_{i1}(b_1)}{b} + \frac{\eta_{i2}(b_2)}{b^2} + \cdots \in [0, 1).$$

(Here we assume that $\eta_{ij}(b_j) \neq b - 1$ for infinitely many j. See Remark 1 in [LNS96].)

   Any sequence constructed in this way is called digital sequence over R.

Definition 5 *Let $b, s, R, \psi_r, \eta_{ij}$ and $C_1, \ldots, C_s$ be like in Definition 1. For $m \geq 1$ let $C_i^{(m)}$ denote the left upper $m \times m$-submatrix of C_i.*

Let $\mathbf{T} : \mathbb{N} \longrightarrow \mathbb{N}_0$ be such that $\mathbf{T}(m) \leq m$ for all m and such that $C_1^{(m)}, \ldots, C_s^{(m)}$ generate a digital $(\mathbf{T}(m), m, s)$-net over R.

Then the digital sequence based on $C_1, \ldots, C_s$, is called digital $(\mathbf{T}, s)$-sequence over R.

If for all $\mathbf{Q} : \mathbb{N} \longrightarrow \mathbb{N}_0$, with $\mathbf{Q}(m) < \mathbf{T}(m)$ for some m, P is not a digital $(\mathbf{Q}, s)$-sequence, then P is called strict digital $(\mathbf{T}, s)$-sequence .

Of special interest are digital $(\mathbf{T}, s)$-sequences with bounded quality function $\mathbf{T}$, say $\mathbf{T}(m) \leq t$ for all m. We then speak of a digital (t, s)-sequence over R.

Again this definition fits the notion of ordinary $(\mathbf{T}, s)$- sequences (and (t, s)-sequences). We recall the definitions:

Definition 6 *Let $b \geq 2$ and $s \geq 1$ be integers and let $\mathbf{T} : \mathbb{N} \longrightarrow \mathbb{N}_0$ be a function with $\mathbf{T}(m) \leq m$ for all $m \in \mathbb{N}$.*

Then a sequence

$$x_0, x_1, \ldots$$

of points in I^s is a $(\mathbf{T}, s)$-sequence in base b, if for all $k, m \in \mathbb{N}_0$ the point set consisting of the x_n with

$$kb^m \leq n < (k+1)b^m,$$

forms a $(\mathbf{T}(m), m, s)$-net in base b.

Furthermore, a $(\mathbf{T}, s)$-sequence S in base b is a strict $(\mathbf{T}, s)$-sequence in base b, if there does not exist a function $\mathbf{U} : \mathbb{N} \longrightarrow \mathbb{N}_0$, with $\mathbf{U}(m) \leq m$ for all $m \in \mathbb{N}$, and $\mathbf{U}(m) < \mathbf{T}(m)$ for at least one $m \in \mathbb{N}$, such that S is a $(\mathbf{U}, s)$-sequence in base b.

Definition 7 *Let $b \geq 2, s \geq 1$, and $t \geq 0$ be integers. Then a sequence P of points in I^s is a (t, s)-sequence in base b if it is a $(\mathbf{T}, s)$-sequence in base b with $\mathbf{T}(m) \leq t$ for all $m \in \mathbb{N}$.*

Furthermore, a (t, s)-sequence S in base b is a strict (t, s)-sequence in base b, if there does not exist a u with $0 \leq u < t$, such that S is a (u, s)-sequence in base b.

Again, a digital (strict) $(\mathbf{T}, s)$-sequence over a ring R of order b is a (strict) $(\mathbf{T}, s)$-sequence in base b.

To show (once more) that the described concept for the construction of digital nets and of digital sequences is quite simple, we give the classical examples of a digital net in dimension 3, and of a digital sequence in dimension 2.

These examples were already provided by Sobol and by Faure, and can also be found in the pioneering work [Nie87a] of Niederreiter.

Take as basic ring the finite field $\mathbb{Z}_2$, and for all bijections use the natural identification between the digits 0, 1 and the elements of $\mathbb{Z}_2$.

For C_1 take the $m \times m$-unit matrix.

For C_2 take the matrix

$$
C_2 = \begin{pmatrix}
0 & 0 & \cdots & 0 & 1 \\
0 & 0 & \cdots & 1 & 0 \\
& & \cdots\cdots\cdots & & \\
0 & 1 & \cdots & 0 & 0 \\
1 & 0 & \cdots & 0 & 0
\end{pmatrix},
$$

and for C_3 take the matrix

$$
C_3 = \begin{pmatrix}
\binom{0}{0} & \binom{1}{0} & \cdots & \binom{m-2}{0} & \binom{m-1}{0} \\
0 & \binom{1}{1} & \cdots & \binom{m-2}{1} & \binom{m-1}{1} \\
& & \cdots\cdots\cdots\cdots & & \\
0 & 0 & \cdots & \binom{m-2}{m-2} & \binom{m-1}{m-2} \\
0 & 0 & \cdots & 0 & \binom{m-1}{m-1}
\end{pmatrix}.
$$

For example, if $m = 4$, then we get the sixth point x_5 of the digital $(0, 4, 3)$-net $x_0, x_1, \ldots, x_{15}$ by calculating

for the first coordinate

$$
\begin{pmatrix}
1 & 0 & 0 & 0 \\
0 & 1 & 0 & 0 \\
0 & 0 & 1 & 0 \\
0 & 0 & 0 & 1
\end{pmatrix}
\times
\begin{pmatrix}
1 \\ 0 \\ 1 \\ 0
\end{pmatrix}
=
\begin{pmatrix}
1 \\ 0 \\ 1 \\ 0
\end{pmatrix}
$$

for the second coordinate

$$
\begin{pmatrix}
0 & 0 & 0 & 1 \\
0 & 0 & 1 & 0 \\
0 & 1 & 0 & 0 \\
1 & 0 & 0 & 0
\end{pmatrix}
\times
\begin{pmatrix}
1 \\ 0 \\ 1 \\ 0
\end{pmatrix}
=
\begin{pmatrix}
0 \\ 1 \\ 0 \\ 1
\end{pmatrix}
$$

and for the third coordinate

$$
\begin{pmatrix}
1 & 1 & 1 & 1 \\
0 & 1 & 0 & 1 \\
0 & 0 & 1 & 1 \\
0 & 0 & 0 & 1
\end{pmatrix}
\times
\begin{pmatrix}
1 \\ 0 \\ 1 \\ 0
\end{pmatrix}
=
\begin{pmatrix}
0 \\ 0 \\ 1 \\ 0
\end{pmatrix}.
$$

Therefore we obtain $x_5 = (\frac{5}{8}, \frac{5}{16}, \frac{1}{8})$.

To construct a digital $(0,2)$-sequence over $\mathbb{Z}_2$ we can use the matrices D_1 and D_2, where D_1 is the infinite continuation of C_1, and D_2 is the infinite continuation of C_3.

There already exist a lot of (available) fast implementations of digital nets and sequences. See for example [Fox86] for Faure sequences, or [BFN94] for the Niederreiter sequences. These implementations may be downloaded, at the time of writing, from the server "Collected Algorithms of the ACM" at the netsite "http://www.acm.org/calgo/".

At Columbia University the software system FINDER was developed, which uses refined implementations of digital nets and sequences for pricing financial derivatives. FINDER is available to academic researchers.

A digital (t, m, s)-net over R, or a digital $(\mathbf{T}, s)$-sequence and especially a digital (t, s)-sequence over R, is of high distribution quality if the parameters t, respectively $\mathbf{T}$ are "small" (in the best case $t = 0$). This assertion is supported by the following discrepancy estimates, which are valid even for arbitrary (t, m, s)-nets and $(\mathbf{T}, s)$-sequences in a base b.

By Theorem 4.10 in [Nie92b] we have:

Theorem 1 *The star-discrepancy $D_N^*(P)$ of a (t, m, s)-net P in base b with $m > 0$ satisfies*

$$ND_N^*(P) \le B(s,b) \cdot b^t \cdot (\log N)^{s-1} + O(b^t (\log N)^{s-2}),$$

with a O-constant depending only on b and on s, and with

$$B(s,b) = \left(\frac{b-1}{2 \log b} \right)^{s-1}$$

if either $s = 2$ or $b = 2, s = 3, 4$, and with

$$B(s,b) = \frac{1}{(s-1)!} \left(\frac{\lfloor \frac{b}{2} \rfloor}{\log b} \right)^{s-1}$$

otherwise.

By Theorem 1 in [LN95] we have

Theorem 2 *The star-discrepancy $D_N^*(S)$ of the first N terms of a $(\mathbf{T}, s)$-sequence S in base b satisfies*

$$ND_N^*(S) \leq B(b,s) \cdot \sum_{m=1}^{k} b^{\mathbf{T}(m)} m^{s-1},$$

where k is such that $b^k \leq N < b^{k+1}$ and where $B(b,s)$ is a constant depending only on b and s.

Finally (this is Theorem 4.17 in [Nie92b]):

Theorem 3 *The star-discrepancy $D_N^*(S)$ of the first N terms of a (t, s)-sequence S in base b satisfies*

$$ND_N^*(S) \leq C(s,b) \cdot b^t (\log N)^s + O(b^t (\log N)^{s-1})$$

for $N \geq 2$. Here

$$C(s,b) = \frac{1}{s} \left(\frac{b-1}{2 \log b} \right)^s$$

if either $s = 2$ or $b = 2, s = 3, 4$ and

$$C(s,b) = \frac{1}{s!} \cdot \frac{b-1}{2 \left[\frac{b}{2} \right]} \left(\frac{\left[\frac{b}{2} \right]}{\log b} \right)^s$$

otherwise.

Comparing these results with the classical conjecture in the theory of uniform distribution, stating that for any finite point set $x_1, \ldots, x_N$ in $[0,1)^s$ its discrepancy satisfies

$$ND_N^* \geq c_s \cdot (\log N)^{s-1},$$

and that for any infinite sequence $x_1, x_2, \ldots$ in $[0,1)^s$, for infinitely many N the discrepancy of the first N elements of the sequence satisfies

$$ND_N^* \geq c_s' \cdot (\log N)^s,$$

where c_s and c_s' are constants depending only on s, this means, that digital (t, m, s)-nets resp. digital (t, s)-sequences with small quality parameter t, provide finite point sets, resp. infinite sequences in the s-dimensional unit-cube with the lowest possible order of discrepancy in the number N of points.

Moreover, the constants of the leading term implied in the discrepancy estimates are rather small and are, indeed, rapidly decreasing with growing dimension, which is quite contrary to the situation in the discrepancy estimates for ordinary Halton sequences or good lattice point sets.

The results moreover show the great importance of these point sets for applications in quasi-Monte Carlo methods and above all in quasi-Monte Carlo integration. And the results heavily motivate the search for s-tuples $(C_1, C_2, \ldots, C_s)$ of finite or infinite matrices over a ring R, with quality parameter t as small as possible.

The optimal quality parameter $t = 0$ cannot be obtained for all choices of dimensions s and of underlying rings R.

So the following was shown in [Nie87a]

Theorem 4 *For any base b and any $m \geq 2$ a $(0, m, s)$-net in base b cannot exist if $s > b + 1$ and a $(0, s)$-sequence cannot exist if $s > b$.*

If b is a prime power and if R is the finite field of order b, then there exist digital (t, m, s)-nets over R for dimensions s up to $b + 1$ and digital (t, s)-sequences over R for dimensions s up to b.

More generally it was shown in Theorem 2 in [LNS96]

Theorem 5 *Let R be a commutative ring with identity and of order ≥ 2, and put $q(R) = q := \min_I [R : I]$ where I runs through all proper ideals of R.*

a) *For $m \geq 2$ there exists a digital $(0, m, s)$-net constructed over R if and only if $s \leq q + 1$.*

b) *A digital $(0, s)$-sequence constructed over R exists if and only if $s \leq q$.*

So for example we have Corollary 2 in [LNS96]. We call it:

Theorem 6 *Let $m \geq 2$.*

a) *Let $b = p_1^{\alpha_1} \cdots p_r^{\alpha_r}$ be the canonical factorization of the integer $b \geq 2$ with primes $p_1 < \ldots < p_r$. Then a digital $(0, m, s)$-net constructed over $\mathbb{Z}_b$ exists if and only if $s \leq p_1 + 1$.*

Furthermore, a digital $(0, s)$-sequence constructed over $\mathbb{Z}_b$ exists if and only if $s \leq p_1$.

b) *Let $b = p_1^{\alpha_1} \cdots p_r^{\alpha_r}$ be the canonical factorization of the integer $b \geq 2$ with coprime prime powers $p_1^{\alpha_1} < \ldots < p_r^{\alpha_r}$. Then a digital $(0, m, s)$-net in base b (i.e., over a ring of order b) exists if and only if $s \leq p_1^{\alpha_1} + 1$. Furthermore, there exists a digital $(0, s)$-sequence in base b (i.e., over a ring of order b) if and only if $s \leq p_1^{\alpha_1}$.*

So by increasing the base b and by choosing a suitable ring R, the optimal quality parameter $t = 0$ can be achieved for any dimension s.

However, it turns out, that for values N which are realistic for applications, point sets of length N with a small base b behave much better than sequences with a large base b, even if the quality parameter of the latter is slightly better than the quality parameter of the former.

Thus most investigations concerning the construction of concrete s-tuples of matrices with small quality parameter t concentrate on the case where R is the finite field $\mathbb{Z}_2$. The, in this sense, most powerful construction method known until now, is the method of Niederreiter-Xing. See the article of Niederreiter-Xing in this volume and the references given there.

They show for every prime power base b the existence of digital (t, s)-sequences over the finite field F_b of order b, with

$$t \leq K'(b) \cdot s,$$

where $K'(b)$ is a constant depending only on b.

Since it is known that for all digital (t, s)-sequences over F_b we have

$$t \geq K''(b) \cdot s$$

(again with $K''(b)$ depending only on b), the quality parameter in the construction of Niederreiter-Xing is of the best possible order of magnitude.

For some concrete estimates on the best possible quality parameter t with given parameters s, m and R for digital (t, m, s)-nets and digital (t, s)-sequences, see Section 4.

The discrepancy estimates, given in this section, are estimates valid for arbitrary (t, m, s)-nets and arbitrary (t, s)-sequences. In the next section we give somewhat refined discrepancy estimates, valid only for digital point sets, which give in some metrical applications improved results.

3 Discrepancy bounds for digital point sets

In this section we shall restrict ourselves to digital point sets over the finite field $\mathbb{Z}_b$ of prime order b. However, most of the results could certainly be obtained also for digital point sets over arbitrary finite fields.

Consider the set M_s of all s-tuples $(C_1, \ldots, C_s)$ of $\mathbb{N} \times \mathbb{N}$-matrices over $\mathbb{Z}_b$. We can define a probability measure μ_s on M_s as the product measure induced by a certain probability measure μ on the set M of all $\mathbb{N} \times \mathbb{N}$-matrices over $\mathbb{Z}_b$. We can view M as the product of denumerably many copies of the sequence space $\mathbb{Z}_b^\infty$ over $\mathbb{Z}_b$, and so we define μ as the product measure induced by a certain probability measure μ^* on $\mathbb{Z}_b^\infty$.

We identify each $c = (c_1, c_2, \ldots) \in \mathbb{Z}_b^\infty$ with the real number $\sum_{k=1}^\infty \frac{c_k}{b^k} \in [0, 1]$. This identification is bijective, up to a subset of $[0, 1]$ of Lebesgue measure zero. Thus we may define μ^* as the measure on $\mathbb{Z}_b^\infty$ which with respect to the above identification coincides with the Lebesgue measure on $[0, 1]$.

The following was shown in Theorem 2 in [LN95]

Theorem 7 *Let $D : \mathbb{N} \longrightarrow [0, \infty)$ be such that*

$$\sum_{m=1}^{\infty} \frac{m^{s-1}}{b^{D(m)}} < \infty.$$

Then μ_s-almost all s-dimensional digital sequences constructed over $\mathbb{Z}_b$ are digital $(\mathbf{T}, s)$-sequences over $\mathbb{Z}_b$, with

$$\mathbf{T}(m) \leq D(m) + O(1)$$

for all $m \geq 1$, where the implied O-constant may depend on the sequence.

For example we obtain: for any $\epsilon > 0$,

$$\mathbf{T}(m) \leq (s + \epsilon) \cdot \log_b m + O(1)$$

for all $m \geq 1$, and for μ_s-almost all s-dimensional digital sequences constructed over $\mathbb{Z}_b$, where $\log_b$ denotes the logarithm to base b.

Using now the discrepancy estimate stated in Section 2, which is valid for arbitrary $(\mathbf{T}, s)$-sequences, from the above example we obtain:

For μ_s-almost all s-dimensional digital sequences constructed over $\mathbb{Z}_b$, the discrepancy of the first N sequence elements satisfies

$$N D_N = O((\log N)^{2s+\epsilon})$$

for all $\epsilon > 0$, with an implied constant depending only on b, s, ϵ and the sequence.

An analogous "metrical" result can also be obtained for digital nets.

For the finite field $\mathbb{Z}_b$ of prime order b, for $m \geq 1$ and $s \geq 1$ let $M_s(m)$ denote the set of all s-tuples $C = (C_1, \ldots, C_s)$ of $m \times m$-matrices over b. By $D_N^*(C)$ we denote the star-discrepancy of the digital (t, m, s)-net based on C.

Combining Lemma 4.32 in [Nie92b] and Theorem 4.33 in [Nie92b] then yields:

$$\frac{1}{card M_s(m)} \sum_{C \in M_s(m)} N D_N^*(C) \leq$$

$$\leq \begin{cases} s - 1 + (\frac{\log N}{\log 4} + 1)^s & if \quad b = 2 \\ s - 1 + ((\frac{2}{\pi} + \frac{7}{5 \log b} - \frac{1}{b \log b}) \log N + \frac{1}{b})^s & if \quad b > 2 \end{cases}$$

$$\leq B(s, b)(\log N)^s,$$

with a constant $B(s, b)$ depending only on s and b.

From this immediately follows that for all δ with $0 < \delta < 1$ the number of s-tuples C providing a digital (t, m, s)-net over $\mathbb{Z}_b$ with star-discrepancy

$$ND_N^* \leq \frac{1}{\delta}B(s, b)(\log N)^s$$

is at least

$$(1 - \delta) \cdot cardM_s(m).$$

These two "metrical" results on the discrepancy of digital point sets are not best possible. But they cannot be improved with the help of the general discrepancy estimates given in Section 2. For this reason it was necessary to provide a more refined discrepancy estimate for digital nets and sequences. This was done in Proposition 1 in [Lar98] and in Lemma 2 in [Lar97].

We first give the discrepancy estimate for digital nets.

Definition 8 Let $m \geq 2, s \geq 1$ and the prime base b be given. Let $C = (C_1, \ldots, C_s)$ be an s-tuple of $m \times m$-matrices over the finite field $\mathbb{Z}_b$, and denote by $c_j^{(i)} \in \mathbb{Z}_b^m$ with $j = 1, \ldots, m$ the rows of C_i for $i = 1, \ldots, s$.
For $0 \leq w \leq s$, a w-tuple $(d_1, \ldots d_w)$ of non-negative integers is called admissible with respect to C if the system

$$\{c_j^{(i)} : j = 1, \ldots d_i; i = 1, \ldots, w\}$$

is linearly independent over $\mathbb{Z}_b$. For $w = 0$ we call the "zero-tuple" () admissible.
For $w \leq s - 1$ and $(d_1, \ldots, d_w)$ admissible we define

$$h(d_1, \ldots, d_w) := \max\{h \geq 0 \mid (d_1, \ldots, d_w, h) \text{ is admissible}\}.$$

Then we have the Proposition 1 in [Lar98]:

Theorem 8 Let D^* denote the star-discrepancy of the digital net defined by $C_1, \ldots, C_s$ over $\mathbb{Z}_b$ (b prime). Then

$$D^* \leq \sum_{w=0}^{s-1}(b - 1)^w \sum_{(d_1, \ldots, d_w) \text{admissible}} b^{-(d_1 + \ldots d_w + h(d_1, \ldots, d_w))}.$$

Based on this estimate we derive the following discrepancy estimate for digital $(\mathbf{T}, s)$-sequences (this is Lemma 2 in [Lar97]):

Theorem 9 Let $D_N^*(S)$ denote the star-discrepancy of the first N elements of the digital sequence S generated by the s-tuple $(C_1, \ldots, C_s)$ of infinite matrices over $\mathbb{Z}_b$ (b prime).

Let $N = \sum_{k=0}^{t-1} a_k b^k$ with $0 \leq a_k < b, a_{t-1} \neq 0$, and let $C_i(m)$ denote the left upper $m \times m$-submatrix of C_i. Then

$$ND_N^*(S) \leq$$

$$ts(b-1) \quad + \quad \sum_{m=1}^{t-1} \sum_{w=0}^{s-1} (b-1)^{w+1} \times$$

$$\times \sum_{(d_1,\ldots,d_w) \text{admissible to } m} b^{-m-(d_1+\cdots+d_w+h(d_1,\ldots,d_w))}.$$

Here "admissible to m" means "admissible to $C_1(m),\ldots,C_s(m)$" and also $h(d_1,\ldots,d_w)$ is determined with respect to these matrices.

We now use these estimates to give improvements of the metrical results on the discrepancy of digital nets and digital sequences. The according result for digital nets in an implicit way was given in some estimates concerning certain pseudo-random-number generators in [Lar98]. (For some details on these results see Section 6.) The result never was explicitly stated and proved. So we give the result and the proof in the following.

Theorem 10 *For given integers $s \geq 1, m \geq 2$ and prime base b we have: for all δ with $0 < \delta < 1$, the number of s-tuples $C = (C_1,\ldots,C_s)$ of $m \times m$-matrices, providing a digital net over $\mathbb{Z}_b$ with star-discrepancy $D_N^*(S)$ satisfying*

$$ND_N^*(C) \leq \frac{1}{\delta} B(s,b)(\log N)^{s-1} \log \log N + O((\log N)^{s-1}),$$

is at least

$$(1 - \delta) \cdot cardM_s(m).$$

(Here $B(s,b)$ is a constant depending only on s and b, whereas the O-constant also depends on δ.)

Proof (This proof essentially follows the lines of the proof of Theorem 1 in [Lar98].)
For a non-negative integer c let

$$M(c) := \{C = (C_1,\ldots,C_s) \in M_s(m) \mid \text{there exist positive integers}$$
$$d_1,\ldots,d_s \text{ with}$$
$$d_1 + \cdots + d_s = m - c$$
$$\text{such that}$$
$$c_j^{(i)} : j = 1,\ldots,d_i; i = 1,\ldots,s$$
$$\text{are linearly dependent in } \mathbb{Z}_b\}.$$

Here again $c_j^{(i)}$ denotes the j-th row of C_i.

We have

$$| M(c) | \leq \sum_{\substack{d:=(d_1,\dots,d_s) \\ d_1+\cdots+d_s=m-c}} \sum_{\substack{\lambda:=(\lambda_1,\dots,\lambda_{m-c})\in \\ Z_b^{m-c}\setminus\{0\}}} | M(\lambda, \mathbf{d}) |$$

with

$$M(\lambda, \mathbf{d}) := \{C \in M_s(m) \mid \lambda_1 c_1^{(1)} + \cdots + \lambda_{d_1} c_{d_1}^{(1)} + \cdots$$
$$\cdots + \lambda_{d_1+\cdots+d_s-1+1} c_1^{(s)} + \cdots \lambda_{m-c} c_{d_s}^{(s)} = 0\}.$$

We now estimate the number of elements of $M(\lambda, \mathbf{d})$:

There is an $i \in \{1,\dots,m-c\}$ with $\lambda_i \neq 0$. So the system

$$\lambda_1 c_1^{(1)} + \cdots + \lambda_{d_1} c_{d_1}^{(1)} + \cdots\cdots \lambda_{d_1+\cdots+d_s-1+1} c_1^{(s)} + \cdots \lambda_{m-c} c_{d_s}^{(s)} = 0$$

in $(m-c)m$ variables has rank m, and therefore $| M(\lambda, \mathbf{d}) |= \frac{1}{b^m} | M_s(m) |$.

Consequently

$$| M(c) | \leq | M_s(m) | \cdot \frac{1}{b^m} \cdot b^{m-c} \sum_{\substack{d_1,\dots,d_s \\ d_1+\dots+d_s=m-c}} 1$$

$$= | M_s(m) | \frac{1}{b^c} \binom{m-c-1}{s-1}.$$

Let $\bar{M}(c) := M_s(m)\setminus M(c)$.

Then

$$| \bar{M}(c) | \geq | M_s(m) | \cdot (1 - R(c))$$

with

$$R(c) := \frac{1}{b^c} \binom{m-c-1}{s-1}.$$

For a positive integer c we consider now

$$\sum := \frac{1}{| \bar{M}(c) |} \sum_{C \in \bar{M}(c)} D_N^*(C).$$

By the above discrepancy estimate for digital nets we obtain

$$\sum \leq \frac{1}{| \bar{M}(c) |} \sum_{C\in\bar{M}(c)} \sum_{w=0}^{s-1} (b-1)^w \sum_{\substack{d_1,\dots,d_w \\ admissible\ to\ C}} b^{-(d_1+\cdots+d_w+h(d_1,\dots,d_w))}$$

$$\leq \frac{1}{| \bar{M}(c) |} \sum_{C\in\bar{M}(c)} \sum_{w=0}^{s-1} (b-1)^w \sum_{\substack{d_1,\dots,d_w \\ admissible\ to\ C}} b^{-(d_1+\cdots+d_w)} \times$$

$$\times \left(\left(\sum_{i=m-(d_1+\cdots+d_w)-c+1}^{m-(d_1+\cdots+d_w)} \sum_{\lambda}^{*} \frac{b}{b-1} \cdot \frac{1}{b^i} \right) + \frac{1}{b^{m-(d_1+\cdots+d_w)}} \right).$$

Here $\sum_{\lambda}^{*}$ means summation over all

$$\lambda := (\lambda_1, \ldots, \lambda_{d_1+\cdots+d_w+i}) \in \mathbb{Z}_b^{d_1+\cdots+d_w+i}\backslash\{0\},$$

for which

$$\lambda_1 c_1^{(1)} + \cdots + \lambda_{d_1} c_{d_1}^{(1)} +$$
$$+ \ldots +$$
$$+ \lambda_{d_1+\cdots+d_{w-1}+1} c_1^{(w)} + \cdots + \lambda_{d_1+\cdots+d_w} c_{d_w}^{(w)} +$$
$$+ \lambda_{d_1+\cdots+d_w+1} c_1^{(w+1)} + \cdots + \lambda_{d_1+\cdots+d_w+i} c_i^{(w+1)} = 0.$$

The summand $\frac{1}{b^{m-(d_1+\cdots+d_w)}}$ results from the case where $h(d_1, \ldots, d_w) = m - (d_1 + \cdots + d_w)$ and the factor $\frac{b}{b-1}$ derives from the fact that, whenever for given $w, C, (d_1, \ldots, d_w)$ and i there is a possible summand λ, then there are at least $b - 1$ such summands λ.

Therefore

$$\Sigma \leq \frac{1}{b^m} \sum_{w=0}^{s-1} (b-1)^w \binom{m}{w} +$$

$$\frac{1}{|\bar{M}(c)|} \frac{b}{b-1} \sum_{w=0}^{s-1} (b-1)^w \sum_{\substack{d_1,\ldots,d_w>0 \\ d_1+\cdots+d_w\leq m}} b^{-(d_1+\cdots+d_w)} \times$$

$$\sum_{i=\max(0,m-(d_1+\cdots+d_w)-c+1)}^{m-(d_1+\cdots+d_w)} \frac{1}{b^i} \sum_{\lambda \in \mathbb{Z}_b^{d_1+\cdots+d_w+i}\backslash\{0\}} | M(\lambda, \mathbf{d}, w) |,$$

where $M(\lambda, \mathbf{d}, w)$ is defined like $M(\lambda, \mathbf{d})$ above, but with w instead of $s-1$. Estimating $M(\lambda, \mathbf{d}, w)$ in the same way as estimating $M(\lambda, \mathbf{d})$ above, we obtain $M(\lambda, \mathbf{d}, w) = \frac{1}{b^m} | M_s(m) |$, and

$$\Sigma \leq \frac{1}{b^m} \sum_{w=0}^{s-1} (b-1)^w \binom{m}{w} +$$

$$\frac{1}{|\bar{M}(c)|} \frac{b}{b-1} \cdot c \cdot \frac{| M_s(m) |}{b^m} \sum_{w=0}^{s-1} (b-1)^w \binom{m}{w}$$

$$\leq \frac{1}{b^m} \sum_{w=0}^{s-1} (b-1)^w \binom{m}{w} \cdot \left[1 + \frac{bc}{b-1}(1 - R(c)) \right]$$

$$=: A(c).$$

Therefore for $\Gamma \geq 1$ the number of $C \in M_s(m)$ with $D_N^*(C) \leq \Gamma \cdot A(c)$ is at least $(1 - \frac{1}{\Gamma})(1 - R(c)) \cdot | M_s(m) |$.

Let now $\Gamma = \frac{2-\delta}{\delta}$ with $0 < \delta < 1$, and choose $c \geq 1$ such that $R(c) \leq \frac{\delta}{2}$, which is satisfied for

$$c \geq [\log_b \frac{2b}{\delta(b-1)} m^{s-1}] + 1.$$

By inserting the choices for c and Γ we obtain the result. $\diamond$

The improvement for the metrical result for the discrepancy of digital sequences has been given in the Theorem in [Lar97] in the following form:

Theorem 11 *Let $G : \longrightarrow \mathbb{R}^+$ be monotonically increasing and such that*

$$\sum_{m=1}^{\infty} \frac{m^{s-1} \log m}{G(m)} < \infty.$$

Then for μ_s-almost all s-dimensional digital sequences S constructed over $\mathbb{Z}_b$, (b prime), the star- discrepancy $D_N^*(S)$ of the first N elements of S satisfies

$$D_N^*(S) = O\left(\frac{G(\log N)}{N}\right).$$

For example, for μ_s-almost all S we have

$$D_N^*(S) = O\left(\frac{(\log N)^s (\log \log N)^{2+\epsilon}}{N}\right)$$

for all $\epsilon > 0$.

4   Special classes of digital point sets and quality bounds

In this section we describe some general sub-classes of digital point sets and we report on (numerical) search for nets with lowest possible quality parameters t in dependence on the parameters s, m and b, mainly based on these concepts. Further, we give lower estimates for the quality parameters t of digital nets and sequences.

If we search for digital nets with given dimension s, given m and given basic ring R, until now, we can follow two different strategies.
    The first strategy is as follows:
In order to obtain a digital (t, m, s)-net over a ring R, first generate a digital $(t, s-1)$-sequence over the ring R based, say, on the matrices $C_1, \ldots, C_s$.

Let $C_i(m)$ denote the left upper $m \times m$-matrix of C_i.

Let $C_s(m)$ be the $m \times m$-matrix

$$C_s(m) = \begin{pmatrix} 0 & 0 & \cdots & 0 & 1 \\ 0 & 0 & \cdots & 1 & 0 \\ & & \cdots\cdots\cdots & & \\ 0 & 1 & \cdots & 0 & 0 \\ 1 & 0 & \cdots & 0 & 0 \end{pmatrix}.$$

Then $C_1(m), \ldots, C_s(m)$ generate a digital (t, m, s)-net over R.

The, up to now, by far best construction method for digital (t, s)-sequences is the method of Niederreiter-Xing (see their article in this volume). Using their sequences and using the above method for deriving digital (t, m, s)-nets from $(t, s-1)$-sequences, leads in many cases (that means, for many parameters s, m and R) to the digital nets with smallest currently known quality parameters t and even to the best currently known general (t, m, s)-nets. This is particularly the case if the parameter m (i.e. the number of points) is large. Note that the above construction method makes no restriction on the parameter m. Thus for instance the method of Niederreiter-Xing provides digital s-dimensional sequences over $\mathbb{Z}_2$ with the following quality parameters t:

Table 1

s	1	2	3	4	5	6	7	8	9	10	11	12	13	14
t	0	0	1	1	2	3	4	5	6	8	9	10	11	13

From this we obtain digital (t, m, s)-nets over $\mathbb{Z}_2$ for all $m \geq 2$, with:

Table 2

s	2	3	4	5	6	7	8	9	10	11	12	13	14	15
t	0	0	1	1	2	3	4	5	6	8	9	10	11	13

In Table 1 the parameters t for the dimensions $s = 1, 2, 3, 4, 5, 6$ are the best possible values for t at all. This was shown (for $s \geq 3$) in Theorem 1 in [SW97].

For smaller values of m, however, improved existence results for digital (t, m, s)-nets with small quality parameter t can be given. So for example trivially exist digital (m, m, s)-nets over any ring R for any s and m.

The construction of improved digital (t, m, s)-nets (for rather small given parameter m) until now essentially was based on "search by hand" for "very small" parameters s and m, on methods from coding theory (see [EB97], [SW97]) or on restriction to certain suitable sub-classes of digital nets and extensive computer search over these sub-classes.

So for example in the Appendix of[SW97], digital $(2,6,8)$, and $(3,7,11)$-nets over $\mathbb{Z}_2$, constructed "by hand" are given, and it is shown in [SW97], that these t parameters are best possible for the corresponding m and s parameters.

As sub-classes of digital nets, which provide nets with small quality parameters and which are well suitable for computer search, the concepts of shift nets and of optimal polynomials have turned out to be of importance. The latter are also of special interest from a theoretical point of view and therefore this concept will be introduced at the end of this section in a bit more detail.

Shift nets where first introduced and investigated by Schmid in [Sch96] and in [Sch97b]. A shift net originally is a digital (t,s,s)-net. By certain "propagation rules" (see Lemma 3 in [SW97]), however, with a certain small loss of quality, nets with other parameters m and s can be derived from these nets.

A shift net based on the $s \times s$-matrices $C_1, \ldots, C_s$ is characterised by the first matrix C_1.

Let $a_1, \ldots, a_s$ be the columns of C_1. Then for $i = 2, \ldots, s$ the matrix C_i is defined by

$$C_i := (a_i, a_{i+1}, \ldots, a_s, a_1, \ldots, a_{i-1}).$$

Until now there are no theoretical results on the existence of shift nets. By computer search it turned out, that there exist for many dimensions s shift nets of very high quality.

So for example shift nets over $\mathbb{Z}_2$ with the following parameters are given by Schmid in [Sch97b]:

Table 3

s	3	4	5	6	7	8	9	10	11	12	13	14	15	16
t	0	1	1	2	2	3	3	4	4	5	6	6	7	8

In a forthcoming paper of Clayman, Lawrence, Mullen, Niederreiter and Sloane, tables of parameters of (t,m,s)-nets are given. Here for given parameters m, s and given bases b the best known quality parameters t and the best known lower bounds for t are given. Most entries in these tables originate either from shift nets or from Niederreiter-Xing sequences.

The concept of optimal polynomials was introduced in [LLNS96] (but see also [Nie92a]). For simplicity we restrict ourselves in the following to the underlying ring $R = \mathbb{Z}_2$. The concept, in some sense, is an analogue of the classical concept of "good-lattice-point sets" and of "optimal coefficients", introduced by Hlawka [Hla62] and by Korobov [Kor59].

Let $\mathbb{Z}_2((x^{-1}))$ be the field of formal Laurent series over $\mathbb{Z}_2$ in the variable x^{-1}. Thus, the elements of $\mathbb{Z}_2((x^{-1}))$ have the form

$$\sum_{k=w}^{\infty} b_k x^{-k},$$

where w is an arbitrary integer and all $b_k \in \mathbb{Z}_2$.

Note that $\mathbb{Z}_2((x^{-1}))$ contains all rational functions over $\mathbb{Z}_2$.

For an integer $m \geq 1$ let Φ_m be the map from $\mathbb{Z}_2((x^{-1}))$ to the interval $[0,1)$ defined by

$$\Phi_m \left(\sum_{k=w}^{\infty} b_k x^{-k} \right) = \sum_{k=\max(1,w)}^{m} b_k 2^{-k}.$$

For a given dimension $s \geq 2$ we choose $f \in \mathbb{Z}_2[x]$ with $deg(f) = m$ and an s-tuple

$$\mathbf{g} = (g_1, \ldots, g_s) \in G_m^s,$$

where G_m is the set of all polynomials $g \in \mathbb{Z}_2[x]$ with $deg(g) < m$, and where here and later on we use the convention $deg(0) = -1$.

For $n = 0, 1, \ldots, 2^m - 1$ let

$$n = \sum_{r=0}^{m-1} a_r(n) 2^r$$

with all $a_r(n) \in \mathbb{Z}_2$, be the binary expansion of n. With each such n we associate the polynomial

$$n(x) = \sum_{r=0}^{m-1} a_r(n) x^r \in \mathbb{Z}_2[x].$$

Then we define the point set $P(\mathbf{g}, f)$ consisting of the 2^m points

$$\mathbf{x}_n = \left(\Phi_m \left(\frac{n(x) g_1(x)}{f(x)} \right), \ldots, \Phi_m \left(\frac{n(x) g_s(x)}{f(x)} \right) \right) \in [0,1)^s$$

for $n = 0, 1, \ldots, 2^m - 1$.

The following quantity plays a crucial role.

Definition 9 *The figure of merit $\rho(\mathbf{g}, f)$ is given by*

$$\rho(\mathbf{g}, f) = s - 1 + \min \sum_{i=1}^{s} deg(h_i),$$

where the minimum is extended over all non-zero s-tuples $(h_1, \ldots, h_s) \in G_m^s$ for which f divides $\sum_{i=1}^{s} g_i h_i$.

We have (see Theorem A in [LLNS96])

Theorem 12 *The point set $P(\mathbf{g}, f)$ is a digital (t, m, s)-net over $\mathbb{Z}_2$ with $t = m - \rho(\mathbf{g}, f)$.*

Therefore, to obtain by this construction method digital (t, m, s)-nets in base 2 with a small value of t, we have to choose the parameters f and $\mathbf{g}$ in such a way that $\rho(\mathbf{g}, f)$ is large.

For practical purposes the choice $f(x) = x^m$ is motivated. Note that in the calculation of the nodes we need the Laurent series expansion of the rational functions $n(x)g_i(x)/f(x)$ in order to calculate their images under Φ_m. This is comparatively easy for $f(x) = x^m$, since then for any numerator polynomial $p(x) = \sum_{j=0}^{q} p_j x^j \in \mathbb{Z}_2[x]$ the Laurent series expansion of $p(x)/x^m$ is immediately obtained as

$$\frac{p(x)}{x^m} = \sum_{j=0}^{q} p_j x^{j-m}.$$

Another advantage of the choice $f(x) = x^m$ is, that the search procedure for large figures of merit is less expensive in this case. For theoretical investigations, however, irreducible denominators f are much easier to handle.

The fundamental problem of search procedures for good numerators $g_1, \ldots, g_s$ now is, that even for moderately large s and m an exhaustive search through all 2^{ms} possible s-tuples $\mathbf{g} = (g_1, \ldots, g_s) \in G_m^s$ is infeasible. Thus for interesting ranges of s and m only a subset of the possible s-tuples $\mathbf{g} = (g_1, \ldots, g_s) \in G_m^s$ can be searched. So we limit the search domain to what is in a sense a one-parameter subset of G_m^s, namely to all $\mathbf{g}$ of the form $\mathbf{g} = (1, g, g^2, \ldots, g^{s-1}) \bmod f$.

A polynomial g which leads to a large figure of merit $\rho(\mathbf{g}, f)$ for the corresponding s-tuple $\mathbf{g} \in G_m^s$, is then called an "optimal polynomial". This informal notion depends, of course on the dimension s and on the polynomial f.

In [LLNS96] various existence results for "good" s-tuples $\mathbf{g}$ and "optimal polynomials" g are given. So for example we have the following two results.
The first is Corollary 1 in [LLNS96]

Theorem 13 *Let $s \geq 2$ and let m be sufficiently large. Then for every irreducible $f \in \mathbb{Z}_2[x]$ with $\deg(f) = m$ there exists*

a) *an s-tuple*

$$\mathbf{g} = (1, g_2, \ldots, g_s) \in G_m^s$$

   with

$$\rho(\mathbf{g}, f) \geq [m - (s-1)(\log_2 m - 1) + \log_2((s-1)!)];$$

b) *a polynomial* $g \in \mathbb{Z}_2[x]$ *such that*

$$\mathbf{g} = (1, g, g^2, \ldots, g^{s-1}) \bmod f$$

satisfies

$$\rho(\mathbf{g}, f) \geq [m - (s-1)(\log_2 m - 1) + \log_2((s-2)!)].$$

The second result is Corollary 2 in [LLNS96].

Theorem 14 *For* $s \geq 3$ *and every sufficiently large* m *there exists a*

$$\mathbf{g} = (1, g_2, \ldots, g_s) \in G_m^s$$

with

$$\rho(\mathbf{g}, x^m) \geq [m - (s-1)(\log_2 m - 1) + \log_2((s-1)!)].$$

For dimension $s = 2$ the parameter $\rho(\mathbf{g}, f)$ for $\mathbf{g} = (1, g) \in G_m^2$ largely depends on the continued fraction expansion of the rational function $\frac{g}{f}$ over $\mathbb{Z}_2[x]$ (see Chapter 4 in [Nie92b] and [Nie87b]).

Until now it was not possible to give a corresponding reasonable result for the for practical purposes more important case that $s \geq 3, f = x^m$ and $\mathbf{g} = (1, g, g^2, \ldots, g^{s-1})$.

However, for this case, for many parameters of m and s extensive computer search has been carried out. Note that from the computational point of view, it is a rather simple task to compute the figure of merit $\rho(\mathbf{g}, f)$ for the above choices of $\mathbf{g}$ and f. The search has mainly been carried out with the help of a parallelized algorithm using several DEC Alpha stations. For further details and for tables of t values for the corresponding sequences, see [LLNS96].

5   Digital sequences based on formal Laurent series and non-archimedean diophantine approximation

The above concept for the construction of digital (t, m, s)-nets over a finite field F_q, based on rational functions $\frac{g_1}{f}, \ldots, \frac{g_s}{f}$ over F_q can be extended for the construction of digital $(\mathbf{T}, s)$-sequences over f_q. The resulting class of sequences is of utmost interest from the purely theoretical point of view. It in some sense can be regarded as a function field analogue to the classical Kronecker sequences

$$\mathbf{x}_n = (\{n\alpha_1\}, \{n\alpha_2\}, \ldots, \{n\alpha_s\})_{n=1,2,\ldots}$$

in $[0, 1)^s$, where $\alpha_1, \ldots, \alpha_s$ are given reals and where $\{x\}$ denotes the fractional part of x. Therefore the new class of sequences introduced in [Nie92b] is called the "class of Kronecker-type sequences".

As the investigation of classical Kronecker sequences leads to the classical theory of diophantine approximation, the investigation of Kronecker-type sequences leads to the theory of non-archimedean diophantine approximation.

We will give the definition of Kronecker-type sequences, the necessary notions from non-archimedean diophantine approximation as well as the basic results.

Let F_q denote the finite field of order q, where q now is an arbitrary prime power.

For a given dimension $s \geq 1$ we choose an s-tuple $(L_1, \ldots, L_s)$ of formal Laurent series over F_q, that is, from $F_q((z^{-1}))$, say

$$L_i = \sum_{k=w_i}^{\infty} u_k^{(i)} z^{-k} \text{ for } 1 \leq i \leq s,$$

where we can assume that $w_i \leq 1$ for $1 \leq i \leq s$.

Then for the construction of a digital $(\mathbf{T}, s)$-sequence we use the matrices $C_1, \ldots, C_s$ with

$$C_i = \begin{pmatrix} u_1^{(i)} & u_2^{(i)} & u_3^{(i)} & \cdots \\ u_2^{(i)} & u_3^{(i)} & u_4^{(i)} & \cdots \\ u_3^{(i)} & u_4^{(i)} & u_5^{(i)} & \cdots \\ \cdots\cdots\cdots\cdots\cdots \end{pmatrix},$$

for $1 \leq i \leq s$.

We denote the resulting digital sequence by $S(L_1, \ldots, L_s)$.

The case, where $L_1, \ldots, L_s$ are rational corresponds to the construction of (t, m, s)-nets over F_q as considered in Section 4.

First we consider these Kronecker-type sequences not from viewpoint of digital point sets, but merely from the perspective of the theory of uniform distribution. In [LN93], Theorem 1, the following criterion was given, according to which a sequence $S(L_1, \ldots, L_s)$ is uniformly distributed in $[0, 1)^s$.

Theorem 15 *The sequence $S(L_1, \ldots, L_s)$ is uniformly distributed in $[0, 1)^s$ if and only if $1, L_1, \ldots, L_s$ are linearly independent over the rational function field $F_q[z]$.*

This is in accordance with the criterion for the uniform distribution of the classical Kronecker sequences which are uniformly distributed if and only if $1, \alpha_1, \ldots, \alpha_s$ are linearly independent over the rationals (see [KN74]).

Concerning the discrepancy of these sequences, first a metric result was given in [Lar95].

The sequence $S(L_1, \ldots, L_s)$ is (apart from the fixed bijections) determined by the Laurent series $L_1, \ldots, L_s$. Indeed, we can actually restrict ourselves to the set $\bar{F}_q((z^{-1}))$ of Laurent series of the form

$$L_i = \sum_{k=1}^{\infty} u_k^{(i)} z^{-k}$$

for $1 \leq i \leq s$.

Let μ be the normalized Haar-measure on $\bar{F}_q((z^{-1}))$ and μ_s be the product measure on $(\bar{F}_q((z^{-1})))^s$.

Then the Theorem in [Lar95] states:

Theorem 16 *Let q be a prime. Then for μ_s-almost all $(L_1, \ldots, L_s) \in (\bar{F}_q((z^{-1})))^s$ we have for every $\epsilon > 0$ for the discrepancy D_N of the first N elements of $S(L_1, \ldots, L_s)$*

$$N D_N = O((\log N)^s (\log \log N)^{2+\epsilon}).$$

An, in some sense, analogous result on the discrepancy of the finite point sets $S(L_1, \ldots, L_s)$ with $L_1, \ldots, L_s$ of the form $L_i = \frac{g_i}{f}$ (g_i, f polynomials from $F_q[z]$), was given in [Lar93].

If we want to give concrete discrepancy estimates for a single Kronecker-type sequence $S(L)$ in dimension $s = 1$, we are led to the theory of the continued fraction expansions of formal Laurent series. (For the classical one-dimensional $\{n\alpha\}$-sequence it is well known that its star-discrepancy can be bounded quite precisely in terms of the continued fraction parameters of α; see Chapter 2 in [KN74] and the more recent work of Schoissengeier [Sch84].)

Every $L \in F_q((z^{-1}))$ has a unique continued fraction expansion

$$L = A_0 + \cfrac{1}{A_1 + \cfrac{1}{A_2 + \cfrac{1}{A_3 + \cfrac{1}{\ddots}}}}$$

$$= [A_0; A_1, A_2, A_3, \ldots],$$

where $A_h \in F_q[z]$ for all $h \geq 0$ and $deg(A_h) \geq 1$ for all $h \geq 1$. The expansion is finite for rational L and infinite for irrational L.

We then have the Theorem 5 in [LN93]:

Theorem 17 *Let $L = [A_0; A_1, A_2, \ldots]$ be the continued fraction expansion of an irrational $L \in F_q((z^{-1}))$ and put*

$$d_H := \sum_{h=1}^{H} deg(A_h)$$

for $H \geq 0$. Then for all integers N with $q^{d_{H-1}} < N \leq q^{d_H}, H \geq 1$, we have for the star-discrepancy D_N^* of the first N elements of $S(L)$

$$ND_N^* \leq \frac{q+1}{q} + \frac{1}{4} \sum_{h=1}^{H} q^{deg(A_h)}(1 + q^{-deg(A_h)}).$$

If, for example, the irrational L has bounded partial quotients, i.e., if there exists a $K \geq 1$ such that $deg(A_h) \leq K$ for all $h \geq 1$, then it follows that

$$ND_N^* = O(\log N) \text{ for all } N \geq 2,$$

with an implied constant depending only on K and on q.

The lower bound of Schmidt ([Sch72b]) for the discrepancy of arbitrary one-dimensional sequences shows, that this order of magnitude ($N^{-1} \log N$ for the star-discrepancy D_N^*) is best possible.

For $q = 2$ the irrationals $L \in F_q((z^{-1}))$ with $deg(A_h) = 1$ for all $h \geq 1$ have been characterized in terms of their Laurent series expansion by Baum and Sweet in [BS77]:

$L \in F_q((z^{-1}))$ satisfies this property if and only if

$$L = \sum_{k=1}^{\infty} u_k z^{-k}$$

with $u_1 = 1$ and

$$u_{2k+1} = u_{2k} + u_k$$

for all $k \geq 1$.

In order to give more-dimensional concrete discrepancy estimates for Kronecker-type sequences and to describe the connection with digital $(\mathbf{T}, s)$-sequences, we need some concepts from the theory of non-archimedean diophantine approximation as they were developed in [LN95].

For

$$L = \sum_{k=w}^{\infty} u_k z^{-k} \in F_q((z^{-1})),$$

with an integer w such that $u_w \neq 0$, the degree valuation v is defined by $v(0) = -\infty$ and $v(L) = -w$ for $L \neq 0$.

By $Fr(L)$ we denote the fractional part of L, that is

$$Fr(L) := \sum_{max(1,w)}^{\infty} u_k z^{-k}.$$

For an s-tuple $(L_1, \ldots, L_s)$ of elements of $F_q((z^{-1}))$ and for $m \in \mathbb{N}_0$ we define

$$e_m(L_1, \ldots, L_s)$$

to be the least integer d for which there exist $Q_1, \ldots, Q_s \in F_q[z]$ not all 0 with $\sum_{i=1}^{s} deg(Q_i) = d$ and

$$v(Fr(\sum_{i=1}^{s} Q_i L_i)) < -m.$$

Then we have the Theorem 3 in [LN95]:

Theorem 18 *For every s-tuple $(L_1, \ldots, L_s)$ of elements of $F_q((z^{-1}))$, the sequence $S(L_1, \ldots, L_s)$ is a strict digital $(\mathbf{T}, s)$-sequence over F_q with*

$$\mathbf{T}(m) = m - e_m(L_1, \ldots, L_s) - s + 1$$

for all $m \in \mathbb{N}_0$.

More detailed results for characterizing Kronecker-type sequences as digital $(\mathbf{T}, s)$- and (t, s)-sequences can be found in Theorems 4 - 6 in [LN95].

From the above result and from results on non-archimedean diophantine approximation we obtain the following metrical result, which is Theorem 8 in [LN95]:

Theorem 19 *Let $D : \mathbb{N} \longrightarrow [0, \infty)$ be such that*

$$\sum_{m=1}^{\infty} \frac{m^{s-1}}{q^{D(m)}} < \infty.$$

Then for μ_s-almost all $(L_1, \ldots L_s) \in (\bar{F}_q((z^{-1})))^s$ the sequence $S(L_1, \ldots, L_s)$ is a digital $(\mathbf{T}, s)$-sequence over F_q with

$$\mathbf{T}(m) \leq D(m) + O(1) \text{ for all } m \geq 1,$$

where the implied constant may depend on $(L_1, \ldots, L_s)$.

Concerning the discrepancy of more-dimensional Kronecker-type sequences the following could be shown (This is Theorem 2 in [LN95]):

Theorem 20 *If there is a constant $c \in \mathbb{Z}$ such that for all polynomials $Q_1, \ldots, Q_s \in F_q[z]$ (not all zero) we have*

$$v(Fr(\sum_{i=1}^{s} Q_i L_i)) \geq -c - \sum_{i=1}^{s} deg(Q_i),$$

then the sequence $S(L_1, \ldots, L_s)$ is a digital (t, s)-sequence over F_q with

$$t = c - s.$$

In particular, we have for the discrepancy of the first N elements of $S(L_1, \ldots, L_s)$:

$$N D_N(S(L_1, \ldots, L_s)) = O((\log N)^s) \text{ for } N \geq 2,$$

with an implied constant depending only on c, q and s.

This result, however, is quite unsatisfactory for the following reason:

An s-tuple $(L_1, \ldots, L_s) \in (F_q((z^{-1})))^s$ satisfying the condition required in the above theorem, in analogy to the theory of linear forms in classical diophantine approximation, may be called "badly approximable". For $s = 1$, an irrational $L \in F_q((z^{-1}))$ is badly approximable if and only if the degrees of the partial quotients in the continued fraction expansion of L are bounded.

For $s \geq 2$, Armitage (in [Arm69] and [Arm70]) claimed to have constructed badly approximable s-tuples $(L_1, \ldots, L_s)$. But this claim was disproved by Taussat [Tau86]. The question, whether there exist badly approximable s-tuples $(L_1, \ldots, L_s)$ for $s \geq 2$ is still open, as is the corresponding question for s-tuples of reals.

However, firstly, the condition in the above result to get Kronecker-type sequences with $ND_N = O((\log N)^s)$ can be weakened. (This was done in Theorem 7 in [LN95]. We do not give the detailed result here, since we would need a number of further rather technical notions from non-archimedean diophantine approximation.) Further the above result can be generalized by using the notion of an "approximation type".

Definition 10 *Let* $\tau : \{1 - s, 2 - s, \ldots\} \longrightarrow [1, \infty)$ *be a function with* $\lim_{x \to \infty} \tau(x) = \infty$.
Then an s-tuple $(L_1, \ldots, L_s)$ of elements of $F_q((z^{-1}))$ is said to be of "type $\leq \tau$" if

$$v(Fr(\sum_{i=1}^{s} Q_i L_i)) \geq -\tau(\sum_{i=1}^{s} \deg(Q_i))$$

for all $Q_1, \ldots, Q_s \in F_q[z]$ that are not all zero.
If the function τ is of the form $\tau(x) = \beta x + \gamma$ for some real constants $\beta \geq 1$ and γ, then $(L_1, \ldots, L_s)$ is said to be of "finite type $\leq \beta$".

Examples of s-tuples $(L_1, \ldots, L_s)$ of finite type can be constructed in the following way:

Let A with $F_q \subseteq A \subseteq F_q((z^{-1}))$ be a Galois extension of $F_q(z)$ of degree $m \geq s + 1$ and let E be the integral closure of $F_q[z]$ in A. Then just take any $L_1, \ldots, L_s \in E$ for which $1, L_1, \ldots, L_s$ are linearly independent over $F_q(z)$. $(L_1, \ldots, L_s)$ is then of finite type $\leq m - 1$. (See [LN95].)

By Theorem 6 in [LN95] we then have

Theorem 21 *For any real $\beta \geq 1$, an s-tuple $(L_1, \ldots, L_s)$ of elements of $F_q((z^{-1}))$ is of finite type $\leq \beta$, if and only if $S(L_1, \ldots, L_s)$ is a digital $(\mathbf{T}, s)$-sequence over F_q with*

$$\mathbf{T}(m) \leq \frac{\beta - 1}{\beta} m + O(1) \text{ for all } m \in \mathbb{N}_0.$$

In consequence we obtain the Corollary 2 in [LN95]

Theorem 22 *If the s-tuple* $(L_1, \ldots, L_s)$ *of elements of* $F_q((z^{-1}))$ *is of finite type* $\leq \beta$, *then the discrepancy* D_N *of the first* $N \geq 2$ *terms of the sequence* $S(L_1, \ldots, L_s)$ *satisfies*

$$D_N = \begin{cases} O(N^{-\frac{1}{\beta}}(\log N)^{s-1} & if \quad \beta > 1 \\ O(N^{-1}(\log N)^s & if \quad b = 1 \end{cases},$$

where in both cases the implied constant depends only on q, s, β *and* $(L_1, \ldots, L_s)$.

A lot of further results on Kronecker-type sequences and their **T**-parameter can be found in [LN93] and in [LN95].

In the following two sections we shall give special applications of digital nets respectively of investigations on digital nets.

6 Analysis of pseudo-random-number generators by digital nets

By work especially of Niederreiter (see [Nie88], [Nie95], [Nie96]) but also of L'Ecuyer ([L'E96]), it turned out that digital nets also play an important role in the analysis of certain pseudo-random-number generators.

One of the most important tests for the quality of pseudo-random-number generators is the so-called serial test. The serial test is a test for the statistical independence of successive pseudo-random-numbers.

For a pseudo-random-number sequence $x_0, x_1, \ldots, x_{N-1}$ in $[0, 1)$ and a fixed dimension $s \geq 2$ let the "serial set" $(\mathbf{x}_n)_{n=0}^{N-1}$ be defined by

$$\mathbf{x}_n := (x_n, x_{n+1}, \ldots, x_{n+s-1}) \in [0, 1)^s.$$

(Here we consider the sequence $(x_n)_{n \geq 0}$ to be periodic with period N.) We then consider the star-discrepancy D_N^* of this sequence in $[0, 1)^s$. A small discrepancy guarantees good statistical independence properties of the successive elements of the pseudo-random-number sequence.

For a number of pseudo-random-number generators now it turned out, that their serial sets show in some sense a "net property", and even a "digital net property". Examples are the recursive matrix method (combined with the p-adic digit method), the digital multistep method, the generalized feedback shift-register method or combined Tausworthe generators.

Using the discrepancy estimate for digital nets, given in Section 3, in [Lar98] for all these generation methods, the existence of parameters which provide pseudo-random-number sequences with large period and with an extremely small discrepancy for serial sets, could be shown. Thereby results from [Nie88], [Nie95], and from [Nie87a] could be improved.

As an example, we give here the corresponding result for the recursive matrix method. For the proofs and for further results see [Lar98].

The recursive matrix method was introduced in full generality by Niederreiter in [Nie93], and it was studied in detail for example in [Nie95], and in [Nie96].

Here we consider only the case of recursive matrix methods of order one. This is a combination of the classical matrix method for the generation of pseudo-random vectors (see [Nie92b]), and of a p-adic digit method.

The method is as follows: let p be a prime and F_p be the finite field of order p. Let m be a positive integer and A be a nonsingular $m \times m$ matrix over F_p.

A sequence $\mathbf{z}_0, \mathbf{z}_1, \ldots$ of row vectors from F_p^m is generated by choosing an initial vector $\mathbf{z}_0$ different from $\mathbf{0}$ and by

$$\mathbf{z}_{n+1} := \mathbf{z}_n \cdot A \text{ for } n = 0, 1, \ldots.$$

We now derive pseudo-random numbers x_n in $[0,1)$ from the $\mathbf{z}_n := (z_n^{(1)}, \ldots, z_n^{(m)}) \in F_p^m$ in the following way:

we identify the elements $z \in F_p$ in the natural way with digits $z \in \{0, \ldots, p-1\}$. Then

$$x_n := \sum_{j=1}^{m} z_n^{(j)} p^{-j} \text{ for } n = 0, 1, \ldots.$$

The sequence $(\mathbf{z}_n)_{n \geq 0}$ and therefore $(x_n)_{n \geq 0}$ is purely periodic because of the non-singularity of the matrix A, with (least) period at most $p^m - 1$. This maximal (least) period is attained if and only if the polynomial

$$\det(x \cdot I_m - A)$$

of degree m is a primitive polynomial over F_p. (Here I_m is the $m \times m$ identity matrix.) This is shown for example in Theorem 10.2 in [Nie92b]. In the following we shall restrict ourselves to this, for practical purposes most important, case of maximal period.

Let in the following $q := p^m$. In Theorem 2 in [Nie95] it was shown that an arbitrary sequence $(\mathbf{z}_n)_{n \geq 0}$ with $\mathbf{z}_n := (z_n^{(1)}, \ldots, z_n^{(m)}) \in F_p^m$ is a recursive vector sequence of the above defined form of period $T := p^m - 1$, if and only if there is a primitive element σ of F_q and a basis $\beta_1, \ldots, \beta_m$ of F_q over F_p such that $z_n^{(j)} = \mathrm{Tr}(\beta_j \sigma^n)$ for $1 \leq j \leq m$ and $n \geq 0$. Here Tr is the trace function from F_q to F_p.

Concerning the star-discrepancy $D_T^{*(s)}$ of the serial sets of dimension s of these sequences in [Nie95] the following was shown:

let $2 \le s \le m$ and let σ be a fixed primitive element of F_q. Then for $D_T^{*(s)}$ we have on the average

$$D_T^{*(s)} \le c(s) \cdot \frac{(\log T)^s}{T}$$

with an implied constant depending only on s, where the average is taken over all ordered bases of F_q over F_p.

From this we immediately deduce the following: let $2 \le s \le m$, let σ be a fixed primitive element of F_q and let $\mathcal{B}$ be the set of ordered bases of F_q over F_p. Let $0 < \gamma < 1$ be given. Then the number of bases $B \in \mathcal{B}$ for which for the discrepancy $D_T^{*(s)}(B)$ of the s-dimensional serial set of the corresponding sequence we have

$$D_T^{*(s)}(B) \le \frac{1}{1-\gamma} c(s) \cdot \frac{(\log T)^s}{T}$$

is at least $\gamma \mid \mathcal{B} \mid$.

Using the new discrepancy estimate for digital nets we could improve this result (at least for small p) by almost one logarithmic factor in the following way (this is Theorem 1 in [Lar98]):

Theorem 23 *Let $2 \le s \le m$, let σ be a primitive element of F_q and let $\mathcal{B}$ be the set of ordered bases of F_q over F_p. Let $0 < \gamma < 1$ be given. Then the number of bases $B \in \mathcal{B}$ for which for the discrepancy $D_T^{*(s)}(B)$ of the s-dimensional serial set $\mathbf{x}_0, \dots, \mathbf{x}_{T-1}$ of the corresponding sequence we have*

$$
\begin{aligned}
D_T^{*(s)}(B) \;\le\; & \frac{1}{T} + \frac{1}{p^m} \sum_{w=0}^{s-1} (p-1)^w \binom{m}{w} \times \\
& \left[(s-1) \left(\frac{p}{p-1} \right)^2 \frac{2}{1-\gamma} \, p \log m + \right. \\
& \left. \left(\frac{p}{p-1} \right)^2 \frac{2}{1-\gamma} \left(1 + {}_p\log \frac{4}{1-\gamma} \right) + \frac{1+\gamma}{1-\gamma} \right] \\
=\; & \mathcal{O}\left(\frac{(\log T)^{s-1} \cdot \log \log T}{T} \right)
\end{aligned}
$$

is at least $\gamma \mid \mathcal{B} \mid$.
(Here by ${}_p\log$ we denote the logarithm to base p.)

(Note that the constant in the $\mathcal{O}$-result also depends on p.)

For example we have for the case p=2:
for at least one half of the bases B in $\mathcal{B}$,

$$D_T^{*(s)}(B) \;\le\; 68 \frac{1}{2^m} \sum_{w=0}^{s-1} \binom{m}{w} +$$

$$16(s-1)\frac{2\log m}{2^m}\sum_{w=0}^{s-1}\binom{m}{w}.$$

Similar results were obtained for a number of further pseudo-random-number generators.

7   The digital lattice rule

In several papers by the author of this article and by several co-authors a so-called digital lattice rule for the numerical integration of multivariate, in a "digital sense" smooth functions, has been developed. This method is based on the use of digital point sets of high quality.

The basic tool for estimating the integration error for number-theoretic integration methods is the well-known inequality of Koksma and Hlawka. We present this inequality here in short:

For a function $f : [0,1)^s \longrightarrow \mathbb{R}$ of bounded variation $V(f)$ in the sense of Hardy and Krause, and a point set $x_1,\ldots,x_N$ in $[0,1)^s$ with star-discrepancy D_N^* it is

$$\left| \int_{[0,1)^s} f(\mathbf{x})d\mathbf{x} - \frac{1}{N}\sum_{k=1}^{N} f(x_k) \right| \le V(f)\cdot D_N^*.$$

(See for example [Nie92b].)

So the use of low-discrepancy point sets $x_1,\ldots,x_N$ in $[0,1)^s$ for approximating the integral of f by the average value of f at the nodes $x_1,\ldots,x_N$ is heavily motivated. And especially the use of digital point sets of high quality for the integration of functions of bounded variation is motivated.

However, there exist more subtle number-theoretical integration methods, respecting certain properties of the special classes of integrands they are dealing with. One such method is the method of good lattice points, introduced by Hlawka and Korobov in the early sixties (See [Hla62] and [Kor59]). This method is especially designed for the efficient numerical integration of smooth periodic functions, i.e. of functions which are representable by rapidly converging Fourier series. In this method as integration nodes so-called good lattice point sets are used and the integration error is estimated by means of the smoothness of the integrand and by means of the quality of the good lattice point set.

The digital lattice rule now was designed for the efficient numerical integration of "digitally smooth" functions, i.e. of functions which can be represented by rapidly converging multivariate Walsh series.

We will introduce this concept below, and we will give the basic results.

The classical concept of multivariate Walsh functions and Walsh series was given by Walsh and Paley (see for example [Pal32]) for base $b = 2$ and by Chrestensen [Chr55] for an arbitrary integer base $b \geq 2$.

The Walsh function system in a base $b \geq 2$ can be defined in the following way:

Definition 11 *For a non-negative integer n with base b representation*

$$n = n_r b^{r-1} + \ldots + n_1 \; ; \; n_i \in \{0, \ldots, b-1\}$$

we define the function $_b wal_n : [0,1) \longrightarrow \mathbb{R}$ by

$$_b wal_n(x) := (w_b)^{x_1 n_1 + \ldots + x_r n_r}$$

for

$$x = \sum_{i=1}^{\infty} \frac{x_i}{b^i} \in [0,1)$$

and with $w_b = e^{\frac{2\pi i}{b}}$.

The set

$$\{_b wal_n \mid n = 0, 1, 2, \ldots\}$$

is an orthonormal function system in $L^2([0,1))$.

Definition 12 *For dimension $s \geq 2$ and non-negative integers $k_1, \ldots, k_s$ we define*

$$_b wal_{k_1, \ldots, k_s} : [0,1)^s \longrightarrow \mathbb{R}$$

by

$$_b wal_{k_1, \ldots, k_s}(x_1, \ldots, x_s) := \prod_{i=1}^{s} {}_b wal_{k_i}(x_i).$$

We obtain a complete orthonormal function system in $L^2([0,1)^s)$.

There exist various generalizations of this concept in the literature. For example Vilenkin introduced what we call now Vilenkin systems (see [Vil47]) or Onneweer studied his concept of Rademacher and Walsh functions over groups [Onn77].

For the digital lattice rules, as explained below, we used another quite general concept of Walsh functions:
Let G be a finite abelian group of order b. Let

$$G = \mathbb{Z}_{b_1} \times \cdots \times \mathbb{Z}_{b_m}$$

and

$$\phi : \{0, 1, \ldots, b-1\} \longrightarrow G = \mathbb{Z}_{b_1} \times \cdots \times \mathbb{Z}_{b_m}$$

with

$$\phi(k) := (\phi_1(k), \ldots, \phi_m(k))$$

be a bijection with $\phi(0) = 0$.

For $g = (g_1, \ldots, g_m) \in G$ let the character $\chi_g \in \hat{G}$ be defined by

$$\chi_g(y) := \prod_{l=1}^{m} e^{\frac{2\pi i g_l y_l}{b_l}}$$

for $y = (y_1, \ldots, y_m) \in G$.

Of course $\hat{G} = \{\chi_g | g \in G\}$.

Definition 13 *For a non-negative integer n with b-adic representation*

$$n = n_v b^{v-1} + \cdots + n_1$$

we define the function $_{G,\phi}wal_n : [0,1) \longrightarrow \mathbb{C}$ in the following way:

$$_{G,\phi}wal_n(x) := \prod_{j=1}^{v} \chi_{\phi(n_j)}(\phi(x_j)),$$

where

$$x = \sum_{j=1}^{\infty} x_j b^{-j}$$

is the b-adic representation of x (with infinitely many of the x_j different from $b-1$).

Definition 14 *The set*

$$W_{G,\phi} := \{_{G,\phi}wal_n | n = 0, 1, \ldots\}$$

is called the Walsh function system over G with respect to ϕ.

In most cases we will write $_G wal$ instead of $_{G,\phi}wal$ if it is clear which ϕ we use.

Consider the following examples:

a) Let $G = \mathbb{Z}_b$ and ϕ be the "identity" between $\{0, 1, \ldots, b-1\}$ and $\mathbb{Z}_b$. Then we obtain the classical systems of Walsh-Paley and of Chrestenson.

b) In [Onn77] Onneweer defines Walsh functions on the infinite product of groups. A "continuation" of his functions to $[0,1)$ is easily possible in an natural way.

Our Walsh functions are products of characters on one fixed group. The functions of Onneweer are products of characters on possibly different groups. In this sense the concept of Onneweer is more general, however, his groups must be of prime order, and he just uses identities for ϕ. In this sense our concept is more general. (The bijection ϕ indeed strongly influences the structure of $W_{G,\phi}$ (see [Pir95]).)

c) Concerning the connection between our concept and Vilenkin systems (see [Vil47]), we again have the situation, that both concepts have a non-empty intersection, but none is a "sub-concept" of the other.

Definition 15 *By* $W^s_{G,\phi}$ *we now define the system of s-dimensional Walsh functions over G with respect to ϕ:*

$$W^s_{G,\phi} := \{ _{G,\phi}wal_{k_1,\ldots,k_s} \mid k_i = 0,1,2,\ldots; i = 1,\ldots,s \},$$

where $_{G,\phi}wal_{k_1,\ldots,k_s} : [0,1)^s \longrightarrow \mathbb{C}$ *is defined by*

$$_{G,\phi}wal_{k_1,\ldots,k_s}(x_1,\ldots,x_s) := \prod_{i=1}^{s} {}_{G,\phi}wal_{k_i}(x_i).$$

Now we shall consider series of such Walsh functions of a fixed Walsh function system. We divide these series into classes according to their speed of convergence.

Definition 16 *Let a finite group G of order b and a corresponding bijection ϕ be given. Assume, that the function $f : [0,1)^s \longrightarrow \mathbb{C}$ can be represented by a series of Walsh functions from $W_{G,\phi}$, say,*

$$f(\mathbf{x}) = \sum_{k_1,\ldots,k_s=0}^{\infty} \hat{f}(k_1,\ldots,k_s) \cdot {}_{G,\phi}wal_{k_1,\ldots,k_s}(\mathbf{x}).$$

Assume further that this Walsh series shows a certain speed of convergence of the following form:
There are $\alpha > 1$ and $C > 0$, such that

$$|\hat{f}(k_1,\ldots,k_s)| \leq \frac{C}{(\bar{k}_1 \cdots \bar{k}_s)^s}$$

for all $k_i = 0,1,\ldots$; $i = 1,\ldots,s$ (where $\bar{k}_i := \max(1,|k_i|)$).
Then we say "f belongs to $_{G,s}E^\alpha(C)$".

The parameter α of the convergence speed represents in some sense the degree of digital smoothness of the function f. In support of this assertion we consider the so-called digital derivative of a function f. Here we give the details only for dimension $s = 1$. For higher dimensions we obtain analogous results (see [Pir97] and [LP97]).

Definition 17 *For an abelian group G of order b, a corresponding bijection ϕ, a function $f : [0,1) \longrightarrow \mathbb{C}$, for $x \in [0,1)$ and positive integers n let*

$$d_n(f) := \sum_{j=0}^{n} b^{j-1} \sum_{k=0}^{b-1} k \sum_{l=0}^{b-1} {}_{G,\phi}\overline{wal}_l(\frac{k}{b}) \cdot f(x \oplus \frac{l}{b^{j+1}}).$$

(Here by "$\oplus$" we denote "digital addition with respect to G and ϕ", i.e., for $u, v \in \mathbb{R}_0^+$ let $u = \sum_{i=w}^{\infty} u_i b^{-i}$ and $v = \sum_{i=w}^{\infty} v_i b^{-i}$ be the b-adic representations of u and v. Then

$$u \oplus v := \sum_{i=w}^{\infty} z_i b^{-i}$$

with

$$z_i := \phi^{-1}(\phi(u_i) + \phi(v_i))$$

for $i = w, w+1, \ldots$.)

We say, that f is digitally differentiable in x with respect to G and ϕ if

$$f^{[1]}(x) := \lim_{n \to \infty} d_n f(x)$$

exists. $f^{[1]}(x)$ is called the digital derivative of f in x.

Higher derivatives can be defined in the usual way.

A function $f \in L_p([0,1]), 1 \le p < \infty$ is called strongly differentiable with respect to G and ϕ if there exists a $g \in L_p([0,1))$ with

$$\lim_{n \to \infty} \|d_n f - g\|_p = 0.$$

We then denote g by $\mathbf{d}f$.

Next we have Theorem 1 of [LP97]

Theorem 24 a) *Let $\alpha > 1$ be an integer. If a Walsh series f over G and ϕ is α-times strongly differentiable as a function in $L_1([0,1))$, then $f \in {}_{G,\phi}E^{\alpha}(C)$ for some $C > 0$.*

b) *If for some integer $\alpha > 2$ we have $f \in {}_{G,\phi}E^{\alpha}(C)$ for some $C > 0$, then f is at least $\alpha - 2$-times strongly differentiable with respect to G and ϕ.*

A detailed study of the above general concept of generalized Walsh series was given in [Pir95] and in [Pir97]. Examples of concrete functions lying in different classes ${}_{g,\phi}E_s^{\alpha}(C)$ can be found in [LT94], [LS94], and in [LSW95]. (See also the numerical examples at the end of this section.)

For the numerical integration of these functions we now use (t,m,s)-nets and especially digital (t,m,s)-nets.

The first results of the theory were given in [LT94] and in [LSW96] . We state the following Theorem 3 from [LSW96]:

Theorem 25 *Let $b \geq 2$ and let $x_0, \ldots, x_{N-1}$ be a digital (t, m, s)-net constructed over $\mathbb{Z}_b$.*

Then with a positive constant $K(s, b, C, \alpha)$ depending only on s, b, C, and α, we have

$$\left| \int_{[0,1)^s} f(\mathbf{x}) d\mathbf{x} - \frac{1}{N} \sum_{k=1}^{N} f(x_k) \right| \leq K(s, b, C, \alpha) \cdot b^{t\alpha} \cdot \frac{(\log N)^{s-1}}{N^{\alpha}}$$

for all $f \in {}_{\mathbb{Z}_b} E_s^{\alpha}(C)$.

Theorem 3 in [LT94] is a special case of this result. The above result is best possible in the quite general sense given in Theorem 4 in [LT94]:

Theorem 26 *For all $b \geq 2, C > 0, \alpha > 1$ and $s \in \mathbb{N}$, there is a constant $L(s, b, C, \alpha) > 0$ such that: for all N and every point set $x_0, \ldots, x_{N-1}$ in $[0,1)^s$ there is a $f \in {}_{\mathbb{Z}_b} E_s^{\alpha}(C)$ with*

$$f(x_k) = 0 \text{ for } k = 0, \ldots, N-1$$

and

$$\int_{[0,1)^s} f(\mathbf{x}) d\mathbf{x} > L(s, b, C, \alpha) \cdot \frac{(\log N)^{s-1}}{N^{\alpha}}.$$

In order to obtain the above upper integration error estimate, it was essential that we used digital nets and not only ordinary nets. For the integration with general nets we have the following upper and corresponding lower bound (this is Theorem 2 in [LSW96]):

Theorem 27 a) *If $x_0, \ldots, x_{N-1}$ is a (t, m, s)-net in base b, then with a constant $K(s, b, C, \alpha)$ depending only on s, b, C, and α, we have*

$$\left| \int_{[0,1)^s} f(\mathbf{x}) d\mathbf{x} - \frac{1}{N} \sum_{k=1}^{N} f(x_k) \right| \leq K(s, b, C, \alpha) \cdot b^{t(\alpha - \frac{1}{2})} \cdot \frac{(\log N)^{s-1}}{N^{\alpha - \frac{1}{2}}}$$

for all $f \in {}_G E_s^{\alpha}(C)$, where G is an arbitrary abelian group of order b.

b) *For all $\alpha > 1, C > 0$ and all b and $s \geq 2$ there is a constant $L(s, b, C, \alpha) > 0$ such that:*

for all t and m for which there exists a (t, m, s)-net in base b, there is a (t, m, s)-net in base b and a $f \in {}_{\mathbb{Z}_b} E_s^{\alpha}(C)$ with

$$\left| \int_{[0,1)^s} f(\mathbf{x}) d\mathbf{x} - \frac{1}{N} \sum_{k=1}^{N} f(x_k) \right| \geq L(s, b, C, \alpha) \cdot b^{t(\alpha - \frac{1}{2})} \cdot \frac{1}{N^{\alpha - \frac{1}{2}}}.$$

A more general version of the upper integration error estimate based on digital nets was given in Theorem 1 in [LNS96]:

Theorem 28 *Let R be a commutative ring of order b with additive group G. Let $\phi : \{0, 1, \ldots, b - 1\} \longrightarrow R$ be a fixed bijection with $\phi(0) = 0$, and let $x_0, \ldots, x_{N-1}$ be a digital (t, m, s)-net constructed over the ring R with respect to the bijections $\psi_r = \phi, \eta_{i,j} = \phi^{-1}$ for all $r, i,$ and j. Then*

$$\left| \int_{[0,1)^s} f(\mathbf{x}) d\mathbf{x} - \frac{1}{N} \sum_{k=1}^{N} f(x_k) \right| \leq K(s, b, C, \alpha) \cdot b^{t\alpha} \cdot \frac{(\log N)^{s-1}}{N^\alpha}$$

for all $f \in {}_G E_s^\alpha(C)$, where K is a constant depending only on s, b, C and α.

It seems that this result completely solves the problem of the numerical integration even for the generalized class of Walsh series. However, for practical applications we have the following problem:

Assume, for example, that we want to integrate ordinary base-b-Walsh series. We would need digital nets of high quality constructed over $\mathbb{Z}_b$. As already mentioned in Section 3 we can generate digital $(0, m, s)$-nets over $\mathbb{Z}_b$ only for $s \leq p + 1$ where p is the least prime divisor of b.

If $b = q_1 \cdot q_2 \cdots q_s$ with pairwise coprime prime powers $q_1, \ldots, q_s$, and if we use the commutative ring $R = F_{q_1} \times \cdots \times F_{q_s}$, then we can generate a $(0, m, s)$-net over R up to dimension $s \leq q+1$, where $q = \min_i q_i$. Especially, if b is a large power of a small prime, this makes a large difference.

However, the above error estimate gives no answer to the question, how well we can integrate a Walsh series over a group G_1 by using a digital net, constructed over a ring R whose additive group is another group G_2.

Some result in this direction first was given in [LSW94].

Definition 18 *For a non-negative integer h let the non-negative real number β_h be defined in the following way:*

$$\beta_h := \frac{h-1}{2h} + \frac{\sum_{k=0}^{h-2} \log \sin \left(\frac{\pi}{4} + \frac{\pi}{2} \left\{ \frac{4^{\lfloor \frac{k}{2} \rfloor} - 1}{3 \cdot 2^{k+1}} \right\} \right)}{h \cdot \log 2}.$$

As is easily checked, the sequence $(\beta_h)_{h \geq 0}$ is monotonically increasing and

$$\beta := \lim_{h \to \infty} \beta_h = \frac{1}{2} + \frac{\log \sin \frac{5\pi}{12}}{\log 2} = 0.4499 \ldots.$$

Then we have

Theorem 29 *If $x_0, \ldots, x_{N-1}$ is a digital $(0, m, s)$-net constructed over the finite field of order h with respect to the bijections $\psi_r = \phi, \eta_{i,j} = \phi^{-1}$ for all $r, i,$ and j, then*

$$\left| \int_{[0,1)^s} f(\mathbf{x}) d\mathbf{x} - \frac{1}{N} \sum_{k=1}^{N} f(x_k) \right| \leq K(s, C, \alpha, h) \cdot \frac{(\log N)^{s-1}}{N^{\alpha - \beta_h}},$$

for all $f \in {}_{\mathbb{Z}_2} E_s^\alpha(C)$, with $\alpha > 1 + \beta_h$.

The investigation of the general question (Walsh-series over a group G_1 integrated by digital nets over a ring with additive group G_2) led to very interesting studies in group theory and to a very interesting notion of a distance between finite abelian groups. This distance notion was studied in detail in [Wol98] (see also [LPW97]), and it played a crucial role in the, until now, most general answer to our general numerical integration problem, an answer which was given in [LP97].

In this paper originally the following question has been considered:

Let a finite abelian group G of order b and a corresponding bijection ϕ be given. Assume, that the function $f : [0,1) \longrightarrow \mathbb{C}$ can be represented by a series of Walsh functions from $W_{G,\phi}$, say,

$$f(x) = \sum_{n=0}^{\infty} \hat{f}(n) \cdot_{G,\phi} wal_n(x).$$

Assume further, that this Walsh series is an element of $_{G,\phi}E^{\alpha}(C)$.

Let now H be another finite abelian group with, say, order c and ϕ a corresponding bijection.

We now ask: is there a $\beta > 1$ and a $\bar{C} > 0$, such that $f \in {}_{H,\psi}E^{\beta}(\bar{C})$?

That is: we are looking for the so-called "base change coefficient":

Definition 19 *For given finite abelian groups G and H and corresponding bijections ϕ and ψ, and for $\alpha > 1$ let the base change coefficient $\beta(G,\phi,H,\psi,\alpha)$ be defined by*

$$\beta(G,\phi,H,\psi,\alpha) := \sup\{\beta > 1| \quad \text{for all } C > 0 \text{ there is a } \bar{C} > 0$$
$$\text{with } {}_{G,\phi}E^{\alpha}(C) \subseteq {}_{H,\psi}E^{\beta}(\bar{C})\}.$$

We set $\beta(G,\phi,H,\psi,\alpha) = 1$ if there is no such β.

Partial results to this questions were given in [LSW94] for the case $G = \mathbb{Z}_2, H = \mathbb{Z}_{2^h}$, and ϕ and ψ identities. It was shown that

$$\beta(\mathbb{Z}_2, id, \mathbb{Z}_{2^h}, id, \alpha)) = \alpha - \beta_h$$

with β_h as in Definition 18 .

Partial results for the "converse" problem: $G = \mathbb{Z}_{2^h}$ and $H = \mathbb{Z}_2$ (the problem is not "commutative ") were given in [LW96].

The general question to some extent then was solved in [LP97] and it gave the basis for the numerical integration results which will be presented below after giving three notations.

Definition 20 *Let G and H be abelian groups with orders b and c. Let ϕ and ψ be corresponding bijections for the generation of the Walsh function systems $W_{G,\phi}$ and $W_{H,\psi}$.*

For non-negative integers j and k let

$$\gamma(j,k) := \int_0^1 {}_G wal_j(x) \cdot {}_H \overline{wal}_k(x) dx.$$

Definition 21 *Assume there are positive integers M and N such that $b^M = c^N =: d$. Then*

$$\beta_{G,H,\phi,\psi} := \log_d \left(\max_{k=0,\ldots,d-1} \sum_{j=0}^{d-1} |\gamma(j,k)| \right).$$

($\log_d$ denotes the logarithm to base d.)

This quantity β is a slight generalization of the notion of a distance of groups, given by Wolf in [Wol98], and has been studied for the first time in [LPW97].

Definition 22 *Assume that b divides c^L for some positive integer L, then with the canonical prime factorizations*

$$c = \prod_{i=1}^r p_i^{\nu_i} \text{ with } \nu_i \geq 1 \text{ for } i = 1,\ldots,r$$

and

$$b = \prod_{i=1}^r p_i^{\mu_i} \text{ with } \mu_i \geq 0 \text{ for } i = 1,\ldots,r$$

let

$$\rho = \rho(b,c) := \min\{\frac{\nu_i}{\mu_i}|i = 1,\ldots,r\}.$$

With these notions we have the Theorem 7 in [LP97]:

Theorem 30 *Let G and H be finite abelian groups of orders b and c, and let ϕ and ψ be corresponding bijections. Let R be a commutative ring with additive group H. Assume, that there exists a positive integer L such that b divides c^L. Let $\rho(b,c)$ be defined like in Definition 22 and let*

$$\theta := \rho(b,c) \cdot \frac{\log b}{\log c}.$$

Let $x_0,\ldots,x_{N-1}$ be a digital (t,m,s)-net constructed over R with respect to ψ.

a) *For all $\alpha > \frac{1}{\theta} + \min(\frac{1}{2}, 2 - \frac{1}{\theta})$ and all $C > 0$ we have*

$$|\int_{[0,1)^s} f(\mathbf{x})d\mathbf{x} - \frac{1}{N}\sum_{k=0}^{N-1} f(x_k)| \leq$$

$$\leq K \cdot c^{t(\alpha\theta - \min(\theta/2, 2\theta - 1))} \cdot \frac{(\log N)^{s-1}}{N^{\alpha\theta - \min(\theta/2, 2\theta - 1)}},$$

for all $f \in {}_{G,\phi}E_s^\alpha(C)$.

b) *Assume, that for some positive integers M and L even $b^M = c^L$ holds. Let $\beta_{G,H,\phi,\psi}$ be defined like in Definition 21 .*

For all $\alpha > 1 + \beta_{G,H,\phi,\psi}$ and all $C > 0$ we have

$$|\int_{[0,1)^s} f(\mathbf{x})d\mathbf{x} - \tfrac{1}{N}\sum_{k=0}^{N-1} f(x_k)| \leq$$

$$\leq K \cdot c^{t(\alpha - \beta_{G,H,\phi,\psi})} \cdot \frac{(\log N)^{s-1}}{N^{\alpha - \beta_{G,H,\phi,\psi}}},$$

for all $f \in {}_{G,\phi}E_s^\alpha(C)$.

In both cases again the K denote constants depending only on s, b, c, C and α.

A lot of open problems remain in this field (see Section 8).

At the end of this section we shall give a small sample of numerical integration results. More numerical results can be found in [LLS95], [LS94], [LS95a] or [LS95b].

Examples of functions in the classes ${}_{z_2}E_s^\alpha(C)$ are given by the following family of suitable test-functions (compare [LSW94]) for details):

For $\beta > \frac{\log 3}{\log 2} - 1$ let

$$g_\beta : [0,1) \longrightarrow \mathbb{R}$$

with

$$g_\beta(x) := \begin{cases} 0 & if \quad x = 0 \\ \frac{1}{2^{(k-1)\beta}} & if \quad \frac{1}{2^k} \leq x < \frac{1}{2^{k-1}} \end{cases} .$$

Let

$$G_\beta : [0,1)^s \longrightarrow \mathbb{R}$$

with

$$G_\beta(x_1, \ldots, x_s) := g_\beta(x_1) \cdots g_\beta(x_s).$$

For $\gamma > 0$ let

$$h_\beta^\gamma : [0,1)^s \longrightarrow \mathbb{R}$$

with

$$h_\beta^\gamma(x_1, \ldots, x_s) := (g_\beta(x_1 \oplus x_2 \oplus \cdots \oplus x_s))^\gamma,$$

where $\oplus$ denotes digit-wise addition in base 2 modulo 2.
Then we use the test-function

$$f_\beta : [0,1)^s \longrightarrow \mathbb{R}$$

with

$$f_\beta(x_1, \ldots, x_s) := G_\beta(x_1, \ldots, x_s) - h_\beta^{s + \frac{s-1}{\beta}}(x_1, \ldots, x_s).$$

It is

$$f_\beta \in {}_{\mathbf{z}_2} E_s^{\beta+1}(C)$$

with

$$C = \left(2^\beta \frac{2^{\beta+1} - 2}{2^{\beta+1} - 1} \right)^s + c_2(\beta, s + \frac{s - 1}{\beta}),$$

where

$$c_2(\beta, \gamma) := 2^{\beta\gamma} \cdot (2^{\beta\gamma+1} - 2)/(2^{\beta\gamma+1} - 1).$$

f is a suitable testing-function because the variables cannot be separated and because only a small number of Walsh coefficients are essentially smaller as it is indicated by the parameters α and C.

We have

$$\int_{[0,1)^s} f_\beta(\mathbf{x}) d\mathbf{x} = \left(\frac{2^\beta}{2^{\beta+1} - 1} \right)^s - \left(\frac{2^{\beta s+s-1}}{2^{\beta s+s} - 1} \right).$$

In the following tables we show the integration error by integrating our test function for $\beta = 3$, in dimensions $3 \le s \le 10$, by using 2^m sample points for $18 \le m \le 24$.

We use digital (t, m, s)-nets over $\mathbb{Z}_2$ with the values of t given in these tables (these nets are provided by the tables of optimal polynomials in [LLNS96]).

We compare these results with the results obtained by using good-lattice-point-sets and Halton sequences.

For all parameters the digital base 2 nets give essentially smaller errors than good-lattice-point-sets and Halton sequences.

Table 4a

s	number of nodes	good lattice error	Halton error	(t, m, s)-net, base 2	
				t	error
3	262144	3.5921562e-05	3.4954750e-07	3	0.0000000e+00
3	524288	7.5182341e-06	5.5828906e-07	2	0.0000000e+00
3	1048576	4.4529897e-06	5.4475822e-06	3	0.0000000e+00
3	2097152	8.6172526e-07	8.4125176e-07	3	0.0000000e+00
3	4194304	2.0849650e-07	1.0106131e-06	3	0.0000000e+00
3	8388608	4.4793381e-07	3.9952123e-07	3	0.0000000e+00
3	16777216	3.2784086e-07	6.4671632e-08	3	0.0000000e+00
4	262144	1.3027725e-04	1.6386438e-04	4	3.2751579e-15
4	524288	2.2080340e-04	5.0877323e-05	4	5.5511151e-17
4	1048576	5.0320786e-05	1.0958468e-05	4	0.0000000e+00
4	2097152	5.0631140e-05	4.0262176e-06	4	0.0000000e+00
4	4194304	2.7676063e-06	1.6404962e-06	3	0.0000000e+00
4	8388608	3.6761937e-06	1.1193337e-06	4	0.0000000e+00
4	16777216	5.5986042e-06	3.8665640e-06	4	0.0000000e+00
5	262144	2.7730587e-05	1.0336330e-04	5	9.1043839e-13
5	524288	1.0152600e-05	1.6758269e-05	5	6.9722005e-14
5	1048576	4.7361956e-05	4.3689596e-05	6	3.5582647e-14
5	2097152	2.4254550e-05	4.2307932e-05	6	1.1657341e-15
5	4194304	3.3457341e-06	2.0441928e-05	6	2.2204460e-16
5	8388608	4.8355707e-06	9.9471858e-06	6	0.0000000e+00
5	16777216	1.1764564e-06	5.1937494e-06	7	5.5511151e-17
6	262144	1.5903416e-04	3.5131274e-03	6	6.4331318e-12
6	524288	1.9330628e-04	1.5324514e-03	6	3.9815373e-12
6	1048576	2.6632950e-04	5.4866633e-04	7	2.6449398e-12
6	2097152	2.9199191e-05	2.4788842e-04	7	5.0404125e-14
6	4194304	4.9700161e-05	9.4332527e-05	7	1.4432899e-15
6	8388608	4.9681390e-04	1.5327245e-05	7	4.9960036e-16
6	16777216	2.8663203e-05	6.1565792e-05	8	0.0000000e+00

Table 4b

s	number of nodes	good lattice error	Halton error	(t, m, s)-net, base 2 t	error
7	262144	7.1847135e-05	2.9296690e-04	7	6.1239319e-10
7	524288	9.3144416e-05	1.2186894e-04	8	5.0859205e-11
7	1048576	1.5392400e-05	2.7168719e-04	8	2.1715962e-11
7	2097152	1.7515935e-05	5.2438885e-05	7	6.5064620e-13
7	4194304	4.7665038e-06	7.5918397e-05	7	1.3405943e-13
7	8388608	1.0427638e-05	4.4513814e-05	8	1.0880185e-14
7	16777216	3.3998087e-06	2.8094389e-05	8	1.1102230e-15
8	262144	1.8209749e-03	5.1281675e-04	8	9.3633312e-10
8	524288	1.2707841e-03	4.2436852e-04	8	2.0319423e-10
8	1048576	1.0953086e-03	4.2716826e-04	8	7.5932038e-12
8	2097152	4.8566452e-05	2.4882064e-04	8	2.7386704e-11
8	4194304	1.2603024e-04	1.6392513e-04	7	1.1632894e-10
8	8388608	1.1412178e-04	8.5103615e-05	8	2.1038726e-14
8	16777216	4.0328154e-05	6.1519340e-05	8	2.3869795e-15
9	262144	8.6234787e-05	6.6091199e-04	9	1.5934146e-09
9	524288	3.9352103e-05	4.1643358e-04	10	2.9273966e-09
9	1048576	1.9483701e-05	2.2708768e-04	11	2.9447552e-09
9	2097152	2.1427557e-05	1.5685086e-04	10	3.8510394e-10
9	4194304	4.0961074e-06	4.7410372e-05	10	7.2016281e-11
9	8388608	8.4314002e-07	1.4556830e-05	10	5.2745030e-12
9	16777216	5.8033041e-06	2.9893953e-05	11	5.5083715e-13
10	262144	1.5661014e-03	1.8029472e-03	9	3.6818831e-09
10	524288	6.7021959e-04	1.2652366e-04	10	8.3319304e-09
10	1048576	2.6825055e-04	7.3354448e-04	11	1.9998303e-10
10	2097152	3.0443267e-04	3.5434624e-04	12	7.8417594e-10
10	4194304	2.7096240e-04	1.2353423e-04	11	5.5515592e-12
10	8388608	2.3489566e-04	4.1152916e-05	11	7.5073447e-11
10	16777216	2.3050129e-04	1.8118245e-05	11	8.3618667e-12

8   Outlook and open research topics

There are still a lot of open problems concerning the analysis and the application of digital point sets. At this point we give just a few problems which are of considerable interest for future research.

A challenging and larger project of general interest would be the following:

Using a digital (t, m, s)-net of optimal quality for quasi-Monte Carlo integration and using the inequality of Koksma and Hlawka for estimating the maximal integration error, then by the discrepancy estimate for digital nets we obtain an integration error of at most

$$W(f, s) \cdot \frac{(\log N)^{s-1}}{N},$$

where $W(f, s)$ is a constant depending only on the integrand f and on the dimension s.

Let us compare the above with the probabilistic integration error estimate provided by the pure Monte Carlo methods, which essentially are based on a result of the following form:

for a given function $f \in L^2([0, 1)^s)$ and a given positive integer N consider $(\mathbf{a}_1, \ldots, \mathbf{a}_N)$ with $\mathbf{a}_j \in [0, 1)^s$ as a random variable in $[0, 1)^{sN}$. Then the expected integration error

$$\left| \int_{[0,1)^s)} f(\mathbf{x}) d\mathbf{x} - \frac{1}{N} \sum_{n=1}^{N} f(\mathbf{a}_n) \right|$$

equals

$$\frac{\sigma(f)}{\sqrt{N}},$$

where $\sigma(f)$ is the variance of f on the unit cube (see [Nie92b], Chapter 1).

Now the "deterministic" quasi-Monte Carlo error estimate

$$W(f, s) \cdot \frac{(\log N)^{s-1}}{N}$$

is in the order of magnitude in N much smaller (and therefore better) than the probabilistic Monte Carlo error estimate

$$\frac{\sigma(f)}{\sqrt{N}}.$$

However, in practice, if the dimension s of our integration problem is rather large, and if the number N of sample points is not extremely large, then the dimension-independent Monte Carlo error estimate in absolute value is much smaller than the quasi-Monte Carlo bound given above. As a consequence we ask the following question:

Is there a chance to find for every N (or at least for "sufficiently many N") a point set in $[0, 1)^s$, whose star discrepancy is at most, say $\frac{1}{\sqrt{N}}$ or $c(s) \cdot \frac{1}{\sqrt{N}}$ with a constant $c(s)$ growing "not too much" in s ?

As candidates for such point sets we naturally consider digital nets. For us in a first step the most favourable result in the light of the above explanations would be the following:

for every $N = 2^m$ there is a digital net in base 2 whose star discrepancy D_N^* satisfies

$$D_N^* \leq \min(\frac{1}{\sqrt{N}}, K(s) \cdot \frac{(\log N)^{s-1}}{N}),$$

with a constant $K(s)$ depending only on the dimension s.

A result of this form would mean a very important step towards further justification of quasi-Monte Carlo methods.

The first investigation that has to be carried out in this direction is the following:

the above discrepancy estimates for digital nets are essentially based on the following discrepancy estimate of Niederreiter for general (and not necessarily digital) nets (see [Nie92b, Chapter 4]):

The star discrepancy D_N^* of a (t, m, s)-net in base 2 satisfies

$$ND_N^* \leq 2^t \cdot \sum_{i=0}^{s-1} \binom{m-t}{i}.$$

This estimate however is certainly too weak for our purposes, since it is a bad estimate if s is rather large, and if m and consequently N is rather small (take for example $m \leq s - 1$, then the above result just gives the trivial estimate $D_N^* \leq 1$).

So in a first step we have to improve for (at least certain classes of) digital nets the above discrepancy estimate of Niederreiter for the values $m \leq s - 1$.

Another important field of investigation would be the following:

As explained before, a digital point set of quality t is generated by an s-tuple of matrices. If we choose any $u \leq s$ of these matrices, then these matrices clearly form a $u-$dimensional digital point set with quality parameter at most t, in many cases, however, with a quality parameter t', essentially smaller than t.

At first glance it seems to be of no special interest to learn about better estimates (or even the exact values) of the quantities t', since for example the discrepancy estimate for (t, m, s)-nets of Niederreiter only depends on t.

However, there are at least three reasons for considering also the "sub-quality parameters t'", that means: there are at least three reasons to introduce, to study in general, and to study for concrete already used examples of digital point sets, a more subtle concept of quality parameter. This quality parameter should consist of the $2^s - 1$ sub-quality parameters t', or at least of the s parameters, say $t_1, \ldots, t_s$, where t_j is the maximum over all t' which belong to j−sub-tuples of the s−tuple of generating matrices. Also a certain weighted mean of the values t_j may be a useful quality parameter. It should turn out by the investigations what is the most valuable concept of parameter.

What are the three reasons for the investigation of such a more general concept ?

First of all:

Various practitioners using digital nets for concrete applications demand that all the projections to lower dimensional facets of the digital net in use should be as well distributed as possible. This distribution of projections is represented by the values t'. So far, by investigating already known construction methods only little interest has been concentrated in the sub-quality values t'. It seems to be of great interest for practitioners to know also the behaviour of the values t' for the commonly used digital nets and sequences.

A second reason:

There is certainly a possibility to improve the discrepancy estimate

$$ND_N^* \leq 2^t \cdot \sum_{i=0}^{s-1} \binom{m-t}{i}$$

of Niederreiter slightly for digital nets (and consequently an analogous discrepancy estimate for digital sequences) by using information on the sub-quality parameters t'. The idea for such an improvement is essentially based on the method with which we derived the new discrepancy estimate for digital nets given in Theorem 8 in this article. However certainly some detailed technical work has to be done to achieve a concrete improvement of the discrepancy bound of Niederreiter.

The third reason:

There is a more refined version of the inequality of Koksma and Hlawka than it is stated in Chapter 2 of [Nie92b]. In the more detailed version (see

[KN74] for the technical details) not only the whole variation in the sense of Hardy and Krause of the integrand and the discrepancy of the complete s−dimensional point set of sample points (as it is in the simplified version) but also the variations of the integrand on all the facets of the unit-cube and the discrepancy of all projections of the sample point set to the facets of the unit-cube are involved.

Thus, using this detailed version together with knowledge on the variation of the integrand on facets (a knowledge which for example can be obtained by the ANOVA decomposition of the integrand; see for example [Hic96]) and together with knowledge on the discrepancy of the projections of the sample point set (a knowledge which for example can be obtained for digital nets by the studying the new concept of a quality parameter as it is suggested) the integration error estimates can be improved in many cases.

Carrying out the investigation would involve the following steps:

- Defining the new quality parameter and establishing its basic properties.

- Establishing new (certainly only slightly) improved discrepancy estimates for digital point sets, based on the new quality parameter.

- Considering possibilities of the application of the new parameter for numerical integration, e.g. via the detailed form of the Koksma-Hlawka inequality.

- Estimating the new quality parameter for digital point sets already in use, both by theoretical methods and also by concrete computer calculations, e.g. by a refined version of an algorithm of Schmid [Sch97a] for testing the quality parameter of digital nets.

- Investigation of the following question: for a given dimension s, how "small" can the new quality parameter be for a digital sequence in dimension s ?

  Let us explain this by an example: as we know, there exist digital $(0, 1)$-sequences and $(0, 2)$-sequences in base 2, but no $(0, 3)$-sequences in base 2. However there exist $(1, 3)$-sequences and $(1, 4)$-sequences, but again no $(1, 5)$-sequences in base 2. There exist digital $(2, 5)$-sequences in base 2. (For these results see [SW97].) So we may ask: is there a digital 5-dimensional sequence of quality 2, such that all its three- and four-dimensional projections have quality 1 and such that all its one- and two-dimensional projections have quality 0 ?

The results concerning this topic would be of interest not only for theoreticians but also for practitioners.

Another important field of investigations which is of high theoretical interest, especially from the number-theoretical point of view, would be the "problem of bounded remainders":

The "problem of bounded remainders" has been solved for various classes of sequences, such as the one-dimensional Kronecker-sequence ($\{n\alpha\}; n = 1, 2, \dots$) in [Kes66], the s-dimensional Kronecker-sequence ($(\{n\alpha_1\}, \dots, \{n\alpha_s\}); n = 1, 2, \dots$) in [Lia87], the one-dimensional Van der Corput sequence in [Sch72a] as well as a slightly generalized form of the s-dimensional Halton sequence in [Hel84].

The problem is as follows:

- for a given sequence $\mathbf{x}_1, \mathbf{x}_2, \dots$ in the s-dimensional unit-cube, and for a given subinterval of the form $B = \prod_{i=1}^{s} [0, b_i)$ of $[0, 1)^s$ with $0 \leq b_i < 1$ for $i = 1, \dots, s$ we define the discrepancy function

$$T_B(N) := | \{1 \leq n \leq N \mid \mathbf{x}_n \in B\} | - N \cdot \lambda(B) .$$

By the lower bound of Roth for the star discrepancy of infinite sequences (see [KN74]), this function cannot be bounded in N for all subintervals B.

So the following question arises: for which subintervals B is $T_B(N)$ bounded in N ? We call these B "sets (subintervals) of bounded remainder".

This problem should be studied for the general class of digital sequences. The idea for this investigation and the prospect for success in studying this problem are motivated by the results and the method given by Hellekalek in [Hel84], where he investigates the s-dimensional Halton sequence.

The Halton sequence may be considered as a certain digital sequence when we allow in different dimensions different underlying bases for the construction of the digital sequence. So the classical Halton sequence is obtained by the digital net concept by using in each coordinate the infinite unit matrix and by taking in the j-th coordinate the j-th prime p_j as base. In the digital point set concept now, the base is fixed and the generating matrices are varied.

In his paper, Hellekalek uses methods from ergodic theory to solve the problem. He constructs a dynamical system on the "group of q-adic integers", underlying the generation procedure of the Halton sequence. Then, using a sort of "coboundary-result" in the theory of dynamical systems, a necessary condition for an interval B to be of bounded remainder is obtained. The sufficient condition is obtained by elementary counting arguments.

We think that in great parts quite similar methods can be used to study the "problem of bounded remainder" for digital (t, s)-sequences.

Of course, we conjecture that just the "elementary intervals"

$$B = \prod_{i=1}^{s} [0, \frac{a_i}{q^l}) \; ; \; 0 \le a_i < q^l \; ; \; l \in \mathbb{N}$$

are the subintervals of bounded remainder for a digital (t, s)-sequence in base q.

We shall now state some further open problems in less detail:

The notion of "shift nets" provides for many parameters the best quality parameters t known so far. These concrete shift nets were found by computer search. Until now there are almost no theoretical results on shift nets. So it would be for example of utmost interest to have information on the "average behaviour" of shift nets, or to give strong necessary properties for the "leading matrix" of a shift net of high quality.

In the best case one could give a concrete generation method for shift nets of high quality.

We have given a lot of upper estimates for the discrepancy and for the quality parameter of digital nets and digital sequences. Many of these estimates are of a "metrical nature".

It seems to be much harder to give sharp lower bounds (concrete bounds or metrical bounds) for the discrepancy or for the quality parameter of digital point sets.

In [Fau97] a best possible lower bound for the discrepancy of a special class of digital sequences was given. An extension of these results would be of great interest.

Further, an answer to the question, to which extent the results of Theorems 7, 10, 11, 16 and 19 are best possible, would be very interesting.

A challenging conjecture on the relation between ordinary nets and digital nets was stated in [SW97] and in [Sch95] and remains still unsolved. We think that it would be a hard task to give a proof for this conjecture if it is true, however, it should be possible to construct a counter-example if it is not true.

It is conjectured that: for a prime-power q, a (t, m, s)-net in base q exists, if and only if there exists a digital (t, m, s)-net over the finite field F_q.

In Theorem 20 and 21 it was shown, that for an s-tuple $(L_1, \ldots, L_s)$ of Laurent series over $F_q[z]$ of type $\beta = 1$ we obtain a Kronecker-type sequence $S(L_1, \ldots, L_s)$ with discrepancy of order

$$ND_N(S(L_1, \ldots, L_s)) = O((\log N)^s).$$

Until now it remains unknown, if such s-tuples exist. (The existence of such an s-tuple would assure the existence of a Kronecker-type sequence which is a digital (t, s)-sequence.) An answer to this question seems hard to find.

However, in Theorem 1b in [LN95] a weaker condition for a digital $(\mathbf{T}, s)$-sequence, to satisfy a discrepancy estimate of the form

$$ND_N = O((\log N)^s)$$

was given. The condition is:

If for a digital $(\mathbf{T}, s)$-sequence S in base b the sequence

$$\left(\frac{1}{k} \sum_{m=1}^{k} b^{\mathbf{T}(m)} \right)_{k=1,2,\ldots}$$

is bounded, then we have

$$D_N(S) = O(N^{-1}(\log N)^s).$$

The question is: do Kronecker-type sequences satisfying this weaker condition exist ?

Out of many open problems in the field of digital lattice rules and of base-change-coefficients we shall just mention two concrete examples:

Close the gap between the upper and the lower bounds in the integration error estimates given in Theorem 27 in this article.

It is known that for the base-change-coefficient $\beta_{G,H,\phi,\psi}$ (see the Definition 19 in this article) we always have

$$0 \leq \beta_{G,H,\phi,\psi} < \frac{1}{2}.$$

(See [LP97].) Is it true that

$$\sup_{G,H} \min_{\phi,\psi} \beta_{G,H,\phi,\psi} < \frac{1}{2} ?$$

Is it even true that

$$\sup_{G,H} \min_{\phi,\psi} \beta_{G,H,\phi,\psi} = \lim_{h \to \infty} \beta_{\mathbf{Z}_2, \mathbf{Z}_{2^h}, id, id} ?$$

It is known (see [LP97]) that

$$\lim_{h\to\infty} \beta_{Z_2.Z_{2^h},id,id} = \frac{1}{2} + \frac{\log \sin \frac{5\pi}{12}}{\log 2} = 0.4499\ldots.$$

Various further problems concerning this topic are given in Section 8 in [LP97].

9 References

[Arm69] J. V. Armitage. An analogue of a problem of Littlewood. *Mathematika*, **16**:101–105, 1969.

[Arm70] J. V. Armitage. Corrigendum and addendum: An analogue of a problem of Littlewood. *Mathematika*, **17**:173–178, 1970.

[BFN94] P. Bratley, B. L. Fox, and H. Niederreiter. Algorithm 738: Programs to generate Niederreiter's low-discrepancy sequences. *ACM Trans. Math. Software*, **20**:494–495, 1994.

[BS77] L. E. Baum and M. M. Sweet. Badly approximable power series in characteristic 2. *Ann. of Math.*, **105**:573–580. 1977.

[Chr55] H. E. Chrestenson. A class of generalized Walsh functions. *Pacific J. Math.*, **5**:17–31, 1955.

[EB97] Y. Edel and J. Bierbrauer. Construction of digital nets from bch-codes. In H. Niederreiter et al., editor, *Monte Carlo and Quasi-Monte Carlo Methods 1996*, volume 127 of *Lecture Notes in Statistics*, pages 221–231. Springer, New York, 1997.

[Fau82] H. Faure. Discrépance de suites associées à un système de numération (en dimension s). *Acta Arith.*, **41**:337–351, 1982.

[Fau97] H. Faure. Discrepancy lower bounds for special quasi-random sequences. In H. Niederreiter et al., editor, *Monte Carlo and Quasi-Monte Carlo Methods 1996*, volume 127 of *Lecture Notes in Statistics*, pages 232–237. Springer, New York, 1997.

[Fox86] B. L. Fox. Algorithm 647: Implementation and relative efficiency of quasirandom sequence generator. *ACM Trans. Math. Software*, **12**:362–376, 1986.

[Hel84] P. Hellekalek. Regularities in the distribution of special sequences. *J. Number Theory*, **18**:41–55, 1984.

[Hic96] F. J. Hickernell. Quadrature error bounds with applications to lattice rules. *SIAM J. Numer. Anal.*, **33**:1995–2016, 1996.

218   Gerhard Larcher, Salzburg

[Hla62]   E. Hlawka. Zur angenäherten Berechnung mehrfacher Integrale. *Monatsh. Math.*, **66**:140–151, 1962.

[Kes66]   H. Kesten. On a conjecture of Erdös and Szüsz related to uniform distribution mod 1. *Acta Arith.*, **12**:193–212, 1966.

[KN74]   L. Kuipers and H. Niederreiter. *Uniform Distribution of Sequences.* John Wiley, New York, 1974.

[Kor59]   N. M. Korobov. The approximate computation of multiple integrals. *Dokl. Akad. Nauk SSSR*, **124**:1207–1210, 1959. (Russian).

[Lar93]   G. Larcher. Nets obtained from rational functions over finite fields. *Acta Arith.*, **63**:1–13, 1993.

[Lar95]   G. Larcher. On the distribution of an analog to classical Kronecker-sequences. *J. Number Theory*, **52**:198–215, 1995.

[Lar97]   G. Larcher. On the distribution of digital sequences. In H. Niederreiter et al., editor, *Monte Carlo and Quasi-Monte Carlo Methods 1996*, volume 127 of *Lecture Notes in Statistics*, pages 109–123. Springer, New York, 1997.

[Lar98]   G. Larcher. A bound for the discrepancy of digital nets and its application to the analysis of certain pseudo-random number generators. *Acta Arith.*, **83**:1–15, 1998.

[L'E96]   P. L'Ecuyer. Maximally equidistributed combined Tausworthe generators. *Math. Comp.*, **65**:203–213, 1996.

[Lia87]   P. Liardet. Regularities of distribution. *Compos. Math.*, **61**:267–293, 1987.

[LLNS96]   G. Larcher, A. Lauß, H. Niederreiter, and W. Ch. Schmid. Optimal polynomials for (t, m, s)-nets and numerical integration of multivariate Walsh series. *SIAM J. Numer. Analysis*, **33**:2239–2253, 1996.

[LLS95]   G. Larcher, A. Lauß, and W. Ch. Schmid. Classical number-theoretical integration methods and the Walsh series lattice rule: a comparison. In G. De Pietro et al., editor, *Proc. International Workshop "Parallel Numerics '95" (Sorrent)*, pages 47–66, 1995.

[LN93]   G. Larcher and H. Niederreiter. Kronecker-type sequences and nonarchimedean diophantine approximations. *Acta Arith.*, **63**:379–396, 1993.

[LN95]   G. Larcher and H. Niederreiter. Generalized (t,s)-sequences, Kronecker-type sequences, and diophantine approximations of formal Laurent series. *Trans. Amer. Math. Soc.*, **347**:2051–2073, 1995.

[LNS96]   G. Larcher, H. Niederreiter, and W. Ch. Schmid. Digital nets and sequences constructed over finite rings and their application to quasi-Monte Carlo integration. *Monatsh. Math.*, **121**:231–253, 1996.

[LP97]   G. Larcher and G. Pirsic. Base change problems for generalized Walsh series, digital derivatives, and multivariate numerical integration. Preprint, University of Salzburg, 1997.

[LPW97] G. Larcher, G. Pirsic, and R. Wolf. Quasi-Monte Carlo integration of digitally smooth functions by digital nets. In H. Niederreiter et al., editor, *Monte Carlo and Quasi-Monte Carlo Methods 1996*, volume 127 of *Lecture Notes in Statistics*, pages 321–329. Springer, New York, 1997.

[LS94]   G. Larcher and W. Ch. Schmid. Numerical integration of multivariate Walsh series by means of different (t,m,s)-nets. In M. Vajteršic and P. Zinterhof, editors, *Proc. International Workshop "Parallel Numerics '94" (Smolenice, Slov.)*, pages 24–43, 1994.

[LS95a]   G. Larcher and W. Ch. Schmid. Multivariate Walsh series, digital nets and quasi-Monte Carlo integration. In H. Niederreiter and P. J.-S. Shiue, editors, *Monte Carlo and Quasi-Monte Carlo Methods in Scientific Computing*, volume 106 of *Lecture Notes in Statistics*, pages 252–262. Springer, New York, 1995.

[LS95b]   G. Larcher and W. Ch. Schmid. On the numerical integration of high-dimensional Walsh series by quasi-Monte Carlo methods. In J. Lacroix and R. Riganti, editors, *Colloque Probabilités Numériques (Paris, 1993)*, volume 38 of *Math. Comput. Simulation*, pages 127–134. Elsevier, North-Holland, 1995.

[LSW94] G. Larcher, W. Ch. Schmid, and R. Wolf. Representation of functions as Walsh series to different bases and an application to the numerical integration of high-dimensional Walsh series. *Math. Comp.*, **63**:701–716, 1994.

[LSW95] G. Larcher, W. Ch. Schmid, and R. Wolf. Digital (t,m,s)-nets, digital (T,s)-sequences, and numerical integration of multivariate Walsh series. In P. Hellekalek, G. Larcher, and P. Zinterhof, editors, *Proceedings of the 1st Salzburg Minisymposium on Pseudorandom Number Generation and Quasi-Monte Carlo Methods,*

Salzburg, Nov. 18, 1994, volume 95–4 of *Technical Report Series*, pages 75–107. ACPC – Austrian Center for Parallel Computation, 1995.

[LSW96] G. Larcher, W. Ch. Schmid, and R. Wolf. Quasi-Monte Carlo methods for the numerical integration of multivariate Walsh series. *Math. and Computer Modelling*, **23**(8/9):55–67, 1996.

[LT94] G. Larcher and C. Traunfellner. On the numerical integration of Walsh series by number-theoretic methods. *Math. Comp.*, **63**:277–291, 1994.

[LW96] G. Larcher and R. Wolf. Nets constructed over finite fields and the numerical integration of multivariate Walsh series. *Finite Fields Appl.*, **2**:304–320, 1996.

[Nie87a] H. Niederreiter. Point sets and sequences with small discrepancy. *Monatsh. Math.*, **104**:273–337, 1987.

[Nie87b] H. Niederreiter. Rational functions with partial quotients of small degree in their contiued fraction expansion. *Monatsh. Math.*, **103**:269–288, 1987.

[Nie88] H. Niederreiter. The serial test for digital k-step pseudorandom numbers. *Math. J. Okayama Univ.*, **30**:93–119, 1988.

[Nie92a] H. Niederreiter. Low-discrepancy point sets obtained by digital constructions over finite fields. *Czechoslovak Math. J.*, **42**:143–166, 1992.

[Nie92b] H. Niederreiter. *Random Number Generation and Quasi-Monte Carlo Methods*. Number **63** in CBMS–NSF Series in Applied Mathematics. SIAM, Philadelphia, 1992.

[Nie93] H. Niederreiter. Factorization of polynomials and some linear-algebra problems over finite fields. *Linear Algebra Appl.*, **192**:301–328, 1993.

[Nie95] H. Niederreiter. The multiple recursive matrix method for pseudorandom number generation. *Finite Fields Appl.*, **1**:3–30, 1995.

[Nie96] H. Niederreiter. Improved bounds in the multiple-recursive matrix method for pseudorandom numbers and vector generation. *Finite Fields Appl.*, **2**:225–240, 1996.

[NX96] H. Niederreiter and C. P. Xing. Quasirandom points and global function fields. In S. Cohen and H. Niederreiter, editors, *Finite Fields and Applications (Glasgow, 1995)*, volume 233 of *Lect. Note Series of the London Math. Soc.*, pages 269–296. Camb. Univ. Press, Cambridge, 1996.

[Onn77]   C. W. Onneweer. Differentiability for Rademacher series on groups. *Acta Sci. Math.*, **39**:121–128, 1977.

[Pal32]   R. E. A. C. Paley. A remarkable system of orthogonal functions. *Proc. Lond. Math. Soc.*, **34**:241–279, 1932.

[Pir95]   G. Pirsic. Schnell konvergierende Walshreihen über Gruppen. Masters Thesis, University of Salzburg, 1995.

[Pir97]   G. Pirsic. *Base change problems for generalized Walsh series and digital differentiation.* PhD thesis, Institute for Mathematics, University of Salzburg, 1997.

[Sch72a]   W. M. Schmidt. Irregularities of distribution, VI. *Comput. Math.*, **24**:63–74, 1972.

[Sch72b]   W. M. Schmidt. Irregularities of distribution, VII. *Acta Arith.*, **21**:45–50, 1972.

[Sch84]   J. Schoißengeier. On the discrepancy of $(n\alpha)$. *Acta Arith.*, **44**:241–279, 1984.

[Sch95]   W. Ch. Schmid. (t, m, s)-*nets: digital construction and combinatorial aspects.* PhD thesis, Institute for Mathematics, University of Salzburg, May 1995.

[Sch96]   W. Ch. Schmid. An algorithm to determine the quality parameter of binary nets, and the new shift-method. In R. Trobek et al., editor, *Proc. International Workshop "Parallel Numerics '96" (Gozd Martuljek, Slovenia)*, pages 51–63, 1996.

[Sch97a]   W. Ch. Schmid. An algorithm to determine the quality parameter of digital nets in base 2. Preprint, University of Salzburg, 1997.

[Sch97b]   W. Ch. Schmid. Shift–nets: a new class of binary digital (t, m, s)-nets. In H. Niederreiter et al., editor, *Monte Carlo and Quasi-Monte Carlo Methods 1996*, volume 127 of *Lecture Notes in Statistics*, pages 369–381. Springer, New York, 1997.

[Sob67]   I. M. Sobol'. The distribution of points in a cube and the approximate evaluation of integrals. *Ž. Vyčisl. Mat. i Mat Fiz.*, **7**:784–802, 1967. (Russian).

[SW97]   W. Ch. Schmid and R. Wolf. Bounds for digital nets and sequences. *Acta Arith.*, **78**:377–399, 1997.

[Tau86]   Y. Taussat. *Approximation Diophantienne Dans un Corps de Séries Formelles.* PhD thesis, Université de Bordeaux, 1986.

[Vil47]   N. Ya. Vilenkin.   A class of complete orthonormal series. *Izv.Akad.Nauk SSSR*, **11**:363–400, 1947.

[Wol98]   R. Wolf. A distance measure on finite abelian groups and an application to quasi-Monte Carlo integration. *Acta Math. Hungar.*, **78**:25–37, 1998.

Institut für Mathematik
Universität Salzburg
Hellbrunnerstraße 34
A–5020 Salzburg
Austria
e-mail address: GERHARD.LARCHER@SBG.AC.AT

Random Number Generators: Selection Criteria and Testing

Pierre L'Ecuyer and Peter Hellekalek

1 Introduction

A random number generator produces a periodic sequence of numbers on a computer. The starting point can be random, but after it is chosen, everything else is deterministic. The goal is to "fake" the realization of a sequence of independent and identically distributed (i.i.d.) random variables uniformly distributed over some set U, which is most often the unit interval $[0, 1]$. In the latter case, the random variables are denoted $U(0, 1)$.

In a more precise way, we define a *random number generator* (RNG) as a structure (S, U, μ, f, g), where S is a finite set of *states* called the state space, U is a set of *output* symbols called the output space, μ is a probability distribution over S, $f : S \to S$ is a mapping called the *transition function*, and $g : S \to U$ is another mapping called the *output function*. The *initial state* (or *seed*) s_0 is a random variable that follows the probability distribution μ. After s_0 has been chosen, the state evolves as $s_n = f(s_{n-1})$, $n \geq 1$, and the generator outputs $u_n = g(s_n)$ at step n. These u_n are the so-called *random numbers* produced by the generator. The sequence $\{s_n\}$ is eventually periodic with period length $\rho \leq |S|$, where $|S|$ denotes the cardinality of the set S. If the state is represented over b bits of memory, an upper bound is $\rho \leq 2^b$ and, for obvious reasons, we prefer ρ to be near 2^b. In this paper, we shall assume that the sequence is *purely periodic*, in the sense that the initial state s_0 is always revisited. In other words, the sequence has no transient part.

The goal is to make it hard to distinguish between the output of the generator and a typical realization of an i.i.d. uniform sequence over U. In *principle*, it is always possible to make the distinction, because the output sequence of the RNG is periodic. But if the period is very long (for example, more than 2^{100}) then no computer can exhaust it in our lifetime. So, there is still room for hope that one cannot distinguish the two sequences in *reasonable time*. For Monte Carlo simulation, we ask for less: we want the output sequence to behave like an i.i.d. uniform sequence *for practical purposes*, for the problem at hand.

It is common practice to select the parameters of RNGs according to certain *figures of merit*. For example, linear congruential or multiple recursive generators are usually selected via the so-called *spectral test*, which measures the "equilibrium" of their lattice structure. Tausworthe-type generators are selected by studying the equidistribution of their points in regular partitions of the unit hypercube into cubic cells of equal sizes. This gives rise to figures of merit measuring the equilibrium or equidistribution of all the points that can be produced by the generator over its entire period, and from all possible initial seeds in case it has more than one cycle. We refer to the contribution of the second author in this volume for further information on measures of equidistribution.

These figures of merit can be seen as *predictors* of the good or bad statistical behavior of an RNG. As a *general rule*, generators with long period and good figures of merit should pass the statistical tests, whereas those with bad figures of merit should fail "more easily". The aim of this paper is to examine this from the empirical viewpoint, for certain classes of generators commonly used in practice. For a number of selected empirical tests, we find (approximately) the minimal sample size required for the generator to clearly fail the test as a function of the size (i.e. period) of the generator and for different values of the figure of merit. We give an example of a test and an RNG family where the generators with the best figures of merit do *worse* in the tests than those with a not-so-good figure of merit. Nevertheless, the general rule given above still holds "typically".

In the next section, we discuss some principles underlying the construction of RNGs. The right approach is to design a generator based on theoretical considerations, keeping an eye of course on practical implementation issues, and test the generator empirically afterwards. Section 3 discusses empirical statistical testing for RNGs. Section 4 describes specific classes of tests: Disjoint and overlapping serial tests and tests based on close pairs of points in the unit torus. Section 5 defines certain collections of generators that will be used to make systematic experiments with some tests. Each such collection contains a generator of period length near 2^e for $e = 12, 13, \ldots, 34$. There are two collections of linear congruential generators (good and bad), one of explicit inversive generators, and one of combined cubic generators. Section 6 applies the tests of Section 4 to these collections of generators, in a systematic way. Section 7 summarizes what happens when they are applied to certain widely used generators.

2   Design Principles and Figures of Merit

2.1   A Roulette Wheel

An RNG can be viewed as a giant roulette wheel around which is written a sequence of output numbers. When the generator has more than a single cycle, we can view it as a collection of roulette wheels, one for each cycle. For simplification in the following discussion we assume that there is a single cycle, of length $\rho = |S|$. The arguments generalize easily to the case of several cycles. We also assume that $U = [0,1]$.

Suppose that μ is the uniform distribution over S, and that the initial seed is random and uniformly distributed over the state space. If the generator is used to produce t output values, then this is equivalent to spinning the wheel to select a starting point at random and take the t successive values $u_0, \ldots, u_{t-1}$ that appear on the wheel from this starting point onward. What we want is for the vector $\mathbf{u}_{0,t} = (u_0, \ldots, u_{t-1})$ to be approximately uniformly distributed over the unit hypercube $[0,1]^t$. This requirement captures both the uniformity and independence of the u_i.

Since s_0 is a random variable, $\mathbf{u}_{0,t}$ is a random vector, uniformly distributed over the multiset $\Psi_t = \{\mathbf{u}_{0,t} = (u_0, \ldots, u_{t-1}) : s_0 \in S\} = \{\mathbf{u}_{n,t}, 0 \leq n < \rho\}$, where the last equality holds only under our assumption that the generator has a single cycle of length $\rho = |S|$. Strictly speaking, Ψ_t is a *multiset* in the sense that repeated elements (if any) are counted as many times as they occur in the list. But we often just call it a *set*, for simplification. The aim is that the uniform distribution over Ψ_t be a good approximation of that over $[0,1]^t$, for each t up to some reasonably large value. It is then natural to ask for Ψ_t to be very evenly spread over the unit hypercube, more uniformly than a typical sample of random points (i.e., *super-uniformly* distributed).

2.2   Sampling from Ψ_t

Suppose now that several t-dimensional points are needed. If one uses $\mathbf{u}_{n,t} = (u_n, \ldots, u_{n+t-1})$ for $n = 0, t, \ldots, (q-1)t$, say, then the concatenation of these points can be viewed as a single point in the qt-dimensional unit cube, and what we need in that case is good uniformity in the qt-dimensional space. But as qt gets higher, it becomes increasingly difficult in practice to analyze the uniformity of Ψ_{qt}. The points $\mathbf{u}_{0,t}, \ldots, \mathbf{u}_{(q-1)t,t}$ can be viewed as a "random" sample of q points taken uniformly from the set Ψ_t, without replacement. If ρ is several orders of magnitude larger than qt, there is practically no difference between *with* and *without* replacement. For further discussion about this argument, see [Com95, L'E97b, L'E98a].

A slightly different requirement would be to ask for the set Ψ_t to look like a typical set of i.i.d. $U(0,1)$ random points. But what needs to behave like a set of i.i.d. uniforms is the set of points that are actually taken from Ψ_t by the random number generator, not necessarily the entire set. If the period is really large, as we recommend, the set Ψ_t will be so huge that only a negligible fraction of it can be generated in "feasible" time. Consequently, only a negligible fraction will be used. One could still argue that super-uniform generators tend to produce point sets that are so regular that some statistical tests (or some simulation problems) may be affected by the regularity for sample sizes that are relatively small compared with the period length. Indeed, we give an example later of a test that detects regularities in the "best" linear congruential generators for a sample size near the square root of the period length ρ. To escape the problem in this particular situation, ρ must be large enough so that generating $\sqrt{\rho}$ random numbers would take too much time to be feasible. If Ψ_t behaves more like a typical "random" point set, then a larger fraction of the period can be used and this is a (small) advantage.

In practice, the theoretical analysis of Ψ_t can be performed only for t up to a few dozens or at most a few hundreds, which is relatively small compared with the sample sizes that are commonly used in large simulations. And even if Ψ_t is well-distributed, this does not guarantee a good empirical behavior for the vectors $u_{n,t}$, $n = 0, t, \ldots, (q-1)t$, when $q \ll \rho$. This is where empirical statistical testing comes into play: To assess the quality of the behavior of large segments of the generator's sequence, or of several successive points of the form $u_{n,t}$, or of specific bits taken out of these numbers, or of certain particular transformations, and so on. Statistical tests are discussed in forthcoming sections.

To assess the super-uniformity of Ψ_t for large-period generators, generating the entire sequence of points is impossible, so one needs more sophisticated mathematical techniques. In practice, the techniques that are used are specific to certain classes of generators. The uniformity of Ψ_t is measured by *figures of merit* whose choice depends on practical considerations such as the ease of computing them.

In particular, the quality of linear congruential generators (LCGs) or multiple recursive generators (MRGs) is assessed by examining their lattice structure, most often via the spectral test, whereas linear feedback shift register generators are evaluated by measuring the equidistribution of their points in certain cubic boxes which form a partition of the unit hypercube. We give more details about these two special cases in what follows.

2.3   The Lattice Structure of MRGs

An MRG is defined by a linear recurrence of the form ([LBC93, L'E94, L'E96a, Nie92]):

$$x_n = (a_1 x_{n-1} + \cdots + a_k x_{n-k}) \bmod m; \tag{1.1}$$

$$u_n = x_n/m, \tag{1.2}$$

where m and k are positive integers called the modulus and the order, and $a_1, \ldots, a_k$ are integers between $-m$ and m, called the coefficients of the recurrence. The largest possible period for this recurrence is $\rho = m^k - 1$, attained if and only if m is prime and the characteristic polynomial $P(z) = z^k - a_1 z^{k-1} - \cdots - a_k$ is primitive over $\mathbb{Z}_m$, the finite field of residue classes modulo m. This can in fact be achieved with only two nonzero coefficients a_k and a_r, $1 \le r < k$. MRGs with very large (composite) moduli can be implemented indirectly by combining two or more MRGs with smaller moduli, as explained in [L'E96a, L'E97a]. When $k = 1$, the MRG becomes the classical LCG, in wide use since the fifties.

The set Ψ_t that corresponds to an MRG of the form (1.1–1.2) turns out to be the intersection of a lattice L_t with the half-open unit cube $[0, 1)^t$ [Knu81, LBC93]. This implies that L_t is covered by each of several families of equidistant parallel hyperplanes. Among all such families that cover L_t (and also Ψ_t), choose one for which the distance between the successive hyperplanes is the largest, and let d_t be this distance. Computing this number d_t is often called the *spectral test* [Fis96, Knu81], for historical reasons. For a fixed cardinality of the set Ψ_t, a smaller d_t means that Ψ_t is more uniformly spread over the unit hypercube, and this is what we want.

Absolute lower bounds on d_t are available for general lattices having m^k points in the hypercube $[0, 1)^t$. Let $d_t^*(m^k)$ be such a lower bound (for further details, see [Knu81] for $t \leq 8$ and [CS88, L'E98b, L'E97a] for more general t). The value of d_t can be normalized to

$$S_t = \frac{d_t^*(m^k)}{d_t},$$

which lies between 0 and 1. For any $\tau \geq k + 1$, one can then define the worst-case *figure of merit*:

$$M_\tau = \min_{2 \leq t \leq \tau} S_t. \tag{1.3}$$

Searches for MRGs and LCGs with good figures of merit M_τ have been performed by several authors. Fishman and Moore [FM86] and Fishman [Fis90] found good LCGs with respect to M_6. L'Ecuyer [L'E98b] gives tables of good LCGs with respect to M_8, M_{16}, and M_{32}, for moduli ranging (approximately) from 2^8 to 2^{128}. A list of MRGs of different sizes, good with respect to a related criterion (the Beyer quotient) for all $t \leq 20$, is given is [LBC93]. L'Ecuyer [L'E97a] reports the results of extensive computer searches for good combined MRGs with respect to M_{32}. He also provides specific implementations for 32-bit and 64-bit computers. These generators are reasonably fast and their period lengths range from 2^{185} to 2^{1319} (approximately). We believe that they constitute excellent choices for general-purpose statistical or mathematical libraries, or for simulation software environments.

Other figures of merit can be defined based on the lattice structure analysis. For example, one can use quantities such as the shortest distance between any two points in the lattice, or the minimal number of hyperplanes that cover all the points of Ψ_t, or Beyer quotients, which are the ratio of the lengths of the shortest and longest vectors in a Minkowski-reduced basis of L_t, and so on. See, e.g., [AG85, Fis96]. The quantity d_t turns out to be the most popular.

2.4   Equidistribution for Regular Partitions in Cubic Boxes

Among the alternatives to (1.1–1.2), there is a large class of generators based on arithmetic modulo 2. Take again the recurrence (1.1) but define the output at step n by

$$u_n = \sum_{j=1}^{L} x_{ns+j-1} m^{-j}, \qquad (1.4)$$

where s and L are two positive integers. This works for general m but for more concreteness let us assume henceforth that $m = 2$. This is the most popular case. This generator is called a *Linear Feedback Shift Register* (LFSR) or *Tausworthe* generator [Fis96, L'E94, Nie92, Tau65]. A commonly used and *economic* version is based on the recurrence $x_n = (x_{n-r} + x_{n-k}) \bmod 2$, whose characteristic polynomial is a trinomial. An efficient implementation can be found in [L'E96b, TL91] for the case where $s \leq r$, $2r > k$, and k does not exceed the number of bits in the computer word length. It is widely accepted that LFSR generators should have a characterictic polynomial of large degree (e.g., $k > 100$ or so), whose coefficients behave somewhat like independent random bits (i.i.d. uniforms over the set $\{0, 1\}$) [Com91, Com95, MK96]. So, a rather simple figure of merit that first comes to mind for these generators is the number N_1 of coefficients a_i that are equal to 1. This number should be not too far from $k/2$. One way of obtaining efficient implementations of LFRSs with a lot of nonzero coefficients is by combining several ones by bitwise exclusive-or [TL91, L'E96b, L'E98c, WC93]. Another technique is the *twisted GFSR* (TGFSR) implementation [MK94, MN98].

Of course, the simple figure N_1 is far from sufficient to assess the quality of a generator. We also need some more direct measure of uniformity of the set Ψ_t. For LFSR generators, Ψ_t does not have a lattice structure in the real space as for MRGs. But LFSR generators have a lattice structure in a more abstract space, namely a space of formal series, and it turns out that this structure can be used to analyze the uniformity of Ψ_t in the unit hypercube via the notion of equidistribution. We now briefly explain this notion. For more details about its analysis and the relationship with the lattice structure in formal series, we refer the reader to [CLT93, L'E94, L'E96b, Tez95].

Partition the interval $[0, 1]$ into 2^{ℓ} segments of equal length, where ℓ is a positive integer such that $\ell \leq L$. This determines a partition of the t-dimensional unit hypercube into $2^{t\ell}$ cubic boxes of volume $2^{-t\ell}$ each. Suppose that the multiset Ψ_t contains 2^k points. This multiset is called (t, ℓ)-*equidistributed in base* 2 if each box of the partition contains exactly $2^{k-t\ell}$ points. This happens if and only if the vector $(u_n, \ldots, u_{n+t-1})$, with each of its components truncated to the first ℓ bits, takes each of the $2^{t\ell}$ possible values the same number of times over the set Ψ_t. Of course, this can happen only for

$$\ell \leq \ell_t^* \overset{\text{def}}{=} \min(L, \lfloor k/t \rfloor).$$

In other words, the number of points cannot be less than the number of boxes. If Ψ_t is (t, ℓ_t^*)-equidistributed for all $t \leq k$, we call Ψ_t (and the generator) *maximally equidistributed* (ME). This is the best equidistribution that one can hope for. If Ψ_t satisfies the further requirement that when $\ell > \ell_t^*$ (i.e., more boxes than points) no box contains more than one point, then we call it (and the generator) *collision-free* (CF). ME-CF generators have their points sets (or multisets) Ψ_t very uniformly distributed into cubic partitions of the hypercube.

Based on this setup, one can define various figures of merit measuring the lack of equidistribution. For instance, let ℓ_t denote the largest ℓ for which Ψ_t is (t, ℓ)-equidistributed. This ℓ_t is called the *resolution* in dimension t, and $\Lambda_t \overset{\text{def}}{=} \ell_t^* - \ell_t$ is called the *resolution gap* in dimension t. From the viewpoint of equidistribution, generators with large resolution gaps are considered bad. The *maximal resolution gap*, $\max_{1 \leq t \leq k} \Lambda_t$, or the *total resolution gap*, $\sum_{t=1}^{k} \Lambda_t$, are examples of figures of merit that naturally come to mind in this context. For ME generators, both are equal to zero. Efficient methods for computing ℓ_t for LFSR generators are explained in [CLT93, L'E96b, Tez95].

The ME-CF property appears so strong that some may wonder if ME-CF generators exist at all. It turns out there are plenty of them among the LFSR generators and they are relatively easy to find. L'Ecuyer [L'E96b, L'E98c] gives several examples and concrete implementations, among certain classes of combined Tausworthe generators.

The TGFSR generator, in its original version, has very large resolution gaps. This was discovered by Tezuka [Tez95] and by its creators Matsumoto and Kurita [MK94], who proposed a modification that adds some *tempering* to the output. The idea is to multiply the bit vector of the output by some well-chosen matrix of bits, at each step, in order to improve the equidistribution. With the type of tempering that is used in [MK94, MN98], it has been proved that ME cannot be achieved. Nevertheless, these authors obtain large resolutions by choosing very large values of k (up to $k = 19937$ in [MN98]).

A notion stronger than that of (t, ℓ)-equidistribution can be defined as follows. Consider the family $\mathcal{P}_q$ of *all* partitions of the unit hypercube into 2^{k-q} so-called elementary dyadic boxes of volume 2^{q-k}, with the length of each side of each box equal to a non-negative power of $1/2$, and where all the boxes have the same dimensions and orientation. In other words, these boxes are of the form $\prod_{i=1}^{t}[a_i 2^{-g_i}, (a_i + 1) 2^{-g_i})$, for some integers a_i, $0 \leq a_i < 2^{g_i}$, $g_i \geq 0$, and $g_1 + \ldots + g_t = k - q$. The partition into cubic boxes discussed previously is one member of this family. The multiset Ψ_t is called a (q, k, t)-*net in base* 2 if each rectangular box of the partition contains exactly 2^q points, for each partition that belongs to $\mathcal{P}_q$. (Using a different notation than ours, this is usually called a (t, m, s)-net.) Niederreiter [Nie92] defines a figure of merit $r^{(t)}$ such that for all $t > \lfloor k/L \rfloor$, Ψ_t is a (q, k, t)-net in base 2 with $q = k - r^{(t)}$. Unfortunately, no efficient method is available for computing $r^{(t)}$ when t and k are large.

2.5   Other Measures of Divergence

The lattice structure analysis of MRGs via the spectral test and the test of equidistribution for LFSR generators are ways of assessing the discrepancy between the distribution of the multiset Ψ_t and the uniform distribution. There are of course several other ways of measuring this discrepancy.

Two forms of the rectangular discrepancy, often called simply *the discrepancy*, can be defined as follows. Suppose that Ψ_t contains N points. For any t-dimensional rectangular box of the form $R = \prod_{j=1}^{t} [\alpha_j, \beta_j)$ with $0 \le \alpha_j < \beta_j \le 1$, let $I(R)$ be the number of points u_n in Ψ_t that fall into R, and let $V(R) = \prod_{j=1}^{t} (\beta_j - \alpha_j)$ be the volume of R. Let $\mathcal{R}$ be the collection of all such boxes R, and define

$$D_N^{(t)} = \sup_{R \in \mathcal{R}} |V(R) - I(R)/N|.$$

This quantity is the so-called t-dimensional *extreme discrepancy* of Ψ_t. If we restrict $\mathcal{R}$ to those boxes which have one corner at the origin, the corresponding quantity is the *star discrepancy*, denoted $D_N^{*(t)}$. Other variants also exist, with different definitions of the family $\mathcal{R}$. See [Nie92] for further discussion. In all cases, *high* uniformity means *low* discrepancy.

To obtain a generator for which Ψ_t is super-uniform, one would look for the *smallest possible* discrepancy, Unfortunately, no efficient algorithm is available for computing $D_N^{(t)}$ or $D_N^{*(t)}$ for large N and moderate t. Lower and upper bounds are often available, but these bounds are usually expressed as asymptotic expressions as a function of N, and often as an average over a certain class of generators. On the other hand, the *discrepancy* is defined for all types of generators, not only for MRGs or LFSRs for instance.

Recently, a new family of equidistribution measures has been introduced by Hellekalek [Hel97], the *weighted spectral test*. Two realizations of this general concept are the classical diaphony (see [HN98]) and the dyadic diaphony [HL97]. These two figures of merit allow the same kind of number-theoretic estimates as discrepancy, but are much easier to compute. Furthermore, the asymptotic distribution of diaphony and dyadic diaphony has already been found, see Leeb [Lee97], Leeb and Hellekalek [LH98], and James, Hoogland, and Kleiss [JHK96]. Both quantities are related to mean-square integration errors over certain classes of functions and also to the so-called L^2-discrepancy, see Strauch [Str94]. We refer to the contribution of the second author in this volume for further information.

3   Empirical Statistical Tests

3.1   What are the Good Tests?

After designing a car on the computer, using all those nice mathematical tools, the car maker will try a prototype on the road. For RNGs, the *road test* means *some simulation problems*. A first thing to do is to try some simplified simulation problems, for which the answer is known. This is basically the idea of empirical statistical testing.

The statistical tests view the RNG as a black box. They make the hypothesis that the output is the realization of a sequence of i.i.d. $U(0,1)$ random variables. This is called the *null hypothesis*, denoted $\mathcal{H}_0$. One knows up-front that $\mathcal{H}_0$ is formally false. But the important question is "How can this affect our statistical results"?

Generally speaking, the answer is unknown. Heuristically, statistical tests try to catch up unusual departures from uniformity, or significant defects. The tests simply go fishing. Any finite set of tests can miss certain particular defects. No amount of testing can guarantee that a generator is foolproof.

In fact, any RNG has its own particular structure. As pointed out by Compagner [Com95], the bad things happen when the structure of the simulation problem at hand interacts constructively with the structure of the RNG. The difference between the good and bad RNGs is that the former have their structure *better hidden*, so that the problems that constructively interact with them are *harder to find* and occur less often in practice.

A *statistical test* can be defined by a real-valued function T of a finite (fixed or random) number of output values from the generator. The function T is called the *test statistic*. There is an infinite number of such functions. To apply the test, one needs to know (or have a good approximation of) the distribution function F of the random variable T under the null hypothesis $\mathcal{H}_0$. One can then compute the value t_1 taken by T and the corresponding p-value of the test, defined as $p_1 = P[T > t_1 | \mathcal{H}_0] = 1 - F(t_1)$. For discrete distributions, the ">" is sometimes replaced by "$\geq$", depending on the context. If F is continuous, p_1 is a $U(0,1)$ random variable under $\mathcal{H}_0$.

When p_1 is extremely close to 0 or to 1, one concludes that the generator *fails* the test. In case of doubt, when the outcome is not clear enough, the test can be replicated independently until either failure becomes obvious or one can declare that the suspect p-value was obtained by chance. In some situations, the test may be only *single-sided* in the sense that failure is declared only if p_1 is too close to 0, or only if it is too close to 1. The choice of rejection area depends on what the test aims to detect. We will talk about a more elaborate concept, so-called *two-level tests*, later in this chapter.

Which tests are the good ones? The question has no general answer. It depends on what we want to use the RNG for. The relevant tests are those for which the distribution of the test statistic T under H_0 is very close to that of the random variable of interest in our simulation, or at least for which the structures that are likely to affect this random variable can be detected by the test.

But this is an elusive ideal, except perhaps for a few very important simulation problems for which people are ready to spend time and money to test the RNG they use for their particular problem. For the vast majority of simulation applications, the users cannot afford more than pick a "well-tested" RNG off the shelf. In most cases, they do not even have any control over the RNG, because it is hidden deep in the simulation software environment they use and they cannot change it. Moreover, even if one can test and choose the RNG, it is rarely easy (or even possible) to construct a test T known to be sensitive to the same type of structure as the simulation problem at hand. This is especially true in the case of very complicated stochastic models.

Therefore, we are back to heuristics. If both our simulation model and our RNG have (independently constructed) complicated structures, the chance that they interact constructively can be deemed negligible. The RNGs that are placed in general purpose software packages must be tested without knowing the millions of things they will be used for. These general purpose RNGs should be submitted to a wide variety of tests of different natures, able to detect different types of correlations or other "simple" structures likely to be harmful.

3.2   Two-Level Tests

Instead of just computing a single p-value p_1, one can replicate the test several times over disjoint parts of the sequence. If this is done N times, then one obtains N copies of T, say $T_1, \ldots, T_N$, the N corresponding p-values, say $p_{1,1}, \ldots, p_{1,N}$, and their empirical distribution $\hat{F}_N$. Under $\mathcal{H}_0$ and the additional assumption that T has a *continuous* distribution, $p_{1,1}, \ldots, p_{1,N}$ are i.i.d. $U(0,1)$. Note that this second assumption is important and often not satisfied in practice, because several commonly used test statistics have in fact a discrete distribution [Knu81]. But if this discrete distribution is well-approximated by a continuous one, the $p_{1,i}$ will be approximately $U(0,1)$. The hypothesis $\mathcal{H}_0$ can then be tested by measuring the *goodness-of-fit* between $\hat{F}_N$ and the uniform distribution. Standard goodness-of-fit tests such as those of Kolmogorov-Smirnov (KS), Anderson-Darling (AD), Cramer-von Mises, Watson, etc., can be used for that purpose. We suggest [Dur73, Ste86a] as good references about these types of tests.

3.3   Collections of Empirical Tests

A classical list of statistical tests for RNGs is given by Knuth [Knu81]. More recently, Marsaglia [Mar96] has made available a large battery of tests with specifically chosen parameters, under the name of DIEHARD.

This battery includes tests that are more "stringent" than the classical ones given in [Knu81], in the sense that more popular generators tend to fail them. Another large testing package called TestU01 [L'E00] is in the work and will be made available sometime in the next century.

Certain types of tests simply throw n points in the t-dimensional unit hypercube and examine how they behave. These points are generally vectors u_i that belong to Ψ_t. A natural idea is to test the equidistribution of these points by partitioning the hypercube into boxes and counting how many points fall in each box. These (empirical) *equidistribution tests* have several variants and we examine some in Section 4.1. The name *serial test* often refers to them. Other types of tests are based on the distances between the close pairs of points in the hypercube. We look at some of them in Section 4.2.

There are of course many other types of tests, sensitive to different types of structures. For example, tests based on discrete *random walks* and certain *weight distribution tests* are generally good to detect correlations for certain families of generators such as GFSR or LFSR when their characteristic polynomials have few nonzero coefficients [MK94, SHB97, Tak94, VANK94, VANK95]. Various types of tests based on the notion of *entropy* are studied in [DR81, DMST95, LCC96, L'E97a, L'E97b, PK97]. Some of them are closely related to sum-functions of spacings, which define yet other classes of tests [L'E97c]. Among several additional references proposing specific tests or applying certain tests to random number generators, we mention [L'E92, LW97, Mar85, Ste86b].

4   Examples of Empirical Tests

4.1   Serial Tests of Equidistribution

Define a mapping $\iota : [0,1]^t \to \{0,\ldots,k-1\}$ such that $\mathrm{Vol}(\iota^{-1}(j)) = 1/k$ for $j = 0,\ldots,k-1$. This determines a partition of the unit hypercube into k pieces of equal volume. An easy way of doing the partition is by dividing the interval $[0,1]$ into d equal segments, which determines a partition of the unit hypercube into $k = d^t$ cubic boxes. Here, we shall consider only this type of partition, with the understanding that the techniques apply more generally.

Generate n disjoint vectors $u_{ti,t} = (u_{ti},\ldots,u_{ti+t-1})$, $i = 0,\ldots,n-1$, and let X_j be the number of these vectors that fall into the jth box, for $j = 0,\ldots,k-1$. Under $\mathcal{H}_0$, $(X_0,\ldots,X_{k-1})$ is a random vector having the multinomial distribution with parameters $(n, 1/k, \ldots, 1/k)$. (In this paper, the n serves both as an index for random numbers as in (1.1), and to denote the sample size in certain statistical tests as here. But the context indicates well the correct interpretation, so the reader should not be confused.)

As a measure of the distance (or *divergence*) between the empirical and uniform distributions, we consider test statistics of the form

$$Y = \sum_{j=0}^{k-1} f(X_j), \tag{1.5}$$

where f is a real-valued function. Different choices are available for f. For example:

$$\text{Chi-square:} \quad X^2 = \sum_{j=0}^{k-1} \frac{(X_j - n/k)^2}{n/k},$$

$$\text{Negative entropy:} \quad -H = \sum_{j=0}^{k-1} (X_j/n) \log_2(X_j/n),$$

$$\text{Collisions:} \quad C = \sum_{j=0}^{k-1} (X_j - 1) I[X_j > 1],$$

$$\text{Empty boxes:} \quad V = \sum_{j=0}^{k-1} I[X_j = 0] = k - n + C.$$

For all these examples, a larger value of the test statistic means that the points are *less* uniformly distributed in the boxes. The negative entropy $-H$ is in fact equivalent (via a linear transformation) to the *loglikelihood ratio* statistic G^2 for the multinomial distribution [LCC96]. Both G^2 and the Pearson chi-square X^2 are members of a parameterized class called the *power divergence* test statistics for the multinomial distribution, studied in Read and Cressie [RC88]. All other members of this class are also special cases of (1.5). The test based on X^2, with $n/k > 5$ or so, is often called the *serial test* or *t-tuple test* [Fis96, Knu81]. The Collision test is analyzed by Knuth [Knu81], who gives an algorithm to calculate the exact distribution of C.

Under the hypothesis $\mathcal{H}_0$, and under mild conditions on f (see [Hol72, LSW98, Mor75, RC88] for details), the following holds: If k is fixed and $n \to \infty$,

$$Y_C \overset{\text{def}}{=} \frac{Y - E[Y] + (k-1)\sigma_C}{\sigma_C} \Rightarrow \chi^2(k-1), \tag{1.6}$$

where $\sigma_C^2 = \text{Var}[Y]/(2(k-1))$, $\Rightarrow$ means convergence in distribution, and $\chi^2(k-1)$ is the chi-square distribution with $k-1$ degrees of freedom. We call this the *dense case*. If $k \to \infty$, $n \to \infty$, and $n/k \to \delta$ where $0 < \delta < \infty$, then

$$Y_N \overset{\text{def}}{=} \frac{Y - E[Y]}{\sigma_N} \Rightarrow N(0,1), \tag{1.7}$$

where $\sigma_N^2 = \text{Var}[Y]$ and $N(0,1)$ is the standard normal distribution. We call this the *sparse case*. The statistics X^2, $-H$, G^2, C, and V, mentioned above all satisfy the conditions in the sparse case. However, C and V *do not* satisfy the conditions for the dense case: In that case, V converges to 0 with probability one as $n \to \infty$. General expressions for the exact mean $E[Y]$ and the exact variance $\text{Var}[Y]$ are given in [LSW98]. The expression for the variance involves $O(n^2)$ terms, but in the sparse case, only a few of these terms are non-negligible, so one can easily obtain a very close numerical approximation. In other cases, asymptotic values are available.

Since C and V are integer-valued random variables, the normal approximation is not so good for them when their expectation is too small (say, less than 50 or so). In that case, they are approximately Poisson. When $\lambda' = n/k \ll 1$, the probability that any given box is empty is approximately $e^{-\lambda'}$, so $E[V] \approx ke^{-\lambda'}$ and $E[C] \approx n + k(e^{-\lambda'} - 1) \approx n^2/(2k)$.

The vectors $u_{ti,t}$ defined above are independent under $\mathcal{H}_0$. The *overlapping* vectors $u_{i,t}$, $i = 0, 1, 2, \ldots$ are not independent. If we use them to define the X_j, the limit theorems for Y_C and Y_N above no longer apply and the analysis is more difficult than for the disjoint case, but still something can be done for certain properly-defined statistics. Let $X_{t,j}^{\emptyset}$ be the number of overlapping vectors that fall into cell j and define

$$X_{(t)}^2 = \sum_{j=0}^{k-1} \frac{(X_{t,j}^{\emptyset} - n/d^t)^2}{n/d^t}. \tag{1.8}$$

Then

$$\tilde{X}^2 = X_{(t)}^2 - X_{(t-1)}^2 \Rightarrow \chi^2(d^{t-1}(d-1))$$

as $n \to \infty$ for d and t fixed. This was proved by Good [Goo53]. The standard rule of thumb recommends to take $n/k \geq 5$ for the chi-square approximation to be good. This is called the *overlapping serial test* or the *overlapping t-tuple test* (often with m instead of t in the notation) in the literature [Alt88, LW97, Mar85, Weg95].

In the sparse case, we believe that $\tilde{X}^2$ converge to a normal with mean $d^{t-1}(d-1)$ and variance $2d^{t-1}(d-1)$. No rigorous proof of this is currently available according to our knowledge but plenty of numerical experimentation supports it. This is also in accordance with the chi-square asymptotic distribution of $\tilde{X}^2$ in the dense case.

When $\lambda' = n/k$ is very small, the overlapping version of C is approximately Poisson with the same mean as in the non-overlapping case, namely $n^2/(2k)$. The overlapping version of V is heuristically discussed in [Mar85, MZ93]. In [MZ93], the authors claim that it is approximately normal with mean $ke^{-\lambda'}$ and variance $ke^{-\lambda'}(1 - 3e^{-\lambda'})$. This approximation is good when λ' is between 2 and 5 (approximately) but it quickly deteriorates when λ' moves away from this range [LSW98].

The interest of the overlapping tests comes from the fact that for a given sample size n, they require approximately t times less random numbers than their non-overlapping counterparts while being almost as effective when testing most commonly used RNGs. We will give examples of this later on in this paper.

4.2   Tests Based on Close Points in Space

We consider again the same set of non-overlapping vectors $\boldsymbol{u}_{ti,t} = (u_{ti}, \ldots, u_{ti+t-1})$, $i = 0, \ldots, n-1$, as in the previous subsection. Let $\| \cdot \|_p$ denote the L_p-norm in $[0,1)^t$ viewed as a *unit torus*, for any $p \in [1, \infty]$, and let V_t be the volume of the unit ball $\{x \in \mathbf{R}^t \mid \|x\|_p \leq 1\}$. (We choose the torus to get rid of boundary effects.) For any $s \geq 0$, let $Y_n(s)$ be the number of pairs (i,j), with $i < j$, such that

$$\|\boldsymbol{u}_{tj,t} - \boldsymbol{u}_{ti,t}\|_p \leq \left(\frac{2V_t s}{n(n-1)} \right)^{1/t}.$$

Under $\mathcal{H}_0$, for each $s_1 > 0$, the stochastic process $\{Y_n(s), 0 \leq s \leq s_1\}$ converges weakly to a Poisson process of unit rate when $n \to \infty$. See [LCS97, RS78, SB78] for details.

The spacings between the first few successive jumps of Y_n should be approximately i.i.d. exponentials with mean one, and this can be the basis of a test. More specifically, if $T_{n,0} = 0$ and if $T_{n,1}, T_{n,2}, \ldots$, are the jump points of Y_n, the spacings are $\Delta_{n,i} = T_{n,i} - T_{n,i-1}$, $i \geq 1$, and the random variables $W_{n,i}^* = 1 - \exp[-\Delta_{n,i}]$, $i = 1, \ldots, M$, are approximately i.i.d. $U(0,1)$ under $\mathcal{H}_0$ if n is large and M is relatively small. As a simple rule of thumb, it is reported in [LCS97] that the approximation is accurate for $n \geq 4M^2$. Values of M in the range from 10 to 100 are reasonable. Computing the M nearest pairs turns out to be faster for larger p and the supremum norm ($p = \infty$) is a very good choice. Values of n up to 10 millions or so can be handled if t is not too large (this depends mainly on the size of the available computer memory).

To test the goodness-of-fit of the empirical distribution of $W_{n,1}^*, \ldots, W_{n,M}^*$ to the uniform distribution, the Anderson-Darling (AD) test statistic happens to be an excellent choice. The reason is that for the generators whose structure is too regular, the jumps of Y_n tend to cluster and this produces several $W_{n,i}^*$ that are very close to zero. When this happens, their empirical distribution goes up quickly near the origin. The AD test statistic is quite appropriate to detect this type of behavior, because it is most sensitive to departures from the uniform distribution near the extremities of the interval $[0,1]$. The AD statistic here is:

$$
\begin{aligned}
A_M^2 &= \int_0^1 \frac{(\hat{F}(u) - u)^2}{u(1-u)} du \\
&= -M - \frac{1}{M} \sum_{j=1}^{M} \left\{ (2j-1)\ln(U_{(j)}) + (2M+1-2j)\ln(1 - U_{(j)}) \right\}
\end{aligned}
$$

where $U_{(1)} \leq \cdots \leq U_{(m)}$ are $W_{n,1}^*, \ldots, W_{n,m}^*$ sorted by increasing order and $\hat{F}$ is their empirical distribution. The hypothesis $\mathcal{H}_0$ is rejected when A_M^2 takes a value a that is too large; i.e., when its p-value $p_1 = P[A_M^2 > a|\mathcal{H}_0]$ is too small. This is the M-nearest-pairs test (M-NP test) proposed by L'Ecuyer, Cordeau, and Simard [LCS97].

A simpler version of this test is proposed and studied by Ripley [Rip77, Rip87], using only $M = 1$. The null hypothesis is rejected when $W_{n,1}^*$ is too close to either 0 or 1. The p-value is simply $p_1 = W_{n,1}^*$. This is the nearest-pair test (NP test).

These two tests can of course be applied in a two-level setup as explained previously. Replicate them N times and compare the empirical distribution of the N p-values to the uniform distribution using a second-level AD test. For the two-level NP test, it is typically much better to apply a spacings transformation or a power ratio transformation to the N values of p_1 before applying the AD test [LCS97]. The reason is that the values of p_1 often tend to cluster, and these two transformations change the small spacings into values that are very close to zero or very close to 1. The empirical distribution of the transformed values then becomes a good prey for the AD test at the second level. See [LCS97] for more details. We give examples later on.

5   Collections of Small RNGs

5.1   Small Linear Congruential Generators

Table 1.1 gives a list of LCGs with prime moduli m of different sizes and period $\rho = m - 1$. These LCGs are defined by

$$x_n = ax_{n-1} \bmod m, \tag{1.9}$$

$$u_n = x_n/m, \tag{1.10}$$

where a is a primitive root modulo m. For each integer e from 12 to 36, we took the largest prime m less than 2^e and found one good and one bad LCG with that modulus. Here, good and bad should be understood to be with respect to the spectral test. The good LCGs are those with the best figure of merit M_8 that were found in [L'E98b], for each prime m. The bad ones have been chosen to have $S_2 = M_2 \approx 0.05$, so they have a mediocre lattice structure in 2 dimensions.

5.2   Explicit Inversive Congruential Generators

We have a collection of explicit inversive congruential generators (EICG) defined by

$$x_n = (an)^{-1} \bmod m, \tag{1.11}$$

TABLE 1.1. Some good and bad (baby) LCGs

m	a (good LCG)	a (bad LCG)
$2^{12} - 3 = 4093$	209	5
$2^{13} - 1 = 8191$	884	2341
$2^{14} - 3 = 16381$	572	2731
$2^{15} - 19 = 32749$	219	10
$2^{16} - 15 = 65521$	17364	17
$2^{17} - 1 = 131071$	43165	68985
$2^{18} - 5 = 262139$	92717	203883
$2^{19} - 1 = 524287$	283741	458756
$2^{20} - 3 = 1048573$	380985	213598
$2^{21} - 9 = 2097143$	360889	202947
$2^{22} - 3 = 4194301$	914334	4079911
$2^{23} - 15 = 8388593$	653276	2696339
$2^{24} - 3 = 16777213$	6423135	486293
$2^{25} - 39 = 33554393$	25907312	5431467
$2^{26} - 5 = 67108859$	26590841	42038579
$2^{27} - 39 = 134217689$	45576512	24990322
$2^{28} - 57 = 268435399$	31792125	31842465
$2^{29} - 3 = 536870909$	16538103	8903330
$2^{30} - 35 = 1073741789$	5122456	3930720
$2^{31} - 1 = 2147483647$	1389796	868723
$2^{32} - 5 = 4294967291$	1588635695	1123161
$2^{33} - 9 = 8589934583$	7425194315	1026767
$2^{34} - 41 = 17179869143$	5295517759	10045
$2^{35} - 31 = 34359738337$	3124199165	10052
$2^{36} - 5 = 68719476731$	49865143810	102254510

$$u_n = x_n/m, \qquad\qquad (1.12)$$

where the inverse of an modulo m can be computed as $x_n = (an)^{m-2}$ mod m. All the generators in our EICG collection have the same prime modulus m as the LCGs, their multiplier is $a = 123$, and their period length is m. This type of nonlinear generator has been proposed by Eichenauer-Herrmann and is discussed in [Eic92, EHW97, Hel95, LW97].

5.3   Compound Cubic Congruential Generators

Our next collection contains a recent type of generator, the *compound cubic congruential generators* (CCCG), in the form proposed by Eichenauer-Herrmann [EH97], with two components. These generators are defined as follows. Let m_1 and m_2 be two distinct primes, a_1 and a_2 two positive integers less than m_1 and m_2, respectively, and put:

$$x_{1,n} = (a_1 x_{1,n-1}^3 + 1) \bmod m_1, \tag{1.13}$$

$$x_{2,n} = (a_2 x_{2,n-1}^3 + 1) \bmod m_2, \tag{1.14}$$

$$u_n = (x_{1,n}/m_1 + x_{2,n}/m_2) \bmod 1. \tag{1.15}$$

For certain values of m_1 and a_1, the recurrence (1.13) has the maximal period length m_1, and similarly for (1.14). These values are not so easy to find, because simple maximum period conditions are not available, and because these values are rare. Fourteen pairs (m_1, a_1) that produce maximum period are given in [EH97], for m_1 near 2^{15}. Table 1.2 gives the pairs that we use to construct our collection of CCCGs. These pairs (m_j, a_j) all yield the maximal period m_j, and they were found by brute force computations. For each e, the period length of the compound generator is equal to $m = m_1 m_2$, which is close to 2^e.

These nonlinear generators tend to have a much less regular structure than the LCGs or MRGs when a significant fraction of the period length is used, just like the inversive ones, but they run much faster than the latter, almost as fast as the LCGs. They seem to offer a good compromise between (slow) inversive generators and (too regular) LCGs.

6   Systematic Testing for Small RNGs

This section reports on empirical experiments made with the tests and the RNG collections described in the two previous sections. For each collection and each type of test, we find (empirically) some function φ and some constant ν^* such that $n \approx 2^{\nu^*} \varphi(e)$ is a crude approximation of the minimal sample size n at which the tests decisively reject the generators. As a general rule, for the two types of LCGs, we take $\varphi(e) = 2^{e/2} \approx \sqrt{\rho}$, whereas for the inversive and the cubic generators, we take $\varphi(e) = 2^e \approx \rho$. These functions turn out to be the right choices for ν^* to be approximately independent of e. The tests were carried out with a preliminary version of the TestU01 package of the second author.

6.1   Serial Tests of Equidistribution for LCGs

Our first set of tests is in 2 dimensions ($t = 2$). For all the generators, we run the tests with sample sizes (the number of points in the hypercube) $n = 2^\nu \varphi(e)$, for $\nu = -5, \ldots, 8$. The number of boxes is $k = (\lfloor (k')^{1/t} + 1/2 \rfloor)^t$, where each of the following four rules is tried for the choice of k': $k' \approx 2^e$, $k' \approx 2^{e+3}$, $k' \approx 2^{e-3}$, and $k' \approx n/8$. The transformation from k' to k is simply to make sure that $d = k^{1/t}$ is an integer. We compute X^2, $-H$, and their p-values with the normal approximation for $k \geq n$, and with the chi-square approximation otherwise. We also compute C and its p-value, using

TABLE 1.2. Full-period compound cubic congruential generators

e	m_1	a_1	m_2	a_2
12	59	15	47	7
13	101	61	59	15
14	101	61	83	9
15	251	135	101	61
16	251	135	233	61
17	503	445	251	135
18	503	445	491	277
19	1019	437	503	445
20	1019	437	1013	31
21	2039	243	1019	437
22	2039	243	2027	349
23	4091	494	2039	243
24	4091	494	4079	3757
25	8147	2410	4091	494
26	8147	2410	8111	2257
27	16361	5595	8147	2410
28	16361	5595	16319	3013
29	32693	1190	16361	5595
30	32693	1190	32687	4535
31	65519	512	32693	1190
32	65519	512	65447	27076
33	131063	110230	65519	512
34	131063	110230	130859	48249

the Poisson approximation when $E[C] < 50$ and the normal approximation otherwise. In fact, when $E[C]$ is too small, the normal approximation gets bad not only for C, but for X^2 and $-H$ as well, as we shall see in a moment. More details about which distribution is most appropriate in which situation can be found in [LSW98].

The results are summarized in a series of tables that follow. In these tables, only the p-values less than 0.01 or larger than 0.99 are given. We call them the *suspect* p-values. The entries where the p-values are not suspect are left blank. The p-values smaller than 10^{-15} are denoted by ϵ. A negative entry $-x$ means a p-value of $1 - x$. In particular, $-\epsilon$ means $> 1 - 10^{-15}$. In general, the columns with a larger ν than those given in a table contain only $\pm\epsilon$ and the columns with a smaller ν than what appears in the table are mostly blank.

When the Poisson approximation is used for C, the p-value printed in the table is defined as follows:

$$p = \begin{cases} P[Y \geq c] & \text{if } P[Y \geq c] < P[Y \leq c]; \\ 1 - P[Y \leq c] & \text{otherwise,} \end{cases}$$

where c is the value of C and Y is a Poisson random variable with mean $\mu = E[C|H_0]$. With this definition, a value of p close to zero [resp., close to one] means that C took a much larger [resp., much smaller] value than expected.

Table 1.3 gives the p-values for X^2, for $k' = 2^e$. The value of ν where the suspect p-values start to appear, as well as the value of ν where the $\pm\epsilon$'s start to appear, are pretty much the same for all e. Let us define $\tilde{\nu}$ and ν^* as the smallest ν for which the majority of the p-values are suspect and the smallest ν for which the majority of the p-values are $\pm\epsilon$, respectively. In Table 1.3, $\tilde{\nu} = 2$ and $\nu^* = 4$. This means that all the good LCGs with period length ρ are definitely rejected by the sparse serial test based on X^2 with $k \approx 2^e$ and $n = 16\sqrt{\rho}$ points. With $n \approx 4\sqrt{\rho}$, the test starts to give a majority of suspect p-values.

TABLE 1.3. The p-values for Pearson's X^2, for $t = 2$ and $k \approx \rho$.

e	$\nu = 1$	$\nu = 2$	$\nu = 3$	$\nu = 4$	$\nu = 5$
12			-4.0E-004	-5.9E-014	$-\epsilon$
13				-3.2E-010	$-\epsilon$
14			-1.0E-004	-9.5E-012	$-\epsilon$
15		-6.7E-003	-5.0E-006	-7.5E-015	$-\epsilon$
16		-2.3E-003	-9.1E-007	$-\epsilon$	$-\epsilon$
17		-6.7E-003	-1.0E-004	$-\epsilon$	$-\epsilon$
18			-1.3E-003	-1.1E-013	$-\epsilon$
19		-6.7E-003	-1.5E-007	$-\epsilon$	$-\epsilon$
20		-2.3E-003	-3.7E-007	$-\epsilon$	$-\epsilon$
21		-6.7E-003	-2.0E-004	-7.6E-013	$-\epsilon$
22			-7.3E-004	-5.4E-010	$-\epsilon$
23		-2.3E-003	-9.1E-007	$-\epsilon$	$-\epsilon$
24		-2.3E-003	-2.1E-008	$-\epsilon$	$-\epsilon$
25			-9.1E-007	$-\epsilon$	$-\epsilon$
26		-6.7E-003	-4.9E-006	$-\epsilon$	$-\epsilon$
27		-6.7E-003	-4.9E-006	$-\epsilon$	$-\epsilon$
28			-4.0E-003	-7.3E-015	$-\epsilon$
29		-2.3E-003	-7.7E-009	$-\epsilon$	$-\epsilon$
30			-5.0E-005	$-\epsilon$	$-\epsilon$
31		-6.7E-003	-1.5E-007	$-\epsilon$	$-\epsilon$
32			-1.0E-004	-1.5E-014	$-\epsilon$
33			-2.4E-005	-3.1E-011	$-\epsilon$
34			-2.4E-005	$-\epsilon$	$-\epsilon$

Table 1.4 gives the p-values of the test statistic C, for the same setup. The expected number of collisions in these tests is approximately $2^{2\nu-1}$. Therefore, the Poisson approximation was used for $\nu \leq 3$ and the normal approximation for $\nu \geq 4$. The entries where the normal approximation was

TABLE 1.4. Some p-values for C (collision test), for $t = 2$ and $k \approx \rho$.

e	$\nu = 1$	$\nu = 2$	$\nu = 3$	$\nu = 4$	$\nu = 5$
12			-2.8E-004	$-\epsilon$	$-\epsilon$
13				-6.8E-011	$-\epsilon$
14			-2.7E-005	-1.9E-012	$-\epsilon$
15		-3.2E-003	-1.6E-007	-1.1E-015	$-\epsilon$
16		-3.5E-004	-5.6E-009	$-\epsilon$	$-\epsilon$
17		-3.1E-003	-2.0E-005	$-\epsilon$	$-\epsilon$
18			-7.3E-004	-6.4E-014	$-\epsilon$
19		-3.1E-003	-8.4E-011	$-\epsilon$	$-\epsilon$
20		-3.4E-004	-6.8E-010	$-\epsilon$	$-\epsilon$
21		-3.0E-003	-4.9E-005	-6.4E-013	$-\epsilon$
22			-3.0E-004	-5.0E-010	$-\epsilon$
23		-3.4E-004	-4.3E-009	$-\epsilon$	$-\epsilon$
24		-3.4E-004	-4.3E-013	$-\epsilon$	$-\epsilon$
25			-4.2E-009	$-\epsilon$	$-\epsilon$
26		-3.0E-003	-1.1E-007	$-\epsilon$	$-\epsilon$
27		-3.0E-003	-1.1E-007	$-\epsilon$	$-\epsilon$
28			-2.8E-003	-7.2E-015	$-\epsilon$
29		-3.4E-004	-1.3E-014	$-\epsilon$	$-\epsilon$
30			-5.6E-006	$-\epsilon$	$-\epsilon$
31		-3.0E-003	-7.6E-011	$-\epsilon$	$-\epsilon$
32			-1.7E-005	-1.5E-014	$-\epsilon$
33			-1.7E-006	-3.1E-011	$-\epsilon$
34			-1.7E-006	$-\epsilon$	$-\epsilon$

used are remarkably similar to the corresponding entries in Table 1.3. In fact, if we had used the normal approximation everywhere, all the corresponding entries in the two tables would be nearly identical. The difference between the two tables in the columns for $\nu = 2$ and 3 is almost entirely due to fact that we use a Poisson (instead of normal) approximation for the distribution of C for these values of ν. The Poisson approximation is quite accurate in this area, so we must conclude that the normal approximation is not very good.

For $-H$ the p-values are almost the same as those of X^2 and C with the normal approximation. Why are these values so similar? Since $n \ll k$ and $E[C]$ is quite small here, most boxes contain 0 or 1 point, a few boxes contain 2 points, and in most cases no box contains more than 2 points. When the latter happens, C is the number of boxes that contain 2 points and there is a one-to-one correspondence between C, X^2, and $-H$, so the p-values of these three statistics should be almost the same. Under $\mathcal{H}_0$, for $n \ll k$, the number of boxes where X_i exceeds 2 is approximately Poisson with mean $\lambda_2 = n(n-1)(n-2)/(6k^2)$ [AGG89].

Considering that the p-values of these three statistics are almost the same when their distribution is approximated by the normal, and that they should effectively be nearly the same under the exact distributions, it follows that when the normal approximation is bad for C because $E[C]$ is too small, it is also bad for X^2 and $-H$. For instance, the correct p-values of X^2 in Table 1.3 should be much closer to the values of Table 1.4. Most of the difference in the two middle columns ($\nu = 2$ and 3) is caused by the lack of normality of X^2 for $\nu \leq 3$.

The $-\epsilon$ entries in the tables mean that X^2, $-H$, and C are much smaller than expected; i.e., that almost all of the points fall into different boxes. This is a consequence of the regular structure of the LCGs.

TABLE 1.5. The p-values of C for the bad LCGs, for $t = 2$ and $k \approx \rho$.

e	$\nu = -3$	$\nu = -2$	$\nu = -1$	$\nu = 0$	$\nu = 1$
14		4.5E-004	7.0E-003	3.1E-009	ϵ
15			9.2E-006	1.6E-010	ϵ
16		4.6E-004	7.1E-003	6.0E-008	ϵ
17		4.8E-004		6.0E-008	ϵ
18			7.1E-003	3.4E-009	ϵ
19		4.7E-004		3.4E-009	ϵ
20				1.7E-004	ϵ
21				3.4E-009	ϵ
22		4.8E-004	7.2E-003	3.2E-013	ϵ
23		4.8E-004		1.7E-004	ϵ
24			7.2E-003	1.7E-004	ϵ
25			3.0E-004	1.0E-006	ϵ
26				1.0E-006	ϵ
27			7.2E-003	1.4E-005	ϵ
28			7.2E-003	6.2E-008	ϵ
29		4.8E-004	7.2E-003	1.0E-006	ϵ
30			3.0E-004	ϵ	ϵ
31		4.8E-004	7.2E-003	1.8E-003	ϵ
32			9.2E-006	6.2E-008	ϵ
33		4.8E-004	7.2E-003	1.8E-003	2.4E-004
34		4.8E-004	9.2E-006		ϵ

Table 1.5 gives the p-values of C for the *bad* LCGs, still for $k = 2^e$. The p-values for X^2 and $-H$ are somewhat different, mainly because they are computed using the normal approximation, which is not so good with these small values of ν (the expected number of collisions is still $2^{2\nu-1}$). Here we have $\tilde{\nu} = -1$ and $\nu^* = 1$, so the strong rejections appear for a sample size n approximately 8 times smaller than for the good LCGs.

The p-values at ν^* are now ϵ instead of $-\epsilon$, which means that C (as well as X^2 and $-H$) is *much larger* than expected. For example, for $\nu = 1$, the number of collisions should be approximately Poisson with mean 2. But it

TABLE 1.6. Some p-values for X^2, for $t = 2$ and $k \approx n/8$, for good LCGs.

e	$\nu = 4$	$\nu = 5$	$\nu = 6$	$\nu = 7$	$\nu = 8$
12		−1.6E-008		ϵ	ϵ
13	−4.2E-003	−3.5E-010		ϵ	ϵ
14		−8.6E-004	−ϵ		ϵ
15		−6.7E-003	−ϵ		ϵ
16	−1.2E-003	−9.4E-003	−1.9E-015	−ϵ	6.2E-003
17			−7.1E-015	−ϵ	−3.0E-011
18			−1.4E-009	−ϵ	−ϵ
19		−4.9E-004	−9.0E-005	−ϵ	−ϵ
20			−3.0E-006	−ϵ	−ϵ
21			−7.8E-005	−ϵ	−ϵ
22		−9.9E-003		−ϵ	−ϵ
23				−9.9E-009	−ϵ
24				−1.6E-006	−ϵ
25			−5.0E-003	−1.1E-007	−ϵ
26			−1.1E-003	−1.0E-005	−ϵ
27					−ϵ
28				−1.7E-004	−1.2E-014

is actually equal to 28 for $e = 14$, 35 for $e = 15$, 25 for $e = 16$, and so on. In other words, the points are concentrated in much fewer boxes than they should. This is due to the fact that they all lie on a small number of lines in the unit square, as indicated by the bad spectral test values.

With $k \approx \rho/8$ and $k \approx 8\rho$, the behavior is similar to that with $k \approx \rho$ and the values of ν^* are the same or slightly larger, both for the good and for the bad LCGs. The p-values are generally not as close to 0 or 1, and they get farther away from 0 or 1 when k gets farther away from ρ.

For the results that we just gave, the number of points is much less than the number of boxes. But in the classical application of the serial test [Knu81, Fis96], the number of points is usually at least 5 to 10 times the number of boxes. This is the dense case. In Tables 1.6 and 1.7, we give the p-values for X^2, for the good and bad LCGs, respectively, for $k' = n/8$. The results for $-H$ are similar. Here, X^2 follows approximately the chi-square distribution. In contrast with the previous tables, ν^* increases with e. To obtain $\pm\epsilon$, one needs a much larger n in this dense case than in the sparse case, especially when ρ is large.

To have a better view of when the $\pm\epsilon$ start to occur in this case, let us change φ. Table 1.8 gives the p-values as in Table 1.6, but with $\varphi(e) = 2^{2e/3} \approx \rho^{2/3}$. The entries marked "—" are cases where we decided not to perform the tests because n was deemed too large. Now, we have $\bar{\nu} = 3$ and $\nu^* = 4$ for the good LCGs. The *rate* of increase of n as a function of ρ to get the small p-values is larger in the dense case than in the sparse case: $O(\rho^{2/3})$ instead of $O(\rho^{1/2})$ (empirically). For the bad LCGs, the behavior is similar and the values of $\tilde{\nu}$ and ν^* are the same.

What we have seen for $t = 2$ can also be observed in a larger number of dimensions, with slightly different values of $\tilde{\nu}$ and ν^*. Table 1.9 gives the p-values of X^2 for the good LCGs, for $k' = 2^e \approx \rho$ and $t = 8$. Here, $\nu^* = 6$, two units higher than for $t = 2$. The results for $-H$ and C are almost the same.

Table 1.10 gives the p-values of X^2 for the same parameters as in Table 1.9, but for the *overlapping* case. Again, the results for $-H$ and C are almost the same. One has the same ν^* as in the *non-overlapping* case, but the overlapping test is more economical because it requires t times less random numbers. A similar behavior can be observed for other values of t and other definitions of k'.

6.2 Serial Tests of Equidistribution for Nonlinear Generators

Table 1.12 and 1.13 show the p-values of X^2 for $t = 2$, $\varphi(e) = 2^e \approx \rho$, and $k' = 2^e$, for the inversive and cubic generators, respectively. Clearly, the sample size n required for rejection increases as $O(\rho)$ in both cases. In fact, suspect p values occur systematically only for n larger than the period length for the inversive generators, and n approximately equal to the period length for the cubic generators. Observe that when $n = \rho$, $2n$ random numbers are used in the test, so the same random numbers are used twice. But the n points are still distinct because ρ is odd. The latter holds in general provided that $\gcd(t, \rho) = 1$. The behavior is similar for $k' = 2^{e-3}$ and $k' = 2^{e+3}$, and also for C, whereas ν^* is sometimes one or two units larger for $-H$.

6.3 A Summary of the Serial Tests Results

Table 1.14 summarizes the values of $\tilde{\nu}$ and ν^* for the serial tests under different setups, for the LCGs. Table 1.15 does the same for the explicit inversive and the cubic generators. The values in parentheses lie in areas where the normal approximation is not very good. In these areas, the tests based on C (with the Poisson approximation) are more reliable. As expected, the bad LCGs fail with much smaller sample sizes than the good LCGs in dimension 2. These LCGs were chosen to be bad in two dimensions, but they turn out to be not so bad in larger dimensions (in general). In 8 dimensions, they do practically as well as the good ones in terms of $\tilde{\nu}$ and ν^*.

TABLE 1.7. Some p-values for X^2, for $t = 2$ and $k \approx n/8$, for bad LCGs.

e	$\nu = 2$	$\nu = 3$	$\nu = 4$	$\nu = 5$	$\nu = 6$
14		ϵ	ϵ	ϵ	ϵ
15		ϵ	ϵ	ϵ	ϵ
16	ϵ	4.6E-005	ϵ	ϵ	ϵ
17	8.9E-007		ϵ	ϵ	ϵ
18			ϵ	ϵ	ϵ
19		ϵ		ϵ	ϵ
20		ϵ	ϵ	ϵ	ϵ
21		ϵ	ϵ		ϵ
22	6.3E-005	ϵ	ϵ	ϵ	ϵ
23				ϵ	−5.4E-005
24				ϵ	ϵ
25					ϵ
26			ϵ	ϵ	ϵ
27			8.3E-003		5.9E-004
28					
29					4.5E-003
30					
31					

TABLE 1.8. Some p-values for X^2, for $t = 2$ and $k \approx n/8$ and $\varphi(e) = 2^{2e/3}$, for good LCGs.

e	$\nu = 0$	$\nu = 1$	$\nu = 2$	$\nu = 3$	$\nu = 4$
12				−1.5E-008	
13				−1.3E-015	
14				−4.8E-011	−4.3E-007
15				−9.3E-010	−ϵ
16				−2.2E-010	−ϵ
17				−1.2E-009	−ϵ
18				−1.4E-009	−ϵ
19			−1.2E-004	−1.6E-006	−ϵ
20		−3.4E-003	−4.1E-003	−2.1E-010	−ϵ
21			−3.3E-003	−1.1E-006	−ϵ
22				−8.9E-008	−ϵ
23			−4.8E-003	−1.2E-004	−ϵ
24				−2.2E-008	−ϵ
25				−3.1E-009	−ϵ
26			−5.5E-004	−4.0E-007	−ϵ
27				−4.0E-006	—
28				−4.3E-007	—

TABLE 1.9. The p-values of X^2 for $t = 8$ and $k \approx \rho$, for good LCGs.

e	$\nu = 2$	$\nu = 3$	$\nu = 4$	$\nu = 5$	$\nu = 6$
12		−7.1E-003	−2.8E-004	ϵ	ϵ
13	5.4E-003			−1.4E-005	ϵ
14				−1.3E-007	ϵ
15		−6.3E-003	−1.9E-008	−5.2E-013	ϵ
16			−1.0E-004	−1.0E-010	ϵ
17			−8.8E-003	−3.1E-010	$-\epsilon$
18			−1.3E-003	−9.0E-012	$-\epsilon$
19		−1.2E-003		−1.2E-010	$-\epsilon$
20				−2.4E-011	$-\epsilon$
21			−9.7E-008	−4.8E-014	$-\epsilon$
22			−3.4E-003	−6.8E-012	$-\epsilon$
23			−4.9E-003	−3.1E-007	$-\epsilon$
24			−9.9E-004	−7.7E-009	$-\epsilon$
25			−1.7E-004	−4.2E-009	$-\epsilon$
26			−1.3E-004	$-\epsilon$	$-\epsilon$
27			−5.0E-003	−1.9E-010	$-\epsilon$
28			−3.1E-004	−5.6E-013	$-\epsilon$
29			−1.0E-005	$-\epsilon$	$-\epsilon$
30			−4.3E-008	−2.3E-011	$-\epsilon$
31			−6.2E-006	−1.5E-013	$-\epsilon$
32			−3.7E-007	−2.7E-015	$-\epsilon$

TABLE 1.10. The p-values of X^2 for $t = 8$ (with overlapping) and $k \approx \rho$, for good LCGs.

e	$\nu = 3$	$\nu = 4$	$\nu = 5$	$\nu = 6$
12		-2.1E-003	-6.6E-004	-8.4E-013
13				-4.8E-008
14			-1.4E-005	$-\epsilon$
15			-5.2E-008	$-\epsilon$
16			-1.2E-005	-4.3E-012
17			-3.8E-005	$-\epsilon$
18		-7.9E-004		-1.1E-011
19		-4.1E-005	-8.2E-008	$-\epsilon$
20		-2.0E-003	-1.7E-007	-1.4E-013
21		-3.6E-004	-3.8E-010	$-\epsilon$
22	-7.7E-004		-2.4E-007	$-\epsilon$
23		-1.1E-003	-4.2E-004	$-\epsilon$
24			-3.3E-010	$-\epsilon$
25		-1.6E-003	-8.4E-011	$-\epsilon$
26		-1.6E-003	-4.8E-012	$-\epsilon$
27			-7.0E-004	$-\epsilon$
28		-1.5E-003	-3.8E-009	$-\epsilon$
29	-1.9E-003	-1.7E-009	-1.2E-013	$-\epsilon$
30		-5.5E-006	-6.2E-011	$-\epsilon$
31		-1.3E-005	-1.4E-014	$-\epsilon$
32		-9.4E-006	-1.8E-009	$-\epsilon$

TABLE 1.11. The p-values of X^2 for $t = 8$ (with overlapping) and $k \approx \rho$, for bad LCGs.

e	$\nu = 3$	$\nu = 4$	$\nu = 5$	$\nu = 6$	$\nu = 7$
12				2.9E-003	ϵ
13		-1.5E-003	-1.1E-005	-3.2E-013	ϵ
14					8.0E-003
15	7.0E-003		1.1E-003	ϵ	ϵ
16					
17		-9.1E-003	-4.6E-013	$-\epsilon$	$-\epsilon$
18					-3.7E-003
19			-1.9E-004	-2.9E-014	$-\epsilon$
20				-1.6E-003	-7.8E-004
21					
22		-7.9E-003	-8.7E-010	$-\epsilon$	$-\epsilon$
23				-3.6E-003	-3.2E-008
24				-1.1E-003	$-\epsilon$
25		-2.6E-004	-4.2E-008	$-\epsilon$	$-\epsilon$
26		-1.4E-003	-3.1E-005	$-\epsilon$	$-\epsilon$
27			-4.9E-004	-1.6E-007	$-\epsilon$
28			1.7E-005	ϵ	ϵ
29				-8.4E-003	-6.9E-008
30			-4.4E-004	-9.3E-010	$-\epsilon$
31		-3.2E-003	-6.6E-005	-1.0E-013	—
32			-2.0E-006	$-\epsilon$	—

TABLE 1.12. The p-values of X^2 for $t = 2$ and $k \approx \rho$, for inversive generators.

e	$\nu = -1$	$\nu = 0$	$\nu = 1$	$\nu = 2$
10			ϵ	ϵ
11			ϵ	ϵ
12			ϵ	ϵ
13	-3.9E-003		ϵ	ϵ
14			ϵ	ϵ
15			ϵ	ϵ
16	-7.1E-003	-2.4E-003	ϵ	ϵ
17		-3.0E-004	ϵ	ϵ
18			ϵ	ϵ
19			ϵ	ϵ
20			ϵ	ϵ
21		1.2E-003	ϵ	—
22			—	—

TABLE 1.13. The p-values of X^2 for $t = 2$ and $k \approx \rho$, for cubic generators.

e	$\nu = -1$	$\nu = 0$	$\nu = 1$
14		ϵ	ϵ
15		ϵ	ϵ
16		ϵ	ϵ
17		ϵ	ϵ
18		ϵ	ϵ
19		ϵ	ϵ
20		ϵ	ϵ
21		ϵ	ϵ
22		ϵ	—

TABLE 1.14. The good and bad LCGs in the serial tests, with $\varphi(e) = 2^{e/2}$.

Overlap	t	k'	Test	Good LCGs $\tilde{\nu}$	ν^*	Bad LCGs $\tilde{\nu}$	ν^*
No	2	2^e	X^2	(2)	4	(−1)	(0)
"	2	2^e	$-H$	(2)	4	(−1)	(0)
"	2	2^e	C	2	4	−1	1
"	2	2^{e+3}	X^2	(4)	5	(0)	(0)
"	2	2^{e+3}	$-H$	(4)	5	(0)	(0)
"	2	2^{e+3}	C	4	5	0	2
"	2	2^{e-3}	X^2	4	6	(−1)	(0)
"	2	2^{e-3}	$-H$	4	6	(−1)	(0)
"	2	2^{e-3}	C	4	6	−1	1
"	2	2^{e-5}	X^2	5	7	(−1)	(0)
"	2	2^{e-5}	$-H$	5	7	(−1)	(0)
"	2	2^{e-5}	C	5	7	−1	1
"	8	2^e	X^2	4	6	5	6
"	8	2^e	$-H$	4	6	5	6
"	8	2^e	C	4	6	5	6
"	8	2^{e+3}	X^2	4	6	5	6
"	8	2^{e+3}	$-H$	4	6	5	6
"	8	2^{e+3}	C	4	6	5	6
"	8	2^{e-3}	X^2	5	6	5	7
"	8	2^{e-3}	$-H$	5	7	5	7
"	8	2^{e-3}	C	5	6	5	7
Yes	8	2^e	X^2	4	6	5	7
"	8	2^e	C	4	6	5	7
"	8	2^{e+3}	X^2	5	6	5	7
"	8	2^{e+3}	C	5	6	5	6
"	8	2^{e-3}	X^2	5	7	6	8

TABLE 1.15. The inversive and cubic generators in the serial tests, with $\varphi(e) = 2^e$.

Overlap	t	k'	Test	Inversive		Cubic	
				$\tilde{\nu}$	ν^*	$\tilde{\nu}$	ν^*
No	2	2^e	X^2	1	1	0	0
"	2	2^e	$-H$	2	3	0	0
"	2	2^e	C	1	1	0	0
"	2	2^{e+3}	X^2	1	1	0	0
"	2	2^{e+3}	$-H$	1	1	0	0
"	2	2^{e+3}	C	1	1	0	0
"	2	2^{e-3}	X^2	1	1	0	0
"	2	2^{e-3}	$-H$	1	1	1	1
"	8	2^e	X^2	1	1	0	0
"	8	2^e	$-H$	2	3	0	0
"	8	2^e	C	1	1	0	0
"	8	2^{e+3}	X^2	0	1	0	0
"	8	2^{e+3}	$-H$	1	1	0	0
"	8	2^{e+3}	C	1	1	0	0
"	8	2^{e-3}	X^2	1	1	0	0
"	8	2^{e-3}	$-H$	1	1	1	1
Yes	8	2^e	X^2	1	1	0	0
"	8	2^{e+3}	X^2	1	1	0	0
"	8	2^{e-3}	X^2	1	1	0	0

TABLE 1.16. NP test with $t = 2$ and $N = 1$ for good LCGs.

e	$\nu = 0$	$\nu = 1$	$\nu = 2$	$\nu = 3$
14			5.5E-007	ϵ
15		4.5E-003	3.9E-010	ϵ
16		7.6E-003	3.3E-009	ϵ
17			4.7E-008	ϵ
18			7.5E-006	ϵ
19		2.9E-003	7.1E-011	ϵ
20		1.8E-003	1.0E-011	ϵ
21			1.9E-008	ϵ
22		9.5E-003	8.1E-009	ϵ
23		1.7E-003	9.2E-012	ϵ
24		7.0E-004	2.3E-013	ϵ
25			1.4E-008	ϵ
26		2.6E-003	4.6E-011	ϵ
27		8.8E-003	5.9E-009	ϵ
28			1.0E-007	ϵ
29		6.6E-004	1.9E-013	ϵ
30			1.8E-007	ϵ
31		1.2E-003	2.1E-012	ϵ
32			6.5E-008	ϵ
33			2.0E-006	ϵ
34		3.0E-003	7.6E-011	ϵ
35		8.1E-003	4.4E-009	ϵ
36		7.2E-003	2.8E-009	ϵ

TABLE 1.17. The 32-NP test with $t = 2$ and $N = 1$ for good LCGs.

e	$\nu = -2$	$\nu = -1$	$\nu = 0$	$\nu = 1$
22	—	—	—	ϵ
24	—	—	ϵ	ϵ
26	—	4.3E–09	ϵ	ϵ
28		2.9E–04	ϵ	ϵ
30	5.9E–05	ϵ	ϵ	ϵ
32	3.4E–04	ϵ	ϵ	ϵ
34	1.6E–03	3.1E–03	ϵ	ϵ
36			ϵ	ϵ
38		6.1E–04	ϵ	ϵ
40		8.8E–12	ϵ	ϵ

TABLE 1.18. The NP test with $t = 2$ and $N = 32$ for good LCGs.

e	$\nu = -2$	$\nu = -1$	$\nu = 0$	$\nu = 1$
14			3.5E–07	ϵ
16			4.4E–09	ϵ
18			2.3E–04	ϵ
20			4.4E–13	ϵ
22			6.1E–10	ϵ
24			ϵ	ϵ
26			9.9E–12	ϵ
28			4.0E–07	ϵ
30			1.9E–06	ϵ
32			5.2E–07	ϵ
34		2.4E–03	2.7E–11	ϵ
36			1.9E–09	ϵ
38			1.3E–11	ϵ
40		3.2E–03	3.0E–13	ϵ

TABLE 1.19. The NP-S test with $t = 2$ and $N = 32$ for good LCGs.

e	$\nu = -4$	$\nu = -3$	$\nu = -2$	$\nu = -1$
14	—	—	ϵ	ϵ
16	—	2.8E–03	ϵ	ϵ
18	7.5E–03	5.4E–12	ϵ	ϵ
20		2.9E–03	ϵ	ϵ
22		1.1E–04	ϵ	ϵ
24		1.8E–15	ϵ	ϵ
26		ϵ	ϵ	ϵ
28			ϵ	ϵ
30		1.5E–04	ϵ	ϵ
32	2.1E–05	5.5E–04	ϵ	ϵ
34		6.2E–09	ϵ	ϵ
36		1.6E–14	ϵ	ϵ
38		8.7E–04	ϵ	ϵ
40	9.5E–03		ϵ	ϵ

TABLE 1.20. NP test with $t = 2$ and $N = 1$ for bad LCGs.

e	$\nu = 4$	$\nu = 5$	$\nu = 6$	$\nu = 7$	$\nu = 8$
14			$-\epsilon$	$-\epsilon$	$-\epsilon$
15		1.9E-003	1.4E-011	$-\epsilon$	$-\epsilon$
16		1.2E-004	ϵ	$-\epsilon$	$-\epsilon$
17		3.6E-003	1.6E-010	ϵ	$-\epsilon$
18		5.1E-003	6.7E-010	ϵ	$-\epsilon$
19		2.6E-003	4.8E-011	ϵ	ϵ
20		3.4E-003	1.3E-010	ϵ	ϵ
21		6.3E-003	1.6E-009	ϵ	ϵ
22		2.7E-003	5.5E-011	ϵ	ϵ
23		2.8E-003	6.5E-011	ϵ	ϵ
24		4.8E-003	5.5E-010	ϵ	ϵ
25		8.9E-003	6.4E-009	ϵ	ϵ
26		2.8E-003	6.1E-011	ϵ	ϵ

6.4   Close-Pairs Tests for LCGs

We applied the close-pairs tests to the good and bad LCGs, with $\varphi(e) = 2^{e/2}$. We now give the results for $t = 2$ and $p = \infty$. Tables 1.16 to 1.19 give the p-values for the NP test with $N = 1$, the 32-NP test with $N = 1$, the NP test with $N = 32$, and the NP-S test with $N = 32$, for the good LCGs. Tables 1.20 gives the results of the NP test with $N = 1$, for the bad LCGs. The entries marked "—" are those for which the test was not run because n was deemed either too small for the Poisson process approximation to be good, or too large. A summary of the other results is given in Table 1.25.

For the NP tests with $t = 2$ and $N = 1$, the good LCGs fail much *sooner* than the bad ones. Why? The LCGs normally fail the NP test because the distance between the two nearest points is *too large*. The reason is that this distance has a lower bound determined by the lattice structure of the point set Ψ_2. And this lower bound turns out to be *smaller* when all the points lie on a set of lines which are *farther* apart, because the successive points on a given line are then closer to each other. The so-called bad LCGs then perform better with respect to the NP test with $N = 1$ since their lower bound is smaller.

For the M-NP tests with $M > 1$ or for the NP-S or NP-PR tests with $N > 1$, the good and bad LCGs fail at about the same sample sizes. For the bad LCGs, the minimal distance tends to be smaller, but a clustering between the values of the few shortest distances occurs anyhow, and this is detected by the AD statistic just as well as for the good LCGs. For the NP test with $N = 32$ (without the spacings or power ratio transformations), the bad LCGs fail significantly sooner than the good ones. The effect of the transformations does not show up in the value of ν^* for the bad LCGs, but the transformations are nevertheless effective (although less spectacularly than for the good LCGs) and this shows up when one looks more closely

TABLE 1.21. The 32-NP test with $t = 8$ and $N = 1$ for good LCGs.

e	$\nu = -2$	$\nu = -1$	$\nu = 0$	$\nu = 1$
24	—	—	ϵ	ϵ
25	—	—	ϵ	ϵ
26	—	5.5E-014	ϵ	ϵ
27	—	5.4E-003	ϵ	ϵ
28		ϵ	ϵ	ϵ
29		ϵ	ϵ	ϵ
30		ϵ	ϵ	ϵ
31	5.3E-003	5.5E-013	ϵ	ϵ
32		7.9E-003	ϵ	ϵ
33	1.8E-006	1.3E-012	ϵ	ϵ
34		2.7E-007	ϵ	ϵ
35		ϵ	ϵ	ϵ
36		6.1E-011	ϵ	ϵ

at the values of the AD statistics.

We also tried close-pairs tests in more than 2 dimensions. As an illustration, Tables 1.21 and 1.22 give the results for the good and bad LCGs, for the 32-NP test in 8 dimensions. These two classes of generators perform about the same with respect to this test. In fact, in 8 dimensions, the bad LCGs are not so bad as in 2 dimensions (they were actually constructed to be bad in 2 dimensions) and this shows up in the results. This remains true in general for the other close-pairs tests and for other values of $t > 2$.

6.5 Close-Pairs Tests for Nonlinear Generators

The behavior of the inversive and cubic generators with the close-pairs tests is similar to the behavior of these generators with the serial tests. The sample size n required for rejection increases as $O(\rho)$, not as $O(\sqrt{\rho})$. As illustrations, Table 1.23 and Table 1.24 give the p-values for the inversive generators, for $t = 2$, $\varphi(e) = 2^e \approx \rho$, for the NP and 32-NP tests, respectively. In both cases, the test is single-level ($N = 1$). The 32-NP test starts detecting something when the sample size n reaches approximately $1/64$ or $1/32$ of the period length, whereas the NP test needs n approximately equal to the period length. The behavior of the compound cubic generators is similar. The results are summarized in the next subsection.

6.6 A Summary of the Close-Pairs Tests Results

Table 1.25 summarizes the values of $\bar{\nu}$ and ν^* for the close-pairs tests in two dimensions.

In two dimensions, for $N = 1$, the NP test suggested by [Rip87] needs $n \approx 8\sqrt{\rho}$ for the good LCGs and $n \approx 128\sqrt{\rho}$ for the bad LCGs, whereas

TABLE 1.22. The 32-NP test with $t = 8$ and $N = 1$ for bad LCGs.

e	$\nu = -2$	$\nu = -1$	$\nu = 0$	$\nu = 1$
24	—	—	ϵ	ϵ
25	—	—	ϵ	ϵ
26	—	ϵ	ϵ	ϵ
27	—	5.9E-005	ϵ	ϵ
28		3.7E-005	ϵ	ϵ
29	1.1E-005	2.4E-008	ϵ	ϵ
30		ϵ	ϵ	ϵ
31		5.3E-006	ϵ	ϵ
32		6.6E-011	ϵ	ϵ
33		4.9E-012	ϵ	ϵ
34		ϵ	ϵ	ϵ
35	1.2E-005	ϵ	ϵ	ϵ
36		1.4E-011	ϵ	ϵ

TABLE 1.23. The p-values for the NP test with $t = 2$, for inversive generators.

e	$\nu = -3$	$\nu = -2$	$\nu = -1$	$\nu = 0$	$\nu = 1$
10				$-\epsilon$	$-\epsilon$
11				$-\epsilon$	$-\epsilon$
12				$-\epsilon$	$-\epsilon$
13				$-\epsilon$	$-\epsilon$
14				$-\epsilon$	$-\epsilon$
15				$-\epsilon$	$-\epsilon$
16				$-\epsilon$	$-\epsilon$
17				$-\epsilon$	$-\epsilon$
18				$-\epsilon$	$-\epsilon$
19				$-\epsilon$	$-\epsilon$
20				$-\epsilon$	$-\epsilon$
21				$-\epsilon$	$-\epsilon$
22				$-\epsilon$	—

TABLE 1.24. The p-values for the 32-NP test with $t = 2$, for inversive generators.

e	$\nu = -8$	$\nu = -7$	$\nu = -6$	$\nu = -5$	$\nu = -4$
17	—	—	—	ϵ	1.5E-009
18	—	—		5.7E-007	7.7E-015
19	—	2.8E-004	1.3E-004	ϵ	ϵ
20			1.7E-007	1.0E-009	8.2E-012
21					ϵ
22		4.8E-005		2.1E-007	ϵ
23			7.5E-006	2.1E-014	ϵ
24				2.5E-013	ϵ
25		1.4E-003		ϵ	ϵ
26			7.8E-006	1.4E-013	ϵ
27				7.9E-008	—

TABLE 1.25. Values of $\bar{\nu}$ and ν^* for the close-pairs tests for $p = \infty$ and $t = 2$.

Generators	Test	N	$\bar{\nu}$	ν^*	$\nu^* + \log_2 N$
Good LCGs	NP	1	2	3	3
"	32-NP	1	-1	0	0
"	NP	16	1	2	6
"	NP-S	16	-2	1	5
"	NP-PR	16	-2	-1	3
"	NP	32	0	1	6
"	NP-S	32	-3	-2	3
"	NP-PR	32	-3	-2	3
"	16-NP	16	-2	-1	3
"	32-NP	32	-2	-1	4
Bad LCGs	NP	1	6	7	7
"	32-NP	1	-1	0	0
"	NP	32	-3	-2	3
"	NP-S	32	-4	-2	3
"	NP-PR	32	-4	-2	3
Inversive	NP	1	0	0	0
"	32-NP	1	-5	-4	-4
"	NP-S	32	-8	-7	-2
"	NP-PR	32	-8	-7	-2
Cubic	NP	1	0	0	0
"	32-NP	1	-5	-3	-3
"	NP-S	32	-7	-6	-1
"	NP-PR	32	-7	-6	-1

TABLE 1.26. Values of $\bar{\nu}$ and ν^* for some close-pairs tests for $p = \infty$ and $t = 8$.

Generators	Test	N	$\bar{\nu}$	ν^*	$\nu^* + \log_2 N$
Good LCGs	NP	1	1	2	2
"	32-NP	1	-1	0	0
Bad LCGs	NP	1	2	3	3
"	32-NP	1	-1	0	0
Inversive	NP	1	0	0	0
"	32-NP	1	-1	0	0
Cubic	NP	1	0	0	0
"	32-NP	1	0	0	0

the 32-NP test needs $n \approx \sqrt{\rho}$ in both cases. With $N = 32$, the NP-S and NP-PR tests need a sample size approximately 8 times smaller than the NP test for the good LCGs. The spacings and power ratio transformations really help. The inversive and cubic generators start failing at sample sizes equal to some value around $\rho/32$ or $\rho/16$ in 2 dimensions.

In 8 dimensions, the good and bad LCGs perform almost equally well. For the inversive and cubic generators, the numbers in the table do not tell the whole story, because there is a lot of variation between the individual generators (i.e., the different values of e) in the results of the 32-NP test. The table gives $\nu^* = 0$ but some p-values around 10^{-15} or less have been observed at $\nu = -3$ or -2, for some values of e.

7   How do Real-Life Generators Fare in These Tests?

The results of the systematic experiments of the previous sections can be used to predict the sample size at which a given generator of the type considered will start to fail a given test. For example, any full-period LCG with modulus 2^{32} or less should fail unequivocally a 32-NP test in two dimensions with sample size $n = 2^{16}$ or more. And any LCG with period length 2^{48} should fail the 32-NP test with $n = 2^{24}$. These expectations are confirmed in [LCS97], where some widely used LCGs have been submitted to several close-pairs tests. Similar experiments with the serial tests are reported in [LSW98, LW97]. We can conclude from this that if an LCG is to be implemented in a general-purpose statistical or simulation package, its period length should then be *at least* larger than the square of the maximal number of output values that is expected to be used in any simulation or statistical experiment.

For an equivalent period length, explicit inversive and cubic generators fare better in the tests than the LCGs, but the inversive ones are much slower, whereas the cubic ones are not yet well-understood from the theoretical viewpoint and few parameters are available for full-period instances. In our opinion, the importance of these generators lies more in the fact that there are no lattice structures than in the irregularity of the full-period point sets with respect to discrepancy.

Other generators described in [L'E96a, L'E97a, MN98] have so far done well in all the statistical tests that have been applied to them. These generators are also reasonably fast and can be recommended for general use.

8 Acknowledgments

This work has been supported by NSERC-Canada grants # OGP0110050 and # SMF0169893, and FCAR-Québec grant # 93-ER-1654 to the first author and by FWF-Austria grants P9285, P11143, and P12654 to the second author. This paper was written while the first author was visiting the pLab group headed by the second author at Salzburg University. Richard Simard helped in the implementations of the tests.

9 REFERENCES

[AG85]   L. Afflerbach and H. Grothe. Calculation of Minkowski-reduced lattice bases. *Computing*, 35:269–276, 1985.

[AGG89]   R. Arratia, L. Goldstein, and L. Gordon. Two moments suffice for Poisson approximation: The Chen-Stein method. *The Annals of Probability*, 17:9–25, 1989.

[Alt88]   N. S. Altman. Bit-wise behavior of random number generators. *SIAM Journal on Scientific and Statistical Computing*, 9(5):941–949, 1988.

[CLT93]   R. Couture, P. L'Ecuyer, and S. Tezuka. On the distribution of k-dimensional vectors for simple and combined Tausworthe sequences. *Mathematics of Computation*, 60(202):749–761, S11–S16, 1993.

[Com91]   A. Compagner. The hierarchy of correlations in random binary sequences. *Journal of Statistical Physics*, 63:883–896, 1991.

[Com95]   A. Compagner. Operational conditions for random number generation. *Physical Review E*, 52(5-B):5634–5645, 1995.

[CS88]   J. H. Conway and N. J. A. Sloane. *Sphere Packings, Lattices and Groups*. Grundlehren der Mathematischen Wissenschaften 290. Springer-Verlag, New York, 1988.

[DR81]   E. J. Dudewicz and T. G. Ralley. *The Handbook of Random Number Generation and Testing with TESTRAND Computer Code*. American Sciences Press, Columbus, Ohio, 1981.

[Dur73]   J. Durbin. *Distribution Theory for Tests Based on the Sample Distribution Function*. SIAM CBMS-NSF Regional Conference Series in Applied Mathematics. SIAM, Philadelphia, 1973.

[DMST95] E. J. Dudewicz, E. C. van der Meulen, M. G. SriRam, and N. K. W. Teoh. Entropy-based random number evaluation. *American Journal of Mathematical and Management Sciences*, 15:115–153, 1995.

[EH97]   J. Eichenauer-Herrmann and E. Herrmann. Compound cubic congruential pseudorandom numbers. *Computing*, 59:85–90, 1997.

[EHW97]   J. Eichenauer-Herrmann, E. Herrmann, and S. Wegenkittl. A survey of quadratic and inversive congruential pseudorandom numbers. In P. Hellekalek, G. Larcher, H. Niederreiter, and P. Zinterhof, editors, *Monte Carlo and Quasi-Monte Carlo Methods in Scientific Computing*, volume 127 of *Lecture Notes in Statistics*, pages 66–97, New York, 1997. Springer-Verlag.

[Eic92]   J. Eichenauer-Herrmann. Inversive congruential pseudorandom numbers: A tutorial. *International Statistical Reviews*, 60:167–176, 1992.

[Fis90]   G. S. Fishman. Multiplicative congruential random number generators with modulus 2^β: An exhaustive analysis for $\beta = 32$ and a partial analysis for $\beta = 48$. *Mathematics of Computation*, 54(189):331–344, Jan 1990.

[Fis96]   G. S. Fishman. *Monte Carlo: Concepts, Algorithms, and Applications*. Springer Series in Operations Research. Springer-Verlag, New York, 1996.

[FM86]   G. S. Fishman and L. S. Moore III. An exhaustive analysis of multiplicative congruential random number generators with modulus $2^{31} - 1$. *SIAM Journal on Scientific and Statistical Computing*, 7(1):24–45, 1986.

[Goo53]   I. J. Good. The serial test for sampling numbers and other tests for randomness. *Proceedings of the Cambridge Philos. Society*, 49:276–284, 1953.

[Hel95]   P. Hellekalek. Inversive pseudorandom number generators: Concepts, results, and links. In C. Alexopoulos, K. Kang, W. R.

Lilegdon, and D. Goldsman, editors, *Proceedings of the 1995 Winter Simulation Conference*, pages 255–262. IEEE Press, 1995.

[Hel97]   P. Hellekalek. On correlation analysis of pseudorandom numbers. In H. Niederreiter, P. Hellekalek, G. Larcher, and P. Zinterhof, editors, *Monte Carlo and Quasi-Monte Carlo Methods 1996*, volume 127 of *Lecture Notes in Statistics*, pages 251–265. Springer-Verlag, New York, 1997.

[HL97]   P. Hellekalek and H. Leeb. Dyadic diaphony. *Acta Arithmetica*, 80:187–196, 1997.

[HN98]   P. Hellekalek and H. Niederreiter. The weighted spectral test: Diaphony. *ACM Transactions on Modeling and Computer Simulation*, 8:43–60, 1998.

[Hol72]   L. Holst. Asymptotic normality and efficiency for certain goodness-of-fit tests. *Biometrika*, 59(1):137–145, 1972.

[JHK96]   F. James, J. Hoogland, and R. Kleiss. Multidimensional sampling for simulation and integration: Measures, discrepancies, and quasi-random numbers. Submitted to Computer Physics Communications, 1996.

[Knu81]   D. E. Knuth. *The Art of Computer Programming, Volume 2: Seminumerical Algorithms*. Addison-Wesley, Reading, Mass., second edition, 1981.

[LBC93]   P. L'Ecuyer, F. Blouin, and R. Couture. A search for good multiple recursive random number generators. *ACM Transactions on Modeling and Computer Simulation*, 3(2):87–98, 1993.

[LCC96]   P. L'Ecuyer, A. Compagner, and J.-F. Cordeau. Entropy tests for random number generators. Submitted, 1996.

[LCS97]   P. L'Ecuyer, J.-F. Cordeau, and R. Simard. Close-point spatial tests and their application to random number generators. Submitted, 1997.

[L'E92]   P. L'Ecuyer. Testing random number generators. In *Proceedings of the 1992 Winter Simulation Conference*, pages 305–313. IEEE Press, Dec 1992.

[L'E94]   P. L'Ecuyer. Uniform random number generation. *Annals of Operations Research*, 53:77–120, 1994.

[L'E96a]   P. L'Ecuyer. Combined multiple recursive random number generators. *Operations Research*, 44(5):816–822, 1996.

[L'E96b]   P. L'Ecuyer. Maximally equidistributed combined Tausworthe generators. *Mathematics of Computation*, 65(213):203–213, 1996.

[L'E97a]   P. L'Ecuyer.   Good parameters and implementations for combined multiple recursive random number generators. Manuscript, 1997.

[L'E97b]   P. L'Ecuyer. Random number generators and empirical tests. In P. Hellekalek, G. Larcher, H. Niederreiter, and P. Zinterhof, editors, *Monte Carlo and Quasi-Monte Carlo Methods in Scientific Computing*, volume 127 of *Lecture Notes in Statistics*, pages 124–138. Springer-Verlag, New York, 1997.

[L'E97c]   P. L'Ecuyer. Tests based on sum-functions of spacings for uniform random numbers. *Journal of Statistical Computation and Simulation*, 59:251–269, 1997.

[L'E98a]   P. L'Ecuyer. Random number generation. In Jerry Banks, editor, *Handbook on Simulation*. Wiley, 1998. To appear. Also GERAD technical report number G-96-38.

[L'E98b]   P. L'Ecuyer. A table of linear congruential generators of different sizes and good lattice structure. *Mathematics of Computation*, 1998. To appear.

[L'E98c]   P. L'Ecuyer. Tables of maximally equidistributed combined LFSR generators. *Mathematics of Computation*, 1998. To appear.

[L'E00]   P. L'Ecuyer. TestU01: Un logiciel pour appliquer des tests statistiques à des générateurs de valeurs aléatoires. In preparation, Circa 2000.

[Lee97]   H. Leeb. *Stochastic Properties of Diaphony*. PhD thesis, Department of Mathematics, University of Salzburg, Austria, November 1997.

[LH98]   H. Leeb and P. Hellekalek. Strong and weak laws for the spectral test and related quantities. In preparation, 1998.

[LSW98]   P. L'Ecuyer, R. Simard, and S. Wegenkittl. Sparse serial tests of randomness. In preparation, 1998.

[LW97]   H. Leeb and S. Wegenkittl. Inversive and linear congruential pseudorandom number generators in empirical tests. *ACM Transactions on Modeling and Computer Simulation*, 7(2):272–286, 1997.

[Mar85]   G. Marsaglia. A current view of random number generators. In *Computer Science and Statistics, Sixteenth Symposium on the Interface*, pages 3–10, North-Holland, Amsterdam, 1985. Elsevier Science Publishers.

[Mar96]   G. Marsaglia. DIEHARD: a battery of tests of randomness. See http://stat.fsu.edu/~geo/diehard.html, 1996.

[MK94]   M. Matsumoto and Y. Kurita.   Twisted GFSR generators II. *ACM Transactions on Modeling and Computer Simulation*, 4(3):254–266, 1994.

[MK96]   M. Matsumoto and Y. Kurita. Strong deviations from randomness in m-sequences based on trinomials. *ACM Transactions on Modeling and Computer Simulation*, 6(2):99–106, 1996.

[MN98]   M. Matsumoto and T. Nishimura. Mersenne twister: A 623-dimensionally equidistributed uniform pseudorandom number generator. *ACM Transactions on Modeling and Computer Simulation*, 8(1):3–30, 1998.

[Mor75]   C. Morris. Central limit theorems for multinomial sums. *The Annals of Statistics*, 3:165–188, 1975.

[MZ93]   G. Marsaglia and A. Zaman. Monkey tests for random number generators. *Computers Math. Applic.*, 26(9):1–10, 1993.

[Nie92]   H. Niederreiter. *Random Number Generation and Quasi-Monte Carlo Methods*, volume 63 of *SIAM CBMS-NSF Regional Conference Series in Applied Mathematics*. SIAM, Philadelphia, 1992.

[PK97]   S. Pincus and R. E. Kalman. Not all (possibly) "Random" sequences are created equal. *Proceedings of the National Academy of Sciences of the USA*, 94:3513–3518, 1997.

[RC88]   T. R. C. Read and N. A. C. Cressie. *Goodness-of-Fit Statistics for Discrete Multivariate Data*. Springer Series in Statistics. Springer-Verlag, New York, 1988.

[Rip77]   B. D. Ripley. Modelling spatial patterns. *Journal of the Royal Statistical Society, Series B*, 39:172–212, 1977.

[Rip87]   B. D. Ripley. *Stochastic Simulation*. Wiley, New York, 1987.

[RS78]   B. D. Ripley and B. W. Silverman. Quick tests for spatial interaction. *Biometrika*, 65(3):641–642, 1978.

[SB78]   B. Silverman and T. Brown. Short distances, flat triangles and Poisson limits. *Journal of Applied Probability*, 15:815–825, 1978.

[SHB97]   L. N. Shchur, J. R. Heringa, and H. W. J. Blöte. Simulation of a directed random-walk model: The effect of pseudo-random number correlations. *Physica A*, 241:579–592, 1997.

[Ste86a]   M. S. Stephens. Tests based on EDF statistics. In R. B. D'Agostino and M. S. Stephens, editors, *Goodness-of-Fit Techniques*. Marcel Dekker, New York and Basel, 1986.

[Ste86b]   M. S. Stephens. Tests for the uniform distribution. In R. B. D'Agostino and M. S. Stephens, editors, *Goodness-of-Fit Techniques*, pages 331–366. Marcel Dekker, New York and Basel, 1986.

[Str94]   O. Strauch. L^2 discrepancy. *Math. Slovaca*, 44(5):601–632, 1994.

[Tak94]   K. Takashima. Sojourn time test for maximum-length linear recurring sequences with characteristic primitive trinomials. *J. Japanese Soc. Comp. Satist.*, 7:77–87, 1994.

[Tau65]   R. C. Tausworthe. Random numbers generated by linear recurrence modulo two. *Mathematics of Computation*, 19:201–209, 1965.

[Tez95]   S. Tezuka. *Uniform Random Numbers: Theory and Practice.* Kluwer Academic Publishers, Norwell, Mass., 1995.

[TL91]   S. Tezuka and P. L'Ecuyer. Efficient and portable combined Tausworthe random number generators. *ACM Transactions on Modeling and Computer Simulation*, 1(2):99–112, 1991.

[VANK94] I. Vattulainen, T. Ala-Nissila, and K. Kankaala. Physical tests for random numbers in simulations. *Physical Review Letters*, 73(19):2513–2516, 11 1994.

[VANK95] I. Vattulainen, T. Ala-Nissila, and K. Kankaala. Physical models as tests of randomness. *Physical Review E*, 52(3):3205–3213, 1995.

[WC93]   D. Wang and A. Compagner. On the use of reducible polynomials as random number generators. *Mathematics of Computation*, 60:363–374, 1993.

[Weg95]   S. Wegenkittl. Empirical testing of pseudorandom number generators. Master's thesis, University of Salzburg, 1995.

Peter Hellekalek
Institute of Mathematics, University of Salzburg
Hellbrunnerstrasse 34, A-5020 Salzburg, Austria
e-mail: peter.hellekalek@sbg.ac.at
web: http://random.mat.sbg.ac.at/team/peter.html

Pierre L'Ecuyer
Institute of Mathematics, University of Salzburg and
Département d'informatique et de recherche opérationnelle
Université de Montréal
C.P. 6128, Succ. Centre-Ville, Montréal, H3C 3J7, Canada
web: http://www.iro.umontreal.ca/~lecuyer

Nets, (t, s)-Sequences, and Algebraic Geometry

Harald Niederreiter and Chaoping Xing

1 Introduction

The star discrepancy is a classical measure for the irregularity of distribution of finite sets and infinite sequences of points in the s-dimensional unit cube $I^s = [0, 1]^s$. Point sets and sequences with small star discrepancy in I^s are informally called *low-discrepancy point sets*, respectively *low-discrepancy sequences*, in I^s. It is also customary to speak of sets, respectively sequences, of *quasirandom points* in I^s. Such point sets and sequences play a crucial role in applications of numerical quasi-Monte Carlo methods. In fact, the efficiency of a quasi-Monte Carlo method depends to a significant extent on the quality of the quasirandom points that are employed, i.e., on how small their star discrepancy is. Therefore, it is a matter of considerable interest to devise techniques for the construction of point sets and sequences with as small a star discrepancy as possible. The reader who desires more background on discrepancy theory and quasi-Monte Carlo methods is referred to the books of Hua and Wang [9], Kuipers and Niederreiter [10], and Niederreiter [21], the survey article of Niederreiter [16], and the recent monograph of Drmota and Tichy [4].

The most powerful known methods for the construction of low-discrepancy point sets and sequences are based on the theory of (t, m, s)-nets and (t, s)-sequences, which are point sets, respectively sequences, satisfying strong uniformity properties with regard to their distribution in I^s. The quality parameter t measures these uniformity properties and should be as small as possible. Various methods for the construction of (t, m, s)-nets and (t, s)-sequences have been developed, and the tools in these constructions stem from number theory, algebra, combinatorics, coding theory, and algebraic geometry. Surveys of these constructions can be found in Niederreiter [20], [21, Chapter 4], Niederreiter and Xing [24], [27], Mullen *et al.* [15], and Clayman *et al.* [2], with the last two papers containing also tables of the quality parameter t.

In this paper we present a survey of constructions of (t, s)-sequences based on algebraic geometry (see Section 5) and several new constructions of (t, m, s)-nets (see Section 6). Methods from algebraic geometry have turned out to be very effective, and the best results were obtained by using algebraic curves over finite fields with many rational points. In Section 7 we list new examples of algebraic curves over finite fields with many rational points and include an updated table of bounds for the number of rational points of such curves. Some background material is given in Sections 2, 3, and 4.

2   Basic Concepts

We review some definitions and basic facts from discrepancy theory. For a subinterval J of I^s and for a point set P consisting of N points $x_0, x_1, \ldots,$ $x_{N-1} \in I^s$ we write $A(J; P)$ for the number of integers n with $0 \leq n \leq N-1$ for which $x_n \in J$. We put

$$R(J; P) = \frac{A(J; P)}{N} - \lambda_s(J),$$

where λ_s is the s-dimensional Lebesgue measure.

Definition 1. The *star discrepancy* $D_N^*(P)$ of the point set P is defined by

$$D_N^*(P) = \sup_J |R(J; P)|,$$

where the supremum is extended over all subintervals J of I^s with one vertex at the origin. For a sequence S of points in I^s, the *star discrepancy* $D_N^*(S)$ is meant to be the star discrepancy of the first N terms of S.

For any $s \geq 1$ and $N \geq 2$, point sets P of N points in I^s with

$$D_N^*(P) = O(N^{-1}(\log N)^{s-1})$$

can be constructed, where the implied constant in the Landau symbol depends only on s, and this order of magnitude is known to be best possible for $s = 1$ and $s = 2$. Similarly, for any $s \geq 1$ there are sequences S of points in I^s with

$$D_N^*(S) = O(N^{-1}(\log N)^s) \quad \text{for all } N \geq 2,$$

where the implied constant in the Landau symbol depends only on s, and this order of magnitude is known to be best possible for $s = 1$. The following definition is fundamental for this paper.

Definition 2. For integers $b \geq 2$ and $0 \leq t \leq m$, a (t, m, s)-*net in base b* is a point set P consisting of b^m points in I^s such that $R(J; P) = 0$ for every subinterval J of I^s of the form

$$J = \prod_{i=1}^{s} [a_i b^{-d_i}, (a_i + 1)b^{-d_i})$$

with integers $d_i \geq 0$ and $0 \leq a_i < b^{d_i}$ for $1 \leq i \leq s$ and $\lambda_s(J) = b^{t-m}$.

Explicit upper bounds on the star discrepancy of (t, m, s)-nets in base b can be found in Niederreiter [17, Section 3], [21, Chapter 4]. We mention here only the following simplified bound: for any (t, m, s)-net P in base b with $m \geq 1$ we have

$$D_N^*(P) \leq B_b(s, t) N^{-1} (\log N)^{s-1} + O(b^t N^{-1} (\log N)^{s-2}),$$

where the implied constant in the Landau symbol depends only on b and s. Here

$$B_b(s, t) = b^t \left(\frac{b-1}{2 \log b} \right)^{s-1}$$

if either $s = 2$ or $b = 2$, $s = 3, 4$; otherwise

$$B_b(s, t) = \frac{b^t}{(s-1)!} \left(\frac{\lfloor b/2 \rfloor}{\log b} \right)^{s-1}.$$

Because of the exponential dependence of $B_b(s, t)$ on the quality parameter t, even a small gain in the value of t leads to a considerable amelioration of the discrepancy bound.

If $m = t$ or $t + 1$, then a (t, m, s)-net in base b always exists, as was noted in [19]. But for $m \geq t + 2$ there are combinatorial obstructions to the general existence of (t, m, s)-nets in base b. This was first pointed out in [17, Section 5] where it was shown, for instance, that for $m \geq 2$ a $(0, m, s)$-net in base b can exist only if $s \leq b + 1$ (by the way, if b is a prime power, then "only if" can be replaced by "if and only if"). Later, these combinatorial obstructions were further exploited to yield lower bounds on the quality parameter t in a (t, m, s)-net in base b. We refer to Clayman et al. [2] for a recent account and tables of such lower bounds.

Next we introduce (t, s)-sequences in the slightly generalized form described in [23], [24] (see [21, Chapter 4] for the original narrower definition). For a base $b \geq 2$ we write $Z_b = \{0, 1, \ldots, b-1\}$ for the set of digits in base b. Given a real number $x \in [0, 1]$, let

$$x = \sum_{j=1}^{\infty} y_j b^{-j} \quad \text{with all } y_j \in Z_b \tag{1}$$

be a b-adic expansion of x, where the case $y_j = b - 1$ for all but finitely many j is allowed. For an integer $m \geq 1$ we define the truncation

$$[x]_{b,m} = \sum_{j=1}^{m} y_j b^{-j}.$$

It should be emphasized that this truncation operates on the *expansion* of x and not on x itself, since it may yield different results depending on which b-adic expansion of x is used. If $\mathbf{x} = (x^{(1)}, \ldots, x^{(s)}) \in I^s$ and the $x^{(i)}, 1 \leq i \leq s$, are given by prescribed b-adic expansions, then we define

$$[\mathbf{x}]_{b,m} = \left([x^{(1)}]_{b,m}, \ldots, [x^{(s)}]_{b,m} \right).$$

Definition 3. Let $b \geq 2$ and $t \geq 0$ be integers. A sequence $\mathbf{x}_0, \mathbf{x}_1, \ldots$ of points in I^s is a (t,s)-*sequence in base* b if for all integers $k \geq 0$ and $m > t$ the points $[\mathbf{x}_n]_{b,m}$ with $kb^m \leq n < (k+1)b^m$ form a (t,m,s)-net in base b. Here the coordinates of all points of the sequence are given by prescribed b-adic expansions of the form (1).

Detailed information on upper bounds for the star discrepancy of (t,s)-sequences is provided in Niederreiter [17, Section 4], [21, Chapter 4]. We mention here only the following simplified bound: for any (t,s)-sequence S in base b we have

$$D_N^*(S) \leq C_b(s,t) N^{-1} (\log N)^s + O\left(b^t N^{-1} (\log N)^{s-1} \right) \quad \text{for all } N \geq 2,$$

where the implied constant in the Landau symbol depends only on b and s. Here

$$C_b(s,t) = \frac{b^t}{s} \left(\frac{b-1}{2 \log b} \right)^s$$

if either $s = 2$ or $b = 2, s = 3, 4$; otherwise

$$C_b(s,t) = \frac{b^t}{s!} \cdot \frac{b-1}{2 \lfloor b/2 \rfloor} \left(\frac{\lfloor b/2 \rfloor}{\log b} \right)^s.$$

As in the case of (t,m,s)-nets, even small improvements in the value of t lead to considerably better discrepancy bounds for (t,s)-sequences.

The following useful principle shows that if we can construct a (t,s)-sequence, then we can construct infinitely many nets in dimension $s+1$. In the form referring to (t,s)-sequences in the sense of Definition 3, this principle was first stated in [23, Section 6]. An earlier form of this principle for a slightly narrower notion of (t,s)-sequence was given in [17, Lemma 5.15].

Lemma 1. *If there exists a* (t,s)-*sequence in base* b, *then for every integer* $m \geq t$ *there exists a* $(t,m,s+1)$-*net in base* b.

If we combine results on the combinatorial obstructions to the existence of (t,m,s)-nets in base b with Lemma 1, then we get necessary conditions for the existence of (t,s)-sequences in base b. For instance, a $(0,s)$-sequence in base b can exist only if $s \leq b$ (if b is a prime power, then "only if" can be replaced by "if and only if"). The following general lower bound on the quality parameter t in a (t,s)-sequence in base b was shown in [24]. We write $\log_b$ for the logarithm to the base b.

Theorem 1. *For every base $b \geq 2$ and every dimension $s \geq 1$, a necessary condition for the existence of a (t, s)-sequence in base b is*

$$t \geq \frac{s}{b} - \log_b \frac{(b-1)s + b + 1}{2}.$$

3   The Digital Method

Most of the known constructions of (t, m, s)-nets and (t, s)-sequences are based on the digital method which was introduced in [17, Section 6]. We fix a commutative ring R with identity and of finite order $b \geq 2$. If the integers $m \geq 1$ and $s \geq 1$ are given, then the construction of (t, m, s)-nets in base b by the digital method proceeds by choosing the following:

(i) bijections $\psi_r : Z_b \to R$ for $0 \leq r \leq m - 1$;

(ii) bijections $\eta_j^{(i)} : R \to Z_b$ for $1 \leq i \leq s$ and $1 \leq j \leq m$;

(iii) elements $c_{j,r}^{(i)} \in R$ for $1 \leq i \leq s$, $1 \leq j \leq m$, and $0 \leq r \leq m - 1$.

For $n = 0, 1, \ldots, b^m - 1$ let

$$n = \sum_{r=0}^{m-1} a_r(n) b^r \quad \text{with all } a_r(n) \in Z_b$$

be the digit expansion of n in base b. We put

$$x_n^{(i)} = \sum_{j=1}^{m} y_{n,j}^{(i)} b^{-j} \quad \text{for } 0 \leq n < b^m \text{ and } 1 \leq i \leq s,$$

with

$$y_{n,j}^{(i)} = \eta_j^{(i)} \left(\sum_{r=0}^{m-1} c_{j,r}^{(i)} \psi_r(a_r(n)) \right) \in Z_b \quad \text{for } 0 \leq n < b^m, 1 \leq i \leq s, 1 \leq j \leq m,$$

and define the point set

$$\mathbf{x}_n = \left(x_n^{(1)}, \ldots, x_n^{(s)} \right) \in I^s \quad \text{for } n = 0, 1, \ldots, b^m - 1. \tag{2}$$

Definition 4. If the point set in (2) is a (t, m, s)-net in base b, then it is called a *digital (t, m, s)-net constructed over R*.

In the digital method for the construction of sequences, we fix a ring R as above and a dimension $s \geq 1$ and then we choose the following:

(i) bijections $\psi_r : Z_b \to R$ for $r \geq 0$, with $\psi_r(0) = 0$ for all sufficiently large r;

(ii) bijections $\eta_j^{(i)} : R \to Z_b$ for $1 \leq i \leq s$ and $j \geq 1$;

(iii) elements $c_{j,r}^{(i)} \in R$ for $1 \leq i \leq s$, $j \geq 1$, and $r \geq 0$.

For $n = 0, 1, \ldots$ let

$$n = \sum_{r=0}^{\infty} a_r(n) b^r$$

be the digit expansion of n in base b, where $a_r(n) \in Z_b$ for $r \geq 0$ and $a_r(n) = 0$ for all sufficiently large r. We put

$$x_n^{(i)} = \sum_{j=1}^{\infty} y_{n,j}^{(i)} b^{-j} \quad \text{for } n \geq 0 \text{ and } 1 \leq i \leq s \tag{3}$$

with

$$y_{n,j}^{(i)} = \eta_j^{(i)} \left(\sum_{r=0}^{\infty} c_{j,r}^{(i)} \psi_r(a_r(n)) \right) \in Z_b \quad \text{for } n \geq 0, 1 \leq i \leq s, \text{ and } j \geq 1.$$

Note that the sum over r is always a finite sum. Now we define the sequence

$$\mathbf{x}_n = (x_n^{(1)}, \ldots, x_n^{(s)}) \in I^s \quad \text{for } n = 0, 1, \ldots. \tag{4}$$

Definition 5. If the sequence in (4) is a (t, s)-sequence in base b, then it is called a *digital (t, s)-sequence constructed over R*. Here the truncations are required to operate on the expansions in (3).

This notion of a digital (t, s)-sequence is slightly more general than the original one in [17] (compare with [23, Remark 1], [24, Remark 3]). The following analog of Lemma 1, which is of similar usefulness, was shown in [23].

Lemma 2. *If there exists a digital (t, s)-sequence constructed over R, then for every integer $m \geq t$ there exists a digital $(t, m, s + 1)$-net constructed over R.*

For practical purposes it suffices to consider the digital method in the special case where the ring R is a finite field $\mathbb{F}_q$ of prime-power order q. The reasons are that: (i) for every base $b \geq 2$ there is a ring R (commutative and with identity) of order b which is given as the direct product of finite fields; (ii) in the passage to ring direct products the behavior of the quality parameter t can be controlled (see [21, Theorem 4.29], [23, Lemma 8]). In the following we will therefore concentrate on the case $R = \mathbb{F}_q$ in the digital method. Information on the general case can be found in Larcher *et al.* [11] and Niederreiter [21, Chapter 4].

The determination of the quality parameter t for point sets and sequences constructed by the digital method with $R = \mathbb{F}_q$ is based on notions of linear algebra. We consider a rectangular array A of vectors from the vector space

$\mathbb{F}_q^m$ for some $m \geq 1$:

$$
\begin{array}{cccc}
\mathbf{a}_1^{(1)} & \mathbf{a}_2^{(1)} & \cdots & \mathbf{a}_m^{(1)} \\
\mathbf{a}_1^{(2)} & \mathbf{a}_2^{(2)} & \cdots & \mathbf{a}_m^{(2)} \\
\vdots & \vdots & & \vdots \\
\mathbf{a}_1^{(s)} & \mathbf{a}_2^{(s)} & \cdots & \mathbf{a}_m^{(s)}
\end{array}
$$

Such an $s \times m$ array is abbreviated as a two-parameter system $A = \{\mathbf{a}_j^{(i)} \in \mathbb{F}_q^m : 1 \leq i \leq s, 1 \leq j \leq m\}$.

Definition 6. Let d be an integer with $0 \leq d \leq m$. The system $A = \{\mathbf{a}_j^{(i)} \in \mathbb{F}_q^m : 1 \leq i \leq s, 1 \leq j \leq m\}$ of vectors is called a (d, m, s)-*system over* $\mathbb{F}_q$ if for any nonnegative integers $d_1, \ldots, d_s$ with $\sum_{i=1}^s d_i = d$ the vectors $\mathbf{a}_j^{(i)}, 1 \leq j \leq d_i, 1 \leq i \leq s$, are linearly independent over $\mathbb{F}_q$ (this property is assumed to be vacuously satisfied for $d = 0$). The largest value of d such that A is a (d, m, s)-system over $\mathbb{F}_q$ is denoted by $\varrho(A)$.

Now we consider the point set in (2) obtained by the digital method with $R = \mathbb{F}_q$. We use the elements $c_{j,r}^{(i)} \in \mathbb{F}_q$ in the digital method to set up the vectors

$$
\mathbf{c}_j^{(i)} = \left(c_{j,0}^{(i)}, \ldots, c_{j,m-1}^{(i)} \right) \in \mathbb{F}_q^m \quad \text{for } 1 \leq i \leq s, 1 \leq j \leq m,
$$

and this leads to the two-parameter system

$$
C = \left\{ \mathbf{c}_j^{(i)} \in \mathbb{F}_q^m : 1 \leq i \leq s, 1 \leq j \leq m \right\}. \tag{5}
$$

Then we arrive at the following result by combining Theorems 6.10 and 6.14 in [17] (note that the value of $\varrho(C)$ according to [17, Definition 6.8] is larger by 1 than that according to Definition 6 above).

Lemma 3. *The point set in (2) obtained by the digital method with $R = \mathbb{F}_q$ is a digital (t, m, s)-net constructed over $\mathbb{F}_q$ if and only if the system C in (5) is a (d, m, s)-system over $\mathbb{F}_q$ with $d = m - t$.*

Next we consider the sequence in (4) obtained by the digital method with $R = \mathbb{F}_q$. If $\mathbb{F}_q^\infty$ is the sequence space over $\mathbb{F}_q$, then we use the elements $c_{j,r}^{(i)} \in \mathbb{F}_q$ in the digital method to set up the sequences

$$
\mathbf{c}_j^{(i)} = \left(c_{j,0}^{(i)}, c_{j,1}^{(i)}, \ldots \right) \in \mathbb{F}_q^\infty \quad \text{for } 1 \leq i \leq s \text{ and } j \geq 1.
$$

These are collected into the two-parameter system

$$
C^{(\infty)} = \left\{ \mathbf{c}_j^{(i)} \in \mathbb{F}_q^\infty : 1 \leq i \leq s \text{ and } j \geq 1 \right\}. \tag{6}
$$

For $m \geq 1$ we define the projection

$$
\pi_m : (c_0, c_1, \ldots) \in \mathbb{F}_q^\infty \mapsto (c_0, \ldots, c_{m-1}) \in \mathbb{F}_q^m,
$$

and we put

$$C^{(m)} = \left\{ \pi_m(\mathbf{c}_j^{(i)}) \in \mathbb{F}_q^m : 1 \leq i \leq s, 1 \leq j \leq m \right\}.$$

Finally, we set

$$\tau(C^{(\infty)}) = \sup_{m \geq 1} \left(m - \varrho(C^{(m)}) \right).$$

Then we can state the following result from [23, Lemma 2].

Lemma 4. *If the system $C^{(\infty)}$ in (6) satisfies $\tau(C^{(\infty)}) < \infty$, then the sequence in (4) obtained by the digital method with $R = \mathbb{F}_q$ is a digital (t, s)-sequence constructed over $\mathbb{F}_q$ with $t = \tau(C^{(\infty)})$.*

4 Background on Algebraic Curves over Finite Fields

Various powerful constructions of digital nets and sequences rely on algebraic curves over finite fields with many rational points, or equivalently on global function fields with many rational places. By an algebraic curve C over $\mathbb{F}_q$ we always mean a smooth, projective, absolutely irreducible algebraic curve defined over $\mathbb{F}_q$. For the number $N(C)$ of $\mathbb{F}_q$-rational points of C we have the Weil-Serre bound

$$N(C) \leq q + 1 + g(C) \left\lfloor 2q^{1/2} \right\rfloor,$$

where $g(C)$ is the genus of C. In particular, the following definition makes sense.

Definition 7. For any prime power q and any integer $g \geq 0$ let

$$N_q(g) = \max N(C),$$

where the maximum is over all algebraic curves C over $\mathbb{F}_q$ with $g(C) = g$.

In most cases we have only bounds for $N_q(g)$, such as the upper bounds obtained from the Weil-Serre bound and its refinements. In an informal way, we say that an algebraic curve C over $\mathbb{F}_q$ of genus g has many rational points if $N(C)$ is reasonably close to $N_q(g)$ or to a known upper bound for $N_q(g)$. We refer to Garcia and Stichtenoth [7] and Niederreiter and Xing [28], [30] for recent surveys of algebraic curves over finite fields with many rational points. Tables of lower and upper bounds for $N_q(g)$ are given in Section 7.

A unifying theme of several methods of obtaining digital (t, s)-sequences constructed over $\mathbb{F}_q$ (see Section 5) is to take an algebraic curve C over $\mathbb{F}_q$ with $N(C) \geq 1$ and to determine the elements $c_{j,r}^{(i)} \in \mathbb{F}_q$ in the digital method from the coefficients in the expansions of certain points of C in local coordinates at a fixed $\mathbb{F}_q$-rational point of C. For the purpose of determining the $c_{j,r}^{(i)}$, it is often more convenient to work with the function field $K/\mathbb{F}_q$ of C. Note that $K/\mathbb{F}_q$ is a global function field, i.e., an algebraic function field with a finite constant field. The notation $K/\mathbb{F}_q$ indicates that $\mathbb{F}_q$ is the full constant field of K, i.e., that $\mathbb{F}_q$ is algebraically closed in K. The genus $g(K/\mathbb{F}_q)$ of $K/\mathbb{F}_q$ is equal to $g(C)$ and the number $N(K/\mathbb{F}_q)$ of rational places of $K/\mathbb{F}_q$, i.e., of places of $K/\mathbb{F}_q$ of degree 1, is equal to $N(C)$. Therefore, the quantity $N_q(g)$ introduced in Definition 7 is also given by

$$N_q(g) = \max N(K/\mathbb{F}_q),$$

where the maximum is over all global function fields $K/\mathbb{F}_q$ with $g(K/\mathbb{F}_q) = g$. By analogy, we say that the global function field $K/\mathbb{F}_q$ of genus g has many rational places if $N(K/\mathbb{F}_q)$ is reasonably close to $N_q(g)$ or to a known upper bound for $N_q(g)$.

For a global function field $K/\mathbb{F}_q$, we write ν_P for the normalized discrete valuation corresponding to the place P of $K/\mathbb{F}_q$. If P is a rational place of $K/\mathbb{F}_q$ and z is a local uniformizing parameter at P, then every $k \in K$ has an expansion

$$k = \sum_{r=v}^{\infty} a_r z^r,$$

where $v \leq \nu_P(k)$ and all $a_r \in \mathbb{F}_q$. For a nonzero $k \in K$ we write (k) for the principal divisor of k. Then for an arbitrary divisor D of $K/\mathbb{F}_q$, the $\mathbb{F}_q$-vector space

$$\mathcal{L}(D) = \{k \in K \backslash \{0\} : (k) + D \geq 0\} \cup \{0\}$$

has a finite dimension which we denote by $l(D)$. A good reference for the theory of global function fields is the book of Stichtenoth [39].

5   Constructions of (t, s)-Sequences

The classical example of a (t, s)-sequence is the one-dimensional van der Corput sequence in base b which is a $(0, 1)$-sequence in base b (see [10, p. 127], [21, pp. 47–48]), and in fact a digital $(0, 1)$-sequence constructed over the ring $R = \mathbf{Z}/b\mathbf{Z}$ (see [21, Remark 4.38]). The first construction of (t, s)-sequences for arbitrary dimensions s was given in base 2. Sobol' [37] showed that there exist digital $(U(s), s)$-sequences constructed over $\mathbb{F}_2$ with

$$U(s) = \sum_{i=2}^{s} (\deg(f_i) - 1), \tag{7}$$

where $f_2, f_3, \ldots$ is the list of all primitive polynomials over $\mathbb{F}_2$ arranged according to nondecreasing degrees. Thus we have $U(s) = 0$ for $s = 1, 2$

and $U(s) = O(s \log s)$ in general. A brief description of the work of Sobol' can also be found in [16, Section 3]. Srinivasan [38] obtained a recursive construction of a $(0, 2)$-sequence in base 2.

The next step was taken by Faure [6] who showed that for every prime p there exists a digital $(0, s)$-sequence constructed over $\mathbb{F}_p$ as long as $s \leq p$. This was generalized by Niederreiter [17] who proved that for every prime power q there exists a digital $(0, s)$-sequence constructed over $\mathbb{F}_q$ provided that $s \leq q$. Note that because of the combinatorial obstructions mentioned in Section 2, a $(0, s)$-sequence in base q cannot exist for $s > q$.

The first construction of (t, s)-sequences for any dimension s and any base b was given by Niederreiter [18]. For a prime-power base q the construction operates in the rational function field over $\mathbb{F}_q$ and uses the Laurent series expansion of rational functions. The method yields digital (t, s)-sequences constructed over $\mathbb{F}_q$ with

$$t = \sum_{i=1}^{s} (\deg(g_i) - 1),$$

where $g_1, \ldots, g_s$ are pairwise coprime nonconstant polynomials over $\mathbb{F}_q$. If $q = 2$ and we choose $g_1(x) = x$ and $g_i = f_i$ for $2 \leq i \leq s$, where the f_i are as in (7), then we obtain the construction of Sobol' [37]. If q is an arbitrary prime power, if $s \leq q$, and if $g_1, \ldots, g_s$ are distinct monic linear polynomials over $\mathbb{F}_q$, then we obtain the construction of Niederreiter [17] (and of Faure [6] in the case where q is prime). The construction is optimized by letting $p_1, p_2, \ldots$ be the list of all monic irreducible polynomials over $\mathbb{F}_q$ arranged according to nondecreasing degrees and choosing $g_i = p_i$ for $1 \leq i \leq s$. This yields a digital $(T_q(s), s)$-sequence constructed over $\mathbb{F}_q$ with

$$T_q(s) = \sum_{i=1}^{s} (\deg(p_i) - 1). \tag{8}$$

If $q = 2$ and $U(s)$ is as in (7), then we have $T_2(s) = U(s)$ for $1 \leq s \leq 7$ and $T_2(s) < U(s)$ for $s \geq 8$. A table comparing the values of $T_2(s)$ and $U(s)$ for $1 \leq s \leq 20$ can be found in [27, Table 1]. For arbitrary prime powers q we have $T_q(s) = O(s \log s)$. By using the digital method with ring direct products of finite fields, the construction can be extended to any base $b \geq 2$. An exposition of the work in Niederreiter [18] is available in [21, Chapter 4]. A related technique of obtaining digital $(T_q(s), s)$-sequences constructed over $\mathbb{F}_q$ was developed by Tezuka [40] and Tezuka and Tokuyama [41].

More recently, a simple way of getting (t, s)-sequences in a prime-power base q with $t = O(s \log s)$ was pointed out by Niederreiter and Xing [23, Remark 8]. For a given dimension $s \geq 2$ put $h = \lceil \log_q s \rceil$, so that $s \leq q^h$. Then by an easy explicit construction in [17, Theorem 6.18] we obtain a $(0, s)$-sequence in base q^h. According to a general principle in [23, Proposition 4], this sequence is also a (t, s)-sequence in base q with

$$t = \left(\lceil \log_q s \rceil - 1 \right) s.$$

The resulting bound $t < s \log_q s$ is better than the bound that is known for the quantity $T_q(s)$ in (8).

An important recent development is the use of algebraic curves over finite fields (or, equivalently, of global function fields) for the construction of (t, s)-sequences. At present, four different construction principles based on algebraic curves are available, and they all rely on the digital method with a finite field $\mathbb{F}_q$. The *first construction* (in the terminology of [24], [27]) is due to Niederreiter and Xing [22] and can be viewed as a direct extension of the method of Niederreiter [18], where the latter method corresponds then to the special case in which the algebraic curve is the projective line over $\mathbb{F}_q$ or, equivalently, the global function field is the rational function field over $\mathbb{F}_q$. Although in principle the first construction works for a wide range of algebraic curves, values of the quality parameter t in digital (t, s)-sequences constructed over $\mathbb{F}_q$ that are smaller than $T_q(s)$ in (8) have been obtained only for certain elliptic curves over $\mathbb{F}_2$ and $\mathbb{F}_3$. Because of this limited interest, the first construction will not be described here, and we refer instead to the original paper [22] and to the brief accounts in [24, Section 5], [27, Section 4].

Dramatic improvements on all previous constructions have been achieved by means of algebraic curves over $\mathbb{F}_q$ with many rational points. The following three constructions use curves of this type, but in each case it will be more convenient to describe the method in terms of the equivalent language of global function fields over $\mathbb{F}_q$. The *second construction* (again in the terminology of [24], [27]) was introduced by Niederreiter and Xing [23]. Let the dimension $s \geq 1$ be given and choose a global function field $K/\mathbb{F}_q$ with $N(K/\mathbb{F}_q) \geq s + 1$. Let $P_\infty, P_1, P_2, \ldots, P_s$ be $s + 1$ distinct rational places of $K/\mathbb{F}_q$. Furthermore, choose a positive divisor D of $K/\mathbb{F}_q$ with $\deg(D) = g(K/\mathbb{F}_q)$ and $l(D) = 1$. It is known that such a divisor D exists if either $g(K/\mathbb{F}_q) = 0$, or $q = 2$ and $N(K/\mathbb{F}_q) \geq 4$, or $q \geq 3$ and $N(K/\mathbb{F}_q) \geq 2$. By the Riemann-Roch theorem it is easily seen that for $1 \leq i \leq s$ and $j \geq 1$ there exist elements $k_j^{(i)} \in \mathcal{L}(D + jP_i)$ with

$$\nu_{P_i}(k_j^{(i)}) = -\nu_{P_i}(D) - j.$$

Since $\nu_{P_\infty}(k_j^{(i)}) \geq -\nu_{P_\infty}(D) =: -v$, we have expansions at P_∞ of the form

$$k_j^{(i)} = z^{-v} \sum_{r=0}^{\infty} b_{j,r}^{(i)} z^r \quad \text{for } 1 \leq i \leq s \text{ and } j \geq 1,$$

where z is a local uniformizing parameter at P_∞ and all $b_{j,r}^{(i)} \in \mathbb{F}_q$. For $1 \leq i \leq s$ and $j \geq 1$ we now define

$$c_{j,r}^{(i)} = \begin{cases} b_{j,r}^{(i)} & \text{for} \quad 0 \leq r \leq v - 1, \\ \\ b_{j,r+1}^{(i)} & \text{for} \quad r \geq v. \end{cases}$$

These elements $c_{j,r}^{(i)} \in \mathbb{F}_q$ are then used in the digital method with $R = \mathbb{F}_q$ (see Section 3). The following result was shown in [23] on the basis of Lemma 4.

Theorem 2. *Let $K/\mathbb{F}_q$ be a global function field with $N(K/\mathbb{F}_q) \geq s + 1$ and suppose that there exists a positive divisor D of $K/\mathbb{F}_q$ with $\deg(D) = g(K/\mathbb{F}_q)$ and $l(D) = 1$. Then the second construction yields a digital (t, s)-sequence constructed over $\mathbb{F}_q$ with $t = g(K/\mathbb{F}_q)$.*

It is known that for any fixed q the function $N_q(g)$ of g in Definition 7 attains arbitrarily large values. Therefore, the following definition is meaningful. The subsequent corollary is an immediate consequence of Theorem 2.

Definition 8. For any prime power q and any dimension $s \geq 1$ let

$$V_q(s) = \min \{g \geq 0 : N_q(g) \geq s + 1\}.$$

Corollary 1. *For every prime power q and every dimension $s \geq 1$ there exists a digital $(V_q(s), s)$-sequence constructed over $\mathbb{F}_q$.*

The *third construction* is due to Xing and Niederreiter [45] and its advantage is that it is more flexible than the second construction in that it permits also the use of places (or closed points in the geometric language) of larger degree. If only rational places are employed in the third construction, then it is of the same quality as the second construction. Another aspect is that, in cases of practical interest, it is trivial to find the auxiliary divisor D that is needed in the third construction, whereas the divisor D for the second construction is sometimes harder to find.

For a given dimension $s \geq 1$ and a prime power q, let $K/\mathbb{F}_q$ be a global function field containing at least one rational place P_∞, and let D be a positive divisor of $K/\mathbb{F}_q$ with $\deg(D) = 2g(K/\mathbb{F}_q)$ and $P_\infty \notin \operatorname{supp}(D)$. We choose s distinct places $P_1, \ldots, P_s$ of $K/\mathbb{F}_q$ with $P_i \neq P_\infty$ for $1 \leq i \leq s$, and we put $e_i = \deg(P_i)$ for $1 \leq i \leq s$. With the abbreviation $g = g(K/\mathbb{F}_q)$ we have $l(D) = g + 1$ by the Riemann-Roch theorem. We choose a basis of $\mathcal{L}(D)$ in the following way. Note that $l(D - P_\infty) = g$ by the Riemann-Roch theorem and $l(D - (2g + 1)P_\infty) = 0$, hence there exist integers $0 = n_0 < n_1 < \ldots < n_g \leq 2g$ such that

$$l(D - n_f P_\infty) = l(D - (n_f + 1)P_\infty) + 1 \quad \text{for } 0 \leq f \leq g.$$

Choose $w_f \in \mathcal{L}(D - n_f P_\infty) \backslash \mathcal{L}(D - (n_f + 1)P_\infty)$, then it is easily seen that $\{w_0, w_1, \ldots, w_g\}$ is a basis of $\mathcal{L}(D)$.

For each $1 \leq i \leq s$ we consider the chain

$$\mathcal{L}(D) \subset \mathcal{L}(D + P_i) \subset \mathcal{L}(D + 2P_i) \subset \ldots$$

of $\mathbb{F}_q$-vector spaces. By starting from the basis $\{w_0, w_1, \ldots, w_g\}$ of $\mathcal{L}(D)$ and successively adding basis vectors at each step of the chain, we obtain for each $n \geq 1$ a basis

$$\left\{ w_0, w_1, \ldots, w_g, k_1^{(i)}, k_2^{(i)}, \ldots, k_{ne_i}^{(i)} \right\}$$

of $\mathcal{L}(D + nP_i)$. Now let z be a local uniformizing parameter at P_∞. For $r = 0, 1, \ldots$ we put

$$z_r = \begin{cases} z^r & \text{if } r \notin \{n_0, n_1, \ldots, n_g\}, \\ w_f & \text{if } r = n_f \text{ for some } f \in \{0, 1, \ldots, g\}. \end{cases}$$

Note that then $\nu_{P_\infty}(z_r) = r$ for all $r \geq 0$. For $1 \leq i \leq s$ and $j \geq 1$ we have $k_j^{(i)} \in \mathcal{L}(D + mP_i)$ for some $m \geq 1$ and also $P_\infty \notin \operatorname{supp}(D + mP_i)$, hence $\nu_{P_\infty}(k_j^{(i)}) \geq 0$. Thus we have expansions at P_∞ of the form

$$k_j^{(i)} = \sum_{r=0}^{\infty} a_{j,r}^{(i)} z_r \quad \text{for } 1 \leq i \leq s \text{ and } j \geq 1,$$

where all coefficients $a_{j,r}^{(i)} \in \mathbb{F}_q$. For $1 \leq i \leq s$ and $j \geq 1$ we now define the sequences

$$\mathbf{c}_j^{(i)} = \left(\widehat{a_{j,n_0}^{(i)}}, a_{j,1}^{(i)}, \ldots, \widehat{a_{j,n_1}^{(i)}}, a_{j,n_1+1}^{(i)}, \ldots, \widehat{a_{j,n_g}^{(i)}}, a_{j,n_g+1}^{(i)}, \ldots \right) \in \mathbb{F}_q^\infty,$$

where the hat indicates that the corresponding term is deleted. If we write

$$\mathbf{c}_j^{(i)} = \left(c_{j,0}^{(i)}, c_{j,1}^{(i)}, \ldots \right),$$

then the elements $c_{j,r}^{(i)} \in \mathbb{F}_q$ are used in the digital method with $R = \mathbb{F}_q$ (see Section 3). On the basis of Lemma 4, the following result was established in [45].

Theorem 3. *The third construction yields a digital (t,s)-sequence constructed over $\mathbb{F}_q$ with*

$$t = g(K/\mathbb{F}_q) + \sum_{i=1}^{s} (\deg(P_i) - 1).$$

The *fourth construction* is due to Niederreiter and Xing [24] and quite a bit simpler, but not as powerful as the second and the third construction. As in the second construction, algebraic curves over $\mathbb{F}_q$ with many rational points are used. In analogy with Corollary 1 the fourth construction yields the result that for every q and s we get a $(W_q(s), s)$-sequence constructed over $\mathbb{F}_q$, where $W_q(s) = 0$ if $1 \leq s \leq q$ and $W_q(s) = 2V_q(s) - 1$ if $s \geq q + 1$.

Now we collect the best results that can be derived from all known methods of obtaining digital (t, s)-sequences constructed over $\mathbb{F}_q$. For this purpose it is convenient to introduce the following definition.

Definition 9. For any prime power q and any dimension $s \geq 1$ let $d_q(s)$ be the least value of t such that there exists a digital (t, s)-sequence constructed over $\mathbb{F}_q$.

From earlier observations in this paper (see also [17]) we deduce that $d_q(s) = 0$ for $1 \leq s \leq q$ and $d_q(s) \geq 1$ for $s \geq q + 1$. We are now interested in upper bounds on $d_q(s)$ that are valid for all dimensions s. Note that in view of Corollary 1, the numbers $V_q(s)$ and $d_q(s)$ introduced in Definitions 8 and 9, respectively, are related by

$$d_q(s) \leq V_q(s) \quad \text{for all } s \geq 1. \tag{9}$$

In the following theorem, the first bound on $d_q(s)$ results from (9) and [23, Theorem 4(i)] and the second one is contained in [45, Theorem 3]. The second bound is asymptotically weaker, but explicit.

Theorem 4. *For all prime powers q and all dimensions $s \geq 1$ we have*

$$d_q(s) \leq \frac{c_1 s}{\log q} + 1$$

with an absolute constant $c_1 > 0$ and

$$d_q(s) \leq \frac{3q - 1}{q - 1}(s - 1) - \frac{(2q + 4)(s - 1)^{1/2}}{(q^2 - 1)^{1/2}} + 2.$$

Stronger bounds can be shown in the special case where q is a square. In the next theorem, the first bound on $d_q(s)$ is obtained from (9) and [23, Theorem 4(ii)] and the second one stems from [45, Theorem 4].

Theorem 5. *If the prime power q is a square, then for all dimensions $s \geq 1$ we have*

$$d_q(s) \leq \frac{c_2 s}{q^{1/4}} + 1$$

with an absolute constant $c_2 > 0$ and

$$d_q(s) \leq \frac{ps}{q^{1/2} - 1},$$

where p is the unique prime factor of q.

There are additional upper bounds on $d_q(s)$ that can be deduced from recent work of the authors on the asymptotics of the number of rational places of global function fields.

Theorem 6. *Let $q = p^e$ with a prime p and an odd integer $e \geq 3$ and let m be the least prime factor of e. If p is odd and $q > 27$, then*

$$d_q(s) \leq \frac{\lceil 2(2q^{1/m} + 3)^{1/2} \rceil + 1}{q^{1/m} + 1} s + 1 \quad \text{for all } s \geq 1.$$

If $p = 2$ and e is composite, then

$$d_q(s) \leq \frac{2\lceil 2(2q^{1/m} + 2)^{1/2} \rceil + 4}{q^{1/m} + 1} s + 1 \quad \text{for all } s \geq 1.$$

Proof. In both cases we apply the bound on $V_q(s)$ in [23, Theorem 3] with $l = 2$. In the first case, according to the proof of [30, Theorem 4] the conditions in [23, Theorem 3] are satisfied for a global function field $K/\mathbb{F}_q$ with

$$g(K/\mathbb{F}_q) = \lceil 2(2q^{1/m} + 3)^{1/2} \rceil + 2$$

and for a set $\mathcal{P}$ of rational places of $K/\mathbb{F}_q$ with $|\mathcal{P}| = 2q^{1/m} + 2$. This yields

$$V_q(s) \leq \frac{\lceil 2(2q^{1/m} + 3)^{1/2} \rceil + 1}{q^{1/m} + 1} s + 1 \quad \text{for all } s \geq 1.$$

In the second case, according to the proof of [29, Theorem 2] the conditions in [23, Theorem 3] are satisfied for a global function field $K/\mathbb{F}_q$ with

$$g(K/\mathbb{F}_q) = 2\lceil 2(2q^{1/m} + 2)^{1/2} \rceil + 5$$

and for a set $\mathcal{P}$ of rational places of $K/\mathbb{F}_q$ with $|\mathcal{P}| = 2q^{1/m} + 2$. This yields

$$V_q(s) \leq \frac{2\lceil 2(2q^{1/m} + 2)^{1/2} \rceil + 4}{q^{1/m} + 1} s + 1 \quad \text{for all } s \geq 1.$$

The desired bounds follow now from (9). $\square$

Theorem 7. *We have*

$$d_5(s) \leq 3s + 1 \quad \text{for all } s \geq 1$$

and

$$d_{27}(s) \leq \frac{12}{5} s + 1 \quad \text{for all } s \geq 1.$$

Proof. In both cases we again apply the bound on $V_q(s)$ in [23, Theorem 3] with $l = 2$. In the first case, according to the proof of [29, Theorem 5] the conditions in [23, Theorem 3] are satisfied for a global function field $K/\mathbb{F}_5$ with $g(K/\mathbb{F}_5) = 13$ and for a set $\mathcal{P}$ of rational places of $K/\mathbb{F}_5$ with $|\mathcal{P}| = 8$. This yields

$$V_5(s) \leq 3s + 1 \quad \text{for all } s \geq 1.$$

In the second case, we use the method in the proof of [30, Theorem 4] with $q = m = 3$ and $n = 8$, and this produces a global function field $K/\mathbb{F}_{27}$ with $g(K/\mathbb{F}_{27}) = 7$ for which we can choose a set $\mathcal{P}$ of rational places of $K/\mathbb{F}_{27}$ with $|\mathcal{P}| = 5$. Then the conditions in [23, Theorem 3] are satisfied, hence we get

$$V_{27}(s) \leq \frac{12}{5}s + 1 \quad \text{for all } s \geq 1.$$

The desired bounds follow now from (9). $\square$

All bounds on $d_q(s)$ in Theorems 4, 5, 6, and 7 are of the form $d_q(s) = O(s)$ with an absolute implied constant, and we can even achieve an implied constant that tends to 0 as $q \to \infty$. According to the current state of affairs, the result $d_q(s) = O(s)$ can be obtained only with the second, the third, and the fourth construction in this section, i.e., with the help of algebraic curves over $\mathbb{F}_q$ with many rational points. It follows from Theorem 1 that

$$d_q(s) \geq \frac{s}{q} - \log_q \frac{(q-1)s + q + 1}{2} \tag{10}$$

for all prime powers q and all dimensions $s \geq 1$, and so the bound $d_q(s) = O(s)$ is best possible as far as the order of magnitude in s is concerned. A different proof of (10) was given by Schmid and Wolf [35], and this paper contains also bounds for related quantities; see also Larcher and Schmid [12] for an earlier lower bound. For the special case $q = 2$, Schmid [34] showed the bound

$$d_2(s) > s \log_2 \frac{3}{2} - 4\log_2(s-2) - 23 \quad \text{for all } s \geq 3, \tag{11}$$

which is better than (10) for all sufficiently large s.

For the convenience of the reader we reprint the table of explicit upper bounds on $d_q(s)$ for $q = 2, 3, 5$ and $1 \leq s \leq 50$ from Niederreiter and Xing [30] at the end of the present paper (see Table 5), with some improvements for $q = 3$ resulting from Table 3.

Now we consider the number

$$L_q := \liminf_{s \to \infty} \frac{d_q(s)}{s}$$

which is defined for any prime power q. It follows from (9) and [23, Proposition 2] that

$$L_q \leq B(q) := \liminf_{s \to \infty} \frac{V_q(s)}{s} \leq \frac{1}{A(q)}, \tag{12}$$

where

$$A(q) := \limsup_{g \to \infty} \frac{N_q(g)}{g}$$

and $N_q(g)$ is as in Definition 7. Serre [36] showed that $A(q)$ is at least of the order of magnitude $\log q$, and an alternative proof and an effective version of this result were given by Niederreiter and Xing [31]. Thus, in view of (12) we obtain

$$L_q \leq \frac{c_3}{\log q}$$

for all prime powers q, with an effective absolute constant $c_3 > 0$. If q is a square, then it is well known that $A(q) = q^{1/2} - 1$ (see the surveys [7], [28]), hence we get

$$L_q \leq \frac{1}{q^{1/2} - 1}.$$

For nonsquares q, further known lower bounds for $A(q)$ yield upper bounds for L_q on account of (12). For instance, the lower bound of Zink [46] for $A(p^3)$ with a prime p implies that

$$L_{p^3} \leq \frac{p + 2}{2(p^2 - 1)}.$$

If $q = p^e$ with an odd prime p and an odd integer $e \geq 3$, then the lower bound for $A(q)$ in [30, Corollary 3] yields

$$L_q \leq \frac{\lceil 2(2q^{1/m} + 3)^{1/2} \rceil + 1}{2q^{1/m} + 2},$$

where m is the least prime factor of e. If $q = 2^e$ with an odd composite integer $e \geq 3$, then the lower bound for $A(q)$ in [29, Theorem 2] shows that

$$L_q \leq \frac{\lceil 2(2q^{1/m} + 2)^{1/2} \rceil + 2}{q^{1/m} + 1},$$

where m is as above. Some slight improvements on the last two bounds can be obtained from the results on $A(q)$ in [31]. A bound for L_q not derived from (12) is

$$L_q \leq \frac{q + 1}{q - 1}$$

for all prime powers q, which was proved in [45] and is based on the third construction in this section. This bound is useful for small nonsquares q, such as $q = 2, 3, 5$. The lower bound $L_q \geq \frac{1}{q}$ for all prime powers q is an immediate consequence of (10). From (11) it follows that $L_2 \geq \log_2 \frac{3}{2}$, as was pointed out by Schmid [34].

Now we turn to results on (t, s)-sequences in an arbitrary base b. It is convenient to introduce the following quantity.

Definition 10. For any integers $b \geq 2$ and $s \geq 1$ let $t_b(s)$ be the least value of t such that there exists a (t, s)-sequence in base b.

As was already mentioned after Lemma 2 in Section 3, we can pass from digital (t, s)-sequences constructed over finite fields to digital (u, s)-sequences constructed over rings of arbitrary order $b \geq 2$, where the quality parameter u can be explicitly determined. Concretely, the relevant result in [23, Lemma 8] leads to the following fact.

Lemma 5. *Let* $b = \prod_{v=1}^{h} q_v$ *be the canonical factorization of the integer* $b \geq 2$ *into pairwise coprime prime powers* $q_1, \dots, q_h$. *Then we have*

$$t_b(s) \leq \max_{1 \leq v \leq h} d_{q_v}(s) \quad \text{for all } s \geq 1.$$

We can now combine Lemma 5 with the various upper bounds for $d_q(s)$ given above to obtain upper bounds for $t_b(s)$. We state only one result of this type which is a consequence of Theorem 4 (see also [23, Corollary 2]).

Theorem 8. *Let* $b = \prod_{v=1}^{h} q_v$ *be the canonical factorization of the integer* $b \geq 2$ *into pairwise coprime prime powers* $q_1 < \cdots < q_h$. *Then we have*

$$t_b(s) \leq \frac{c_1 s}{\log q_1} + 1 \quad \text{for all } s \geq 1$$

with an absolute constant $c_1 > 0$.

It follows from Theorem 8 that we have $t_b(s) = O(s)$ with an absolute implied constant. In view of Theorem 1, this bound is best possible as far as the order of magnitude in s is concerned.

6   New Constructions of (t, m, s)-Nets

We present four new methods of obtaining digital (t, m, s)-nets constructed over a finite field. The first method shows that a digital net constructed over an extension field yields a digital net constructed over the ground field in a higher dimension.

Theorem 9. *Let* q *be a prime power and* r *a positive integer. If there exists a digital* (t, m, s)-net *constructed over* $\mathbb{F}_{q^r}$, *then there exists a digital* $((r - 1)m + t, rm, rs)$-net *constructed over* $\mathbb{F}_q$.

Proof. By Lemma 3 it suffices to show that a (d, m, s)-system over $\mathbb{F}_{q^r}$ yields a (d, rm, rs)-system over $\mathbb{F}_q$. Let

$$A = \left\{ \mathbf{a}_j^{(i)} \in \mathbb{F}_{q^r}^m : 1 \leq i \leq s, 1 \leq j \leq m \right\}$$

be a (d, m, s)-system over $\mathbb{F}_{q^r}$. Choose an ordered basis $\beta_1, \dots, \beta_r$ of $\mathbb{F}_{q^r}$ over $\mathbb{F}_q$ and an $\mathbb{F}_q$-linear isomorphism $\varphi : \mathbb{F}_{q^r}^m \to \mathbb{F}_q^{rm}$. Then we consider the system

$$B = \left\{ \mathbf{b}_j^{(h)} \in \mathbb{F}_q^{rm} : 1 \leq h \leq rs, 1 \leq j \leq rm \right\}$$

such that for any $1 \leq i \leq s$ and $1 \leq k \leq r$,

$$\mathbf{b}_j^{((i-1)r+k)} = \begin{cases} \varphi(\beta_k \mathbf{a}_j^{(i)}) & \text{for} \quad 1 \leq j \leq m, \\ \mathbf{0} & \text{for} \quad m < j \leq rm. \end{cases}$$

We claim that B is a (d, rm, rs)-system over $\mathbb{F}_q$. Choose nonnegative integers $d_{i,k}$, $1 \leq i \leq s$, $1 \leq k \leq r$, with

$$\sum_{i=1}^{s} \sum_{k=1}^{r} d_{i,k} = d$$

and consider a relation

$$\sum_{i=1}^{s} \sum_{k=1}^{r} \sum_{j=1}^{d_{i,k}} c_j^{(i,k)} \mathbf{b}_j^{((i-1)r+k)} = \mathbf{0} \in \mathbb{F}_q^{rm} \tag{13}$$

with all $c_j^{(i,k)} \in \mathbb{F}_q$. Note that all $d_{i,k} \leq m$ since $d \leq m$. For $1 \leq i \leq s$ put

$$d_i = \max_{1 \leq k \leq r} d_{i,k}.$$

Furthermore, for $1 \leq i \leq s$ and $1 \leq k \leq r$ put $e_j^{(i,k)} = 1 \in \mathbb{F}_q$ for $1 \leq j \leq d_{i,k}$ and $e_j^{(i,k)} = 0 \in \mathbb{F}_q$ for $d_{i,k} < j \leq d_i$. Then (13) can be written in the form

$$\sum_{i=1}^{s} \sum_{k=1}^{r} \sum_{j=1}^{d_i} e_j^{(i,k)} c_j^{(i,k)} \varphi(\beta_k \mathbf{a}_j^{(i)}) = \mathbf{0} \in \mathbb{F}_q^{rm}.$$

Since φ is an $\mathbb{F}_q$-linear isomorphism, we conclude that

$$\sum_{i=1}^{s} \sum_{j=1}^{d_i} \gamma_j^{(i)} \mathbf{a}_j^{(i)} = \mathbf{0} \in \mathbb{F}_{q^r}^{m}$$

with

$$\gamma_j^{(i)} = \sum_{k=1}^{r} e_j^{(i,k)} c_j^{(i,k)} \beta_k \in \mathbb{F}_{q^r}. \tag{14}$$

If we note that

$$\sum_{i=1}^{s} d_i \leq \sum_{i=1}^{s} \sum_{k=1}^{r} d_{i,k} = d$$

and that A is a (d, m, s)-system over $\mathbb{F}_{q^r}$, then we get $\gamma_j^{(i)} = 0$ for $1 \leq j \leq d_i$, $1 \leq i \leq s$. Hence (14) implies that $e_j^{(i,k)} c_j^{(i,k)} = 0$ for $1 \leq j \leq d_i$, $1 \leq i \leq s$, $1 \leq k \leq r$, and so by the definition of $e_j^{(i,k)}$ we obtain that all coefficients $c_j^{(i,k)}$ in (13) are 0. $\square$

Corollary 2. *For any prime power q and any integers $m \geq g \geq 0$ and $r \geq 1$ there exists a digital $((r-1)m+g, rm, rN_{q^r}(g))$-net constructed over $\mathbb{F}_q$, where $N_{q^r}(g)$ is given by Definition 7.*

Proof. By Corollary 1 there exists a digital $(g, N_{q^r}(g) - 1)$-sequence constructed over $\mathbb{F}_{q^r}$, and so Lemma 2 shows that there exists a digital $(g, m, N_{q^r}(g))$-net constructed over $\mathbb{F}_{q^r}$. The rest follows from Theorem 9. □

Example 1. With the known lower bounds on $N_4(g)$ for $g = 1, 2, 3, 4, 5, 6, 7, 9$, Corollary 2 yields the following net parameters in base 2:

$(m+1, 2m, 18)$ for $m \geq 1$; $(m+2, 2m, 20)$ for $m \geq 2$; $(m+3, 2m, 28)$ for $m \geq 3$;

$(m+4, 2m, 30)$ for $m \geq 4$; $(m+5, 2m, 34)$ for $m \geq 5$; $(m+6, 2m, 40)$ for $m \geq 6$;

$(m + 7, 2m, 42)$ for $m \geq 7$; $(m + 9, 2m, 52)$ for $m \geq 9$.

Example 2. With the known values of $N_8(g)$ for $g = 0, 1, 2$, Corollary 2 yields the following net parameters in base 2:

$(2m, 3m, 27)$ for $m \geq 1$; $(2m+1, 3m, 42)$ for $m \geq 1$; $(2m+2, 3m, 54)$ for $m \geq 2$.

Example 3. With the known values of $N_9(g)$ for $g = 0, 1, 2, 3$ and $N_{27}(0) = 28$, Corollary 2 yields the following net parameters in base 3:

$(m, 2m, 20)$ for $m \geq 1$; $(m + 1, 2m, 32)$ for $m \geq 1$; $(m + 2, 2m, 40)$ for $m \geq 2$;

$(m + 3, 2m, 56)$ for $m \geq 3$; $(2m, 3m, 84)$ for $m \geq 1$.

The second method allows us to obtain a new digital net constructed over $\mathbb{F}_q$ from two or more known digital nets constructed over $\mathbb{F}_q$ in lower dimensions.

Theorem 10. *If there exists a digital (t_1, m_1, s_1)-net and a digital (t_2, m_2, s_2)-net constructed over $\mathbb{F}_q$, then there exists a digital $(t, m_1 + m_2, s_1 + s_2)$-net constructed over $\mathbb{F}_q$ with*

$$t = \max(m_1 + t_2, m_2 + t_1).$$

Proof. By Lemma 3 it suffices to show that a (d_1, m_1, s_1)-system and a (d_2, m_2, s_2)-system over $\mathbb{F}_q$ yield a $(d, m_1 + m_2, s_1 + s_2)$-system over $\mathbb{F}_q$ with $d = \min(d_1, d_2)$. Let

$$A = \left\{ \mathbf{a}_j^{(i)} \in \mathbb{F}_q^{m_1} : 1 \leq i \leq s_1, 1 \leq j \leq m_1 \right\}$$

be a (d_1, m_1, s_1)-system and

$$B = \left\{ \mathbf{b}_j^{(i)} \in \mathbb{F}_q^{m_2} : 1 \leq i \leq s_2, 1 \leq j \leq m_2 \right\}$$

be a (d_2, m_2, s_2)-system over $\mathbb{F}_q$, where we can assume $m_2 \geq m_1$. Now we consider the system

$$C = \left\{ \mathbf{c}_j^{(i)} \in \mathbb{F}_q^{m_1+m_2} : 1 \leq i \leq s_1 + s_2, 1 \leq j \leq m_1 + m_2 \right\}$$

with

$$\mathbf{c}_j^{(i)} = \begin{cases} (\mathbf{a}_j^{(i)}, \mathbf{0}) & \text{for } 1 \leq i \leq s_1, 1 \leq j \leq m_1, \\ (\mathbf{0}, \mathbf{b}_j^{(i-s_1)}) & \text{for } s_1 < i \leq s_1 + s_2, 1 \leq j \leq m_1, \\ \mathbf{0} & \text{for } 1 \leq i \leq s_1 + s_2, m_1 < j \leq m_1 + m_2. \end{cases}$$

Then it is easy to see that C is a $(d, m_1 + m_2, s_1 + s_2)$-system over $\mathbb{F}_q$. $\square$

Corollary 3. *If there exist n digital (t_k, m_k, s_k)-nets constructed over $\mathbb{F}_q$ for $1 \leq k \leq n$, then there exists a digital $(t, \sum_{k=1}^n m_k, \sum_{k=1}^n s_k)$-net constructed over $\mathbb{F}_q$ with*

$$t = \sum_{k=1}^n m_k - \min_{1 \leq k \leq n} (m_k - t_k).$$

In particular, if there exists a digital (t, m, s)-net constructed over $\mathbb{F}_q$, then for every $n \geq 1$ there exists a digital $((n-1)m + t, nm, ns)$-net constructed over $\mathbb{F}_q$.

Corollary 4. *If there exist n digital (t_k, m_k, s_k)-nets constructed over $\mathbb{F}_{q^{r_k}}$ for $1 \leq k \leq n$, where $r_1, \ldots, r_n$ are positive integers, then there exists a digital $(t, \sum_{k=1}^n r_k m_k, \sum_{k=1}^n r_k s_k)$-net constructed over $\mathbb{F}_q$ with*

$$t = \sum_{k=1}^n r_k m_k - \min_{1 \leq k \leq n} (m_k - t_k).$$

In particular, if there exists a digital (t, m, s)-net constructed over $\mathbb{F}_{q^r}$, where r is a positive integer, then for every $n \geq 1$ there exists a digital $((nr-1)m + t, nrm, nrs)$-net constructed over $\mathbb{F}_q$.

Proof. This is obtained by combining Theorem 9 and Corollary 3. $\square$

The next two methods apply linear codes to the construction of digital nets. Earlier applications of coding theory to the construction of nets can be found in Adams and Shader [1], Clayman and Mullen [3], Edel and Bierbrauer [5], Lawrence [13], Lawrence et al. [14], Schmid [33], and Schmid and Wolf [35]. When we speak of a linear $[n, k, d]$ code, we mean as usual that n is the length, k the dimension, and d the minimum distance of the code. If "k", say, is replaced by "$\geq k$", then this indicates that the dimension of the code is at least k, and similarly for the minimum distance.

The third method uses a linear code over a given finite field and a digital net constructed over an extension field to produce a digital net constructed over the given finite field.

Theorem 11. *Let q be a prime power and d, n, and r be integers with $d \geq 0$ and $1 \leq r \leq n$ such that there exists a linear $[n, \geq n - r, \geq d + 1]$ code over $\mathbb{F}_q$. Then from a digital (t, m, s)-net constructed over $\mathbb{F}_{q^r}$ we can obtain a digital (u, rm, ns)-net constructed over $\mathbb{F}_q$ with*

$$u = rm - \min(d, m - t).$$

Proof. From a linear $[n, \geq n-r, \geq d+1]$ code over $\mathbb{F}_q$ we can derive a linear $[n, n-r, \geq d+1]$ code L over $\mathbb{F}_q$. Let $\mathbf{c}_1, \ldots, \mathbf{c}_n \in \mathbb{F}_q^r$ be the column vectors of a parity-check matrix of L. Furthermore, choose $\mathbb{F}_q$-linear isomorphisms $\varphi : \mathbb{F}_{q^r}^m \to \mathbb{F}_q^{rm}$ and $\pi : \mathbb{F}_q^r \to \mathbb{F}_{q^r}$. By assumption and Lemma 3, there exists an $(m - t, m, s)$-system

$$A = \left\{ \mathbf{a}_j^{(i)} \in \mathbb{F}_{q^r}^m : 1 \leq i \leq s, 1 \leq j \leq m \right\}$$

over $\mathbb{F}_{q^r}$. Now we consider the system

$$B = \left\{ \mathbf{b}_j^{(h)} \in \mathbb{F}_q^{rm} : 1 \leq h \leq ns, 1 \leq j \leq rm \right\}$$

such that for any $1 \leq i \leq s$ and $1 \leq k \leq n$,

$$\mathbf{b}_j^{((i-1)n+k)} = \begin{cases} \varphi(\pi(\mathbf{c}_k)\mathbf{a}_j^{(i)}) & \text{for} \quad 1 \leq j \leq m, \\ \mathbf{0} & \text{for} \quad m < j \leq rm. \end{cases}$$

The proof is completed by showing that B is a (v, rm, ns)-system over $\mathbb{F}_q$ with

$$v = \min(d, m - t).$$

Choose nonnegative integers $d_{i,k}, 1 \leq i \leq s, 1 \leq k \leq n$, with

$$\sum_{i=1}^{s} \sum_{k=1}^{n} d_{i,k} = v$$

and consider a relation

$$\sum_{i=1}^{s} \sum_{k=1}^{n} \sum_{j=1}^{d_{i,k}} f_j^{(i,k)} \mathbf{b}_j^{((i-1)n+k)} = \mathbf{0} \in \mathbb{F}_q^{rm} \tag{15}$$

with all $f_j^{(i,k)} \in \mathbb{F}_q$. Note that all $d_{i,k} \leq m$. For $1 \leq i \leq s$ put

$$d_i = \max_{1 \leq k \leq n} d_{i,k}.$$

Furthermore, for $1 \leq i \leq s$ and $1 \leq k \leq n$ put $e_j^{(i,k)} = 1 \in \mathbb{F}_q$ for $1 \leq j \leq d_{i,k}$ and $e_j^{(i,k)} = 0 \in \mathbb{F}_q$ for $d_{i,k} < j \leq d_i$. Then (15) can be written in the form

$$\sum_{i=1}^{s} \sum_{k=1}^{n} \sum_{j=1}^{d_i} e_j^{(i,k)} f_j^{(i,k)} \varphi(\pi(\mathbf{c}_k)\mathbf{a}_j^{(i)}) = \mathbf{0} \in \mathbb{F}_q^{rm}.$$

Since φ is an $\mathbb{F}_q$-linear isomorphism, we obtain

$$\sum_{i=1}^{s} \sum_{j=1}^{d_i} \gamma_j^{(i)} \mathbf{a}_j^{(i)} = \mathbf{0} \in \mathbb{F}_{q^r}^m$$

with

$$\gamma_j^{(i)} = \sum_{k=1}^{n} e_j^{(i,k)} f_j^{(i,k)} \pi(\mathbf{c}_k) \in \mathbb{F}_{q^r}. \tag{16}$$

If we note that

$$\sum_{i=1}^{s} d_i \leq \sum_{i=1}^{s} \sum_{k=1}^{n} d_{i,k} = v \leq m - t$$

and that A is an $(m - t, m, s)$-system over $\mathbb{F}_{q^r}$, then we get $\gamma_j^{(i)} = 0$ for $1 \leq j \leq d_i, 1 \leq i \leq s$. For fixed j and i with $1 \leq j \leq d_i$ and $1 \leq i \leq s$, the number of terms in (16) with $e_j^{(i,k)} \neq 0$ is

$$\#\{1 \leq k \leq n : e_j^{(i,k)} = 1\} = \#\{1 \leq k \leq n : j \leq d_{i,k}\}$$

$$\leq \sum_{k=1}^{n} d_{i,k} \leq \sum_{i=1}^{s} \sum_{k=1}^{n} d_{i,k} = v \leq d.$$

Since L has minimum distance $\geq d + 1$, any d of the vectors $\mathbf{c}_1, \ldots, \mathbf{c}_n$, and so any d of the elements $\pi(\mathbf{c}_1), \ldots, \pi(\mathbf{c}_n)$, are linearly independent over $\mathbb{F}_q$. Thus we deduce from (16) that $e_j^{(i,k)} f_j^{(i,k)} = 0$ for $1 \leq j \leq d_i, 1 \leq i \leq s, 1 \leq k \leq n$, and so by the definition of $e_j^{(i,k)}$ we obtain that all coefficients $f_j^{(i,k)}$ in (15) are 0. $\square$

Corollary 5. *Let q be a prime power, r a positive integer, and g an integer with $0 \leq g \leq r \leq N_q(g) - 1$, where $N_q(g)$ is as in Definition 7. Then from a digital (t, m, s)-net constructed over $\mathbb{F}_{q^r}$ we can obtain a digital $(u, rm, (N_q(g) - 1)s)$-net constructed over $\mathbb{F}_q$ with*

$$u = rm - \min(r - g, m - t).$$

Proof. Let $K/\mathbb{F}_q$ be a global function field of genus g with $N_q(g)$ rational places. Put $n = N_q(g) - 1$ and let $P, P_1, \ldots, P_n$ be the rational places of $K/\mathbb{F}_q$. Let L be the linear code over $\mathbb{F}_q$ associated with the divisors $\sum_{k=1}^{n} P_k$ and $(n - r + g - 1)P$ of $K/\mathbb{F}_q$ according to [39, Definition II.2.1]. By [39, Corollary II.2.3], L is a linear $[n, \geq n - r, \geq r - g + 1]$ code over $\mathbb{F}_q$. The desired result follows now from Theorem 11. $\square$

Example 4. Since $N_q(0) = q + 1$, Corollary 5 shows that if r is an integer with $1 \leq r \leq q$ and if there exists a digital (t, m, s)-net constructed over $\mathbb{F}_{q^r}$, then there exists a digital (u, rm, qs)-net constructed over $\mathbb{F}_q$ with

$$u = rm - \min(r, m - t).$$

Example 5. Since it is known that $N_q(1) \geq q + \lfloor 2q^{1/2} \rfloor$ (see the surveys [7], [28]), Corollary 5 shows that if r is an integer with $1 \leq r \leq q + \lfloor 2q^{1/2} \rfloor - 1$ and if there exists a digital (t, m, s)-net constructed over $\mathbb{F}_{q^r}$, then there exists a digital $(u, rm, (q + \lfloor 2q^{1/2} \rfloor - 1)s)$-net constructed over $\mathbb{F}_q$ with

$$u = rm - \min(r - 1, m - t).$$

The fourth method uses a digital net constructed over a given finite field and a linear code over an extension field to produce a digital net constructed over the given finite field.

Theorem 12. *Let q be a prime power and $d, m, n,$ and r be integers with $d \geq 0, m \geq 1,$ and $1 \leq r \leq n$ such that there exists a linear $[n, \geq n - r, \geq d + 1]$ code over $\mathbb{F}_{q^m}$. Then from a digital (t, m, s)-net constructed over $\mathbb{F}_q$ we can obtain a digital (u, rm, ns)-net constructed over $\mathbb{F}_q$ with*

$$u = rm - \min(d, m - t).$$

Proof. The proof is similar to that of Theorem 11. We first derive a linear $[n, n - r, \geq d + 1]$ code L over $\mathbb{F}_{q^m}$ and let $\mathbf{c}_1, \ldots, \mathbf{c}_n \in \mathbb{F}_{q^m}^r$ be the column vectors of a parity-check matrix of L. Then we choose $\mathbb{F}_q$-linear isomorphisms $\varphi : \mathbb{F}_{q^m}^r \to \mathbb{F}_q^{rm}$ and $\pi : \mathbb{F}_q^m \to \mathbb{F}_{q^m}$. By assumption and Lemma 3, there exists an $(m - t, m, s)$-system

$$A = \left\{ \mathbf{a}_j^{(i)} \in \mathbb{F}_q^m : 1 \leq i \leq s, 1 \leq j \leq m \right\}$$

over $\mathbb{F}_q$. Now we consider the system

$$B = \left\{ \mathbf{b}_j^{(h)} \in \mathbb{F}_q^{rm} : 1 \leq h \leq ns, 1 \leq j \leq rm \right\}$$

such that for any $1 \leq i \leq s$ and $1 \leq k \leq n$,

$$\mathbf{b}_j^{((i-1)n+k)} = \begin{cases} \varphi(\pi(\mathbf{a}_j^{(i)})\mathbf{c}_k) & \text{for} \quad 1 \leq j \leq m, \\ \mathbf{0} & \text{for} \quad m < j \leq rm. \end{cases}$$

The proof is completed by showing that B is a (v, rm, ns)-system over $\mathbb{F}_q$ with

$$v = \min(d, m - t).$$

Let the $d_{i,k}$ be as in the proof of Theorem 11 and consider a relation of the form (15). Since φ is an $\mathbb{F}_q$-linear isomorphism, we obtain

$$\sum_{k=1}^{n} \gamma_k \mathbf{c}_k = \mathbf{0} \in \mathbb{F}_{q^m}^r \qquad (17)$$

with

$$\gamma_k = \sum_{i=1}^{s} \sum_{j=1}^{d_{i,k}} f_j^{(i,k)} \pi(\mathbf{a}_j^{(i)}) \in \mathbb{F}_{q^m}.$$

The number of terms in (17) with $\gamma_k \neq 0$ is

$$\leq \#\{1 \leq k \leq n : d_{i,k} \geq 1 \text{ for some } 1 \leq i \leq s\} \leq \sum_{k=1}^{n} \sum_{i=1}^{s} d_{i,k} = v \leq d.$$

Since L has minimum distance $\geq d+1$, any d of the vectors $\mathbf{c}_1, \ldots, \mathbf{c}_n$ are linearly independent over $\mathbb{F}_{q^m}$, and so it follows from (17) that $\gamma_k = 0$ for $1 \leq k \leq n$. For fixed k with $1 \leq k \leq n$ we have

$$\sum_{i=1}^{s} d_{i,k} \leq \sum_{i=1}^{s} \sum_{k=1}^{n} d_{i,k} = v \leq m - t.$$

Thus, since A is an $(m-t, m, s)$-system over $\mathbb{F}_q$ and π is an $\mathbb{F}_q$-linear isomorphism, we infer that all coefficients $f_j^{(i,k)}$ in (15) are 0. $\square$

Corollary 6. *Let q be a prime power and $m, r,$ and g be integers with $m \geq 1, r \geq 1,$ and $0 \leq g \leq r \leq N_{q^m}(g) - 1,$ where $N_{q^m}(g)$ is given by Definition 7. Then from a digital (t, m, s)-net constructed over $\mathbb{F}_q$ we can obtain a digital $(u, rm, (N_{q^m}(g) - 1)s)$-net constructed over $\mathbb{F}_q$ with*

$$u = rm - \min(r - g, m - t).$$

Proof. Proceed as in the proof of Corollary 5, but with $\mathbb{F}_{q^m}$ as the constant field, and apply Theorem 12. $\square$

Example 6. Since $N_{q^m}(0) = q^m + 1$, Corollary 6 shows that if r is an integer with $1 \leq r \leq q^m$ and if there exists a digital (t, m, s)-net constructed over $\mathbb{F}_q$, then there exists a digital $(u, rm, q^m s)$-net constructed over $\mathbb{F}_q$ with

$$u = rm - \min(r, m - t).$$

For instance, since for any $m \geq 1$ there exists a digital $(0, m, 3)$-net constructed over $\mathbb{F}_2$, there exists also a digital $(rm - m, rm, 3 \cdot 2^m)$-net constructed over $\mathbb{F}_2$ for $m \leq r \leq 2^m$.

Example 7. Since it is known that $N_{q^m}(1) \geq q^m + \lfloor 2q^{m/2} \rfloor$ (see the surveys [7], [28]), Corollary 6 shows that if r is an integer with $1 \leq r \leq q^m + \lfloor 2q^{m/2} \rfloor - 1$ and if there exists a digital (t, m, s)-net constructed over $\mathbb{F}_q$, then there exists a digital $(u, rm, (q^m + \lfloor 2q^{m/2} \rfloor - 1)s)$-net constructed over $\mathbb{F}_q$ with

$$u = rm - \min(r - 1, m - t).$$

7 New Algebraic Curves with Many Rational Points

The constructions of (t, s)-sequences and (t, m, s)-nets in Sections 5 and 6, respectively, lead to the requirement of finding good lower bounds for the number $N_q(g)$ in Definition 7, or equivalently to the problem of constructing algebraic curves over $\mathbb{F}_q$ of given genus g with many rational points. The new feature of some methods in Section 6 is that data for extension fields of $\mathbb{F}_q$ are now also useful for obtaining digital nets constructed over $\mathbb{F}_q$. In practice, digital nets have been considered mostly for $q = 2, 3, 5$, and so it is of interest to study algebraic curves over extension fields of $\mathbb{F}_2, \mathbb{F}_3$, or $\mathbb{F}_5$ with many rational points.

The most recent table of bounds on $N_q(g)$ in [30] lists data for $q = 2, 3, 4, 8, 16$ and all genera $g \leq 50$, as well as for $q = 5$ and all genera $g \leq 22$. This provides ample information for obtaining good digital nets constructed over $\mathbb{F}_2$ by the methods mentioned above. The situation is less satisfactory for $\mathbb{F}_3$ and $\mathbb{F}_5$. The table of van der Geer and van der Vlugt [42] lists some bounds on $N_q(g)$ for $q = 9, 27, 81$, but many genera $g \leq 50$ are not covered. Recent work of the authors [32] has closed some gaps for $q = 9, 27$. We now provide additional lower bounds on $N_q(g), q = 9, 27$, so that *in toto* all genera $g \leq 50$ are now covered for these two cases as well. In the following we will use the equivalent language of global function fields (compare with Section 4).

We start with the case $q = 9$. Our main tool is [26, Theorem 3] which allows us to obtain global function fields with many rational places from global function fields of smaller genus. Table 1 at the end of the present paper lists the data that are needed in [26, Theorem 3] to get our examples of global function fields $K/\mathbb{F}_9$ with many rational places. The column in Table 1 labeled "$N(K)$" gives a lower bound on $N(K/\mathbb{F}_9)$ and not necessarily the exact value. In Table 1 we abbreviate $g(K/\mathbb{F}_9)$ and $g(F/\mathbb{F}_9)$ by $g(K)$ and $g(F)$, respectively. The following additional examples of global function fields $K/\mathbb{F}_9$ with many rational places are also new.

Example 8. $g(K/\mathbb{F}_9) = 17, N(K/\mathbb{F}_9) \geq 56$. We start from the global function field $L/\mathbb{F}_9$ in [32, Table 4] with $g(L/\mathbb{F}_9) = 49, N(L/\mathbb{F}_9) \geq 168$. The corresponding base field $F/\mathbb{F}_3$ in [32, Theorem 1] satisfies $g(F/\mathbb{F}_3) = 2, N(F/\mathbb{F}_3) = 8$. We consider the constant field extension $F_2 = \mathbb{F}_9 \cdot F$ of F. Then L is an extension field of F_2 of degree 21 such that all 8 rational places of $F/\mathbb{F}_3$ split completely in the extension L/F_2. The only ramified place in L/F_2 is a place Q of F_2 with $\deg(Q) = 3$ and ramification index $e_Q(L/F_2) = 7$. Now let K be a subfield of the abelian extension L/F_2 with $[K : F_2] = 7$. Then $N(K/\mathbb{F}_9) \geq 8 \cdot 7 = 56$. Moreover $e_Q(K/F_2) = 7$, and so the Hurwitz genus formula shows that $g(K/\mathbb{F}_9) = 17$.

Example 9. $g(K/\mathbb{F}_9) = 18, N(K/\mathbb{F}_9) = 46$. Let $K = \mathbb{F}_9(x, y_1, y_2)$ with

$$y_1^4 = x^2 + x - 1, \quad y_2^3 + y_2 = \frac{1}{(x - \alpha)^4},$$

where $\alpha \in \mathbb{F}_9$ with $\alpha^2 + \alpha = 1$. The value of $g(K/\mathbb{F}_9)$ is obtained from the theory of Kummer and generalized Artin-Schreier extensions (see [39, Section III.7]). The following rational places of the rational function field $\mathbb{F}_9(x)$ yield the 46 rational places of K: the pole of x splits into 6 rational places, the places x, $x - 1$, and $x + 1$ all split completely, the place $x - \alpha$ is totally ramified, and the place $x - \alpha^3$ splits into 3 rational places.

Example 10. $g(K/\mathbb{F}_9) = 31, N(K/\mathbb{F}_9) \geq 84$. We use the theory of cyclotomic function fields as developed by Hayes [8]; see also [25] for a convenient summary. Let E_M be the cyclotomic function field over $\mathbb{F}_9$ with modulus $M = (x^3 - x - 1)^2 \in \mathbb{F}_9[x]$. With $F = \mathbb{F}_9(x)$ we have $\text{Gal}(E_M/F) \simeq (\mathbb{F}_9[x]/M)^*$ and

$$[E_M : F] = (9^3 - 1) \cdot 9^3 = 8 \cdot 7 \cdot 13 \cdot 3^6.$$

The order of the subgroup of $(\mathbb{F}_9[x]/M)^*$ generated by the residue classes mod M of $x, x + 1, x - 1$, and all $\gamma \in \mathbb{F}_9^*$ divides $8 \cdot 13 \cdot 3^3$. Thus, there exists a subfield K of the extension E_M/F with $[K : F] = 21$ such that the places $x, x + 1, x - 1$ and the pole of x split completely in K/F. Hence $N(K/\mathbb{F}_9) \geq 4 \cdot 21 = 84$. The only ramified place in K/F is $x^3 - x - 1$ and it is totally ramified. Thus, by [30, Corollary 2] we get $g(K/\mathbb{F}_9) = 1 - 21 + 21 \cdot (1 - \frac{8}{42}) \cdot 3 = 31$.

Example 11. $g(K/\mathbb{F}_9) = 38, N(K/\mathbb{F}_9) \geq 105$. We start from the global function field $L/\mathbb{F}_9$ in [32, Table 4] with $g(L/\mathbb{F}_9) = 112, N(L/\mathbb{F}_9) \geq 315$. Let $F/\mathbb{F}_3$ and its constant field extension F_2 be as in Example 8. Then L is an extension field of F_2 of degree 42. Furthermore, 7 rational places of $F/\mathbb{F}_3$ split completely in the extension L/F_2, and the remaining rational place ∞ of $F/\mathbb{F}_3$ splits into 21 rational places of L with ramification index $e_\infty(L/F_2) = 2$. The only other ramified place in L/F_2 is a place Q of F_2 with $\deg(Q) = 3$ and $e_Q(L/F_2) = 14$. Now let K be a subfield of the abelian extension L/F_2 with $[K : F_2] = 14$. Then 7 rational places of $F/\mathbb{F}_3$ split completely in the extension K/F_2, whereas ∞ splits into 7 rational places of K with $e_\infty(K/F_2) = 2$. Then $N(K/\mathbb{F}_9) \geq 7 \cdot 14 + 7 = 105$. The only other ramified place in K/F_2 is Q and it is totally ramified. Thus, the Hurwitz genus formula yields $g(K/\mathbb{F}_9) = 38$.

Now we turn to the case $q = 27$. Our main tool is again [26, Theorem 3], and Table 2 at the end of the present paper lists the data that are needed in this theorem to get our examples of global function fields $K/\mathbb{F}_{27}$ with many rational places. The column in Table 2 labeled "$N(K)$" gives a lower bound on $N(K/\mathbb{F}_{27})$ and not necessarily the exact value. In Table 2 we abbreviate $g(K/\mathbb{F}_{27})$ and $g(F/\mathbb{F}_{27})$ by $g(K)$ and $g(F)$, respectively. The following additional examples of global function fields $K/\mathbb{F}_{27}$ with many rational places are also new. These examples depend mostly on the theory of Kummer and Artin-Schreier extensions for which we refer to [39, Section III.7]. Let $\beta \in \mathbb{F}_{27}$ denote a root of $x^3 - x + 1 \in \mathbb{F}_3[x]$.

Example 12. $g(K/\mathbb{F}_{27}) = 7, N(K/\mathbb{F}_{27}) = 64$. Let $K = \mathbb{F}_{27}(x, y_1, y_2)$ with

$$y_1^2 = x, \quad y_2^3 - y_2 = -x^2(x^2 - x - 1).$$

The following rational places of the rational function field $\mathbb{F}_{27}(x)$ yield the 64 rational places of K: the pole of x is totally ramified, the place x splits into 3 rational places, and the places $x - \beta^{2n}, 0 \le n \le 11, n \ne 4, 10$, all split completely.

Example 13. $g(K/\mathbb{F}_{27}) = 9, N(K/\mathbb{F}_{27}) = 82$. Let $K = \mathbb{F}_{27}(x, y)$ with

$$y^3 - y = x^{10} - x^4.$$

The pole of x in $\mathbb{F}_{27}(x)$ is totally ramified and all other rational places of $\mathbb{F}_{27}(x)$ split completely in the extension $K/\mathbb{F}_{27}(x)$.

Example 14. $g(K/\mathbb{F}_{27}) = 10, N(K/\mathbb{F}_{27}) = 82$. Let $K = \mathbb{F}_{27}(x, y)$ with

$$y^3 - y = x^{11} - x^7.$$

The pole of x in $\mathbb{F}_{27}(x)$ is totally ramified and all other rational places of $\mathbb{F}_{27}(x)$ split completely in the extension $K/\mathbb{F}_{27}(x)$.

Example 15. $g(K/\mathbb{F}_{27}) = 11, N(K/\mathbb{F}_{27}) = 96$. Let $K = \mathbb{F}_{27}(x, y_1, y_2)$ with

$$y_1^2 = -x(x + 1)(x^2 + 1), \quad y_2^3 - y_2 = -x^2(x^2 - x - 1).$$

The following rational places of $\mathbb{F}_{27}(x)$ yield the 96 rational places of K: the places x and $x + 1$ each split into 3 rational places and the places $x - \beta^n, 1 \le n \le 22, n \ne 5, 8, 13, 15, 17, 19, 20$, all split completely.

Example 16. $g(K/\mathbb{F}_{27}) = 14, N(K/\mathbb{F}_{27}) \ge 84$. We use the theory of cyclotomic function fields (compare with Example 10). Let E_M be the cyclotomic function field over $\mathbb{F}_{27}$ with modulus $M = (x^2 + 1)^2 \in \mathbb{F}_{27}[x]$. With $F = \mathbb{F}_{27}(x)$ we have $\mathrm{Gal}(E_M/F) \simeq (\mathbb{F}_{27}[x]/M)^*$ and

$$[E_M : F] = (27^2 - 1) \cdot 27^2 = 8 \cdot 7 \cdot 13 \cdot 3^6.$$

The order of the subgroup of $(\mathbb{F}_{27}[x]/M)^*$ generated by the residue classes mod M of $x, x + 1, x - 1$, and all $\gamma \in \mathbb{F}_{27}^*$ divides $8 \cdot 13 \cdot 3^3$. Thus, there exists a subfield K of the extension E_M/F with $[K : F] = 21$ such that the places $x, x + 1, x - 1$ and the pole of x split completely in K/F. Hence $N(K/\mathbb{F}_{27}) \ge 4 \cdot 21 = 84$. The only ramified place in K/F is $x^2 + 1$ and it is totally ramified. Thus, by [30, Corollary 2] we get $g(K/\mathbb{F}_{27}) = 1 - 21 + 21 \cdot (1 - \frac{8}{42}) \cdot 2 = 14$.

Example 17. $g(K/\mathbb{F}_{27}) = 15, N(K/\mathbb{F}_{27}) = 82$. Let $K = \mathbb{F}_{27}(x, y)$ with

$$y^3 - y = x^{16} - x^{14}.$$

The pole of x in $\mathbb{F}_{27}(x)$ is totally ramified and all other rational places of $\mathbb{F}_{27}(x)$ split completely in the extension $K/\mathbb{F}_{27}(x)$.

Example 18. $g(K/\mathbb{F}_{27}) = 18, N(K/\mathbb{F}_{27}) = 94$. Let $K = \mathbb{F}_{27}(x, y_1, y_2)$ with

$$y_1^2 = x(x^2 - x - 1), \quad y_2^3 - y_2 = -x^2(x^6 - x^2 - 1).$$

The following rational places of $\mathbb{F}_{27}(x)$ yield the 94 rational places of K: the pole of x is totally ramified, the place x splits into 3 rational places, and the places $x - \beta^n, 1 \leq n \leq 24, n \neq 4, 10, 12, 13, 14, 16, 17, 22, 23$, all split completely.

Example 19. $g(K/\mathbb{F}_{27}) = 22, N(K/\mathbb{F}_{27}) = 112$. Let $K = \mathbb{F}_{27}(x, y_1, y_2)$ with

$$y_1^2 = x(x^2 - x - 1), \quad y_2^3 - y_2 = x^{10} - x^4.$$

The following rational places of $\mathbb{F}_{27}(x)$ yield the 112 rational places of K: the pole of x is totally ramified, the place x splits into 3 rational places, and the places $x - \beta^n, 1 \leq n \leq 24, n \neq 13, 14, 16, 17, 22, 23$, all split completely.

Example 20. $g(K/\mathbb{F}_{27}) = 23, N(K/\mathbb{F}_{27}) = 114$. Let $K = \mathbb{F}_{27}(x, y_1, y_2)$ with

$$y_1^2 = -x(x + 1)(x^2 + 1), \quad y_2^3 - y_2 = x^{10} - x^4.$$

The following rational places of $\mathbb{F}_{27}(x)$ yield the 114 rational places of K: the places x and $x + 1$ each split into 3 rational places and the places $x - \beta^n, 1 \leq n \leq 24, n \neq 5, 13, 15, 17, 19, 23$, all split completely.

At the end of the paper we provide two tables of bounds for $N_q(g)$. Table 3 is for $q = 2, 3, 4, 8, 9, 16, 27$ and $1 \leq g \leq 50$ and for $q = 5$ and $1 \leq g \leq 22$ (for $g = 0$ we trivially have $N_q(0) = q + 1$), and Table 4 extends Table 3 for the important case $q = 2$ to the range $51 \leq g \leq 95$. In each entry of the tables, the first number is a lower bound for $N_q(g)$ and the second is an upper bound for $N_q(g)$. If only one number is given, then this is the exact value of $N_q(g)$. A program for calculating upper bounds for $N_q(g)$, which is based on Weil's explicit formula for the number of rational places in terms of the zeta function and on the trigonometric polynomials of Oesterlé, was kindly supplied to us by Jean-Pierre Serre. The lower bounds in Table 3 are obtained by combining data from the earlier tables of Niederreiter and Xing [30] and van der Geer and van der Vlugt [42] with the new data from the present paper for $q = 9, 27$, from the paper [32] for $q = 4, 8, 9, 16, 27$, from the paper [43] for $q = 3, g = 35, 36, 39, 49$, and $q = 4, g = 39$, and from the paper [44] for $q = 27, g = 21, 39$. Table 4 is a reprinted version of [30, Table 2].

Table 1: Fields over $\mathbb{F}_9$

$g(K)$	$N(K)$	$g(F)$	m	d	l
20	48	2	15	8	1
27	60	3	19	10	1
30	60	2	19	13	1
32	81	0	8	5	2
33	78	3	25	13	1
35	84	3	27	14	1
39	84	3	27	16	1
40	90	0	9	6	2
41	84	3	27	17	1
42	90	4	29	16	1
44	90	4	29	17	1

Table 2: Fields over $\mathbb{F}_{27}$

$g(K)$	$N(K)$	$g(F)$	m	d	l
16	81	0	26	9	1
26	108	0	3	2	3
27	114	1	37	13	1
28	108	2	35	12	1
29	114	1	37	14	1
30	117	2	38	13	1
31	114	1	37	15	1
32	126	0	13	5	2
34	135	2	44	15	1
35	126	3	41	14	1
38	144	2	47	17	1
41	153	3	50	17	1
44	153	4	50	17	1
45	171	3	56	19	1
46	162	4	53	18	1
47	174	3	57	20	1
50	180	4	59	20	1

Table 3: Bounds for $N_q(g)$

$g \backslash q$	2	3	4	5	8	9	16	27
1	5	7	9	10	14	16	25	38
2	6	8	10	12	18	20	33	48
3	7	10	14	16	24	28	38	58
4	8	12	15	18	25-29	30	45-47	64-68
5	9	12-14	17-18	20-22	29-32	32-36	49-55	55-78
6	10	14-15	20	21-25	33-36	35-40	65	76-88
7	10	16-17	21-22	22-27	33-39	39-43	63-70	64-98
8	11	15-18	21-24	22-29	34-43	38-47	61-76	92-108
9	12	19	26	26-32	45-47	40-51	72-81	82-118
10	13	19-21	27-28	27-34	38-50	54-55	81-87	82-128
11	14	20-22	26-30	32-36	48-54	55-59	80-92	96-138
12	14-15	22-24	28-31	30-38	49-57	55-63	68-97	109-148
13	15	24-25	33	36-40	50-61	60-66	97-103	136-156
14	15-16	24-26	32-35	39-43	65	56-70	97-108	84-164
15	17	28	33-37	32-45	54-68	64-74	98-113	82-171
16	16-18	27-29	36-38	40-47	56-71	74-78	93-118	81-178
17	17-18	24-30	40	42-49	61-74	56-82	96-124	128-185
18	18-19	26-31	41-42	32-51	65-77	46-85	93-129	94-192
19	20	27-32	36-43	41-54	58-80	84-88	121-134	126-199
20	19-21	30-34	36-45	30-56	68-83	48-91	121-140	133-207
21	21	32-35	41-47	48-58	72-86	82-95	129-145	163-214
22	21-22	28-36	40-48	51-60	66-89	78-98	129-150	112-221
23	22-23	26-37	40-50		68-92	92-101	126-155	114-228
24	20-23	28-38	42-52		66-95	91-104	129-161	166-235
25	24	36-40	51-53		66-97	64-108	144-166	196-242
26	24-25	36-41	55		72-100	110-111	150-171	108-249
27	22-25	39-42	49-56		96-103	60-114	128-176	114-256
28	24-26	37-43	44-58		97-106	105-117	136-181	108-263
29	25-27	42-44	49-60		97-109	104-120	130-187	114-270
30	24-27	34-46	53-61		80-112	60-123	144-192	117-277
31	27-28	40-47	60-63		72-115	84-127	150-197	114-284
32	26-29	38-48	52-65		72-118	81-130	132-202	126-291
33	28-29	37-49	65-66		92-121	78-133	153-207	220-298
34	27-30	44-50	57-68		80-124	111-136	156-213	135-305
35	28-31	47-51	58-69		106-127	84-139	144-218	126-312
36	30-31	46-52	64-71		105-130	110-142	185-223	244-319
37	28-32	48-54	66-72		121-132	120-145	208-228	162-326
38	28-33	36-55	56-74		129-135	105-149	152-233	144-333
39	33	46-56	65-75		117-138	84-152	160-239	271-340
40	32-34	54-57	75-77		100-141	90-155	162-244	244-346
41	32-35	50-58	65-78		112-144	84-158	216-249	153-353
42	30-35	39-59	66-80		129-147	90-161	162-254	280-360
43	33-36	55-60	72-81		100-150	120-164	226-259	196-367
44	32-37	42-61	68-83		129-153	90-167	162-264	153-374
45	32-37	48-62	80-84		144-156	112-170	242-268	171-381
46	34-38	55-63	81-86		129-158	138-173	243-273	162-388
47	36-38	47-65	68-87		120-161	154-177	176-277	174-395
48	34-39	55-66	77-89		126-164	163-180	184-282	244-402
49	36-40	63-67	81-90		130-167	168-183	192-286	268-409
50	40	56-68	91-92		130-170	182-186	200-291	180-416

Table 4: Bounds for $N_2(g)$

g	51	52	53	54	55	56	57	58	59
$N_2(g)$	36-41	34-42	40-42	42-43	36-43	38-44	40-45	40-45	40-46

g	60	61	62	63	64	65	66	67	68
$N_2(g)$	40-47	40-47	44-48	42-48	42-49	48-50	42-50	44-51	45-51

g	69	70	71	72	73	74	75	76	77
$N_2(g)$	49-52	46-53	44-53	48-54	48-54	48-55	48-56	50-56	52-57

g	78	79	80	81	82	83	84	85	86
$N_2(g)$	48-57	52-58	56-59	48-59	53-60	52-60	57-61	52-62	56-62

g	87	88	89	90	91	92	93	94	95
$N_2(g)$	56-63	56-63	56-64	56-65	54-65	60-66	56-66	56-67	65-68

Table 5: Upper bounds for $d_q(s)$

$q\backslash s$	1	2	3	4	5	6	7	8	9	10	11	12	13
2	0	0	1	1	2	3	4	5	6	8	9	10	11
3	0	0	0	1	1	1	2	3	3	4	4	5	6
5	0	0	0	0	0	1	1	1	1	2	2	3	3

$q\backslash s$	14	15	16	17	18	19	20	21	22	23	24	25	26
2	13	15	15	18	19	19	21	23	25	25	29	31	31
3	7	7	9	9	9	11	12	12	13	13	15	15	15
5	3	3	4	4	5	5	6	7	8	9	9	9	10

$q\backslash s$	27	28	29	30	31	32	33	34	35	36	37	38
2	33	36	36	39	39	39	45	47	47	50	50	50
3	15	20	20	21	21	25	25	25	25	26	27	27
5	11	11	11	11	11	13	13	13	13	14	14	14

$q\backslash s$	39	40	41	42	43	44	45	46	47	48	49	50
2	50	54	54	59	62	65	65	65	65	69	75	77
3	29	29	29	33	34	35	35	35	37	40	40	40
5	16	17	17	18	21	21	21	21	21	22	22	22

8 REFERENCES

[1] M.J. Adams and B.L. Shader, A construction for (t, m, s)-nets in base q, *SIAM J. Discrete Math.* **10**, 460–468 (1997).

[2] A.T. Clayman, K.M. Lawrence, G.L. Mullen, H. Niederreiter, and N.J.A. Sloane, Updated tables of parameters of (t, m, s)-nets, *J. Combinatorial Designs*, to appear.

[3] A.T. Clayman and G.L. Mullen, Improved (t, m, s)-net parameters from the Gilbert-Varshamov bound, *Applicable Algebra Engrg. Comm. Comp.* **8**, 491–496 (1997).

[4] M. Drmota and R.F. Tichy, *Sequences, Discrepancies and Applications*, Lecture Notes in Math., Vol. 1651, Springer, Berlin, 1997.

[5] Y. Edel and J. Bierbrauer, Construction of digital nets from BCH-codes, *Monte Carlo and Quasi-Monte Carlo Methods 1996* (H. Niederreiter *et al.*, eds.), Lecture Notes in Statistics, Vol. 127, pp. 221–231, Springer, New York, 1997.

[6] H. Faure, Discrépance de suites associées à un système de numération (en dimension s), *Acta Arith.* **41**, 337–351 (1982).

[7] A. Garcia and H. Stichtenoth, Algebraic function fields over finite fields with many rational places, *IEEE Trans. Information Theory* **41**, 1548–1563 (1995).

[8] D.R. Hayes, Explicit class field theory for rational function fields, *Trans. Amer. Math. Soc.* **189**, 77–91 (1974).

[9] L.K. Hua and Y. Wang, *Applications of Number Theory to Numerical Analysis*, Springer, Berlin, 1981.

[10] L. Kuipers and H. Niederreiter, *Uniform Distribution of Sequences*, Wiley, New York, 1974.

[11] G. Larcher, H. Niederreiter, and W.Ch. Schmid, Digital nets and sequences constructed over finite rings and their application to quasi-Monte Carlo integration, *Monatsh. Math.* **121**, 231–253 (1996).

[12] G. Larcher and W.Ch. Schmid, Multivariate Walsh series, digital nets and quasi-Monte Carlo integration, *Monte Carlo and Quasi-Monte Carlo Methods in Scientific Computing* (H. Niederreiter and P.J.-S. Shiue, eds.), Lecture Notes in Statistics, Vol. 106, pp. 252–262, Springer, New York, 1995.

[13] K.M. Lawrence, Construction of (t, m, s)-nets and orthogonal arrays from binary codes, *Finite Fields Appl.*, to appear.

[14] K.M. Lawrence, A. Mahalanabis, G.L. Mullen, and W.Ch. Schmid, Construction of digital (t, m, s)-nets from linear codes, *Finite Fields and Applications* (S. Cohen and H. Niederreiter, eds.), London Math. Soc. Lecture Note Series, Vol. 233, pp. 189–208, Cambridge Univ. Press, Cambridge, 1996.

[15] G.L. Mullen, A. Mahalanabis, and H. Niederreiter, Tables of (t, m, s)-net and (t, s)-sequence parameters, *Monte Carlo and Quasi-Monte Carlo Methods in Scientific Computing* (H. Niederreiter and P.J.-S. Shiue, eds.), Lecture Notes in Statistics, Vol. 106, pp. 58–86, Springer, New York, 1995.

[16] H. Niederreiter, Quasi-Monte Carlo methods and pseudo-random numbers, *Bull. Amer. Math. Soc.* **84**, 957–1041 (1978).

[17] H. Niederreiter, Point sets and sequences with small discrepancy, *Monatsh. Math.* **104**, 273–337 (1987).

[18] H. Niederreiter, Low-discrepancy and low-dispersion sequences, *J. Number Theory* **30**, 51–70 (1988).

[19] H. Niederreiter, Orthogonal arrays and other combinatorial aspects in the theory of uniform point distributions in unit cubes, *Discrete Math.* **106/107**, 361–367 (1992).

[20] H. Niederreiter, Constructions of low-discrepancy point sets and sequences, *Sets, Graphs and Numbers* (Budapest, 1991), Colloquia Math. Soc. János Bolyai, Vol. 60, pp. 529–559, North-Holland, Amsterdam, 1992.

[21] H. Niederreiter, *Random Number Generation and Quasi-Monte Carlo Methods*, SIAM, Philadelphia, 1992.

[22] H. Niederreiter and C.P. Xing, Low-discrepancy sequences obtained from algebraic function fields over finite fields, *Acta Arith.* **72**, 281–298 (1995).

[23] H. Niederreiter and C.P. Xing, Low-discrepancy sequences and global function fields with many rational places, *Finite Fields Appl.* **2**, 241–273 (1996).

[24] H. Niederreiter and C.P. Xing, Quasirandom points and global function fields, *Finite Fields and Applications* (S. Cohen and H. Niederreiter, eds.), London Math. Soc. Lecture Note Series, Vol. 233, pp. 269–296, Cambridge Univ. Press, Cambridge, 1996.

[25] H. Niederreiter and C.P. Xing, Cyclotomic function fields, Hilbert class fields, and global function fields with many rational places, *Acta Arith.* **79**, 59–76 (1997).

[26] H. Niederreiter and C.P. Xing, Drinfeld modules of rank 1 and algebraic curves with many rational points. II, *Acta Arith.* **81**, 81–100 (1997).

[27] H. Niederreiter and C.P. Xing, The algebraic-geometry approach to low-discrepancy sequences, *Monte Carlo and Quasi-Monte Carlo Methods 1996* (H. Niederreiter et al., eds.), Lecture Notes in Statistics, Vol. 127, pp. 139–160, Springer, New York, 1997.

[28] H. Niederreiter and C.P. Xing, Algebraic curves over finite fields with many rational points, *Proc. Number Theory Conf.* (Eger, 1996), W. de Gruyter, Berlin, to appear.

[29] H. Niederreiter and C.P. Xing, Towers of global function fields with asymptotically many rational places and an improvement on the Gilbert-Varshamov bound, *Math. Nachr.*, to appear.

[30] H. Niederreiter and C.P. Xing, Global function fields with many rational places and their applications, *Proc. Finite Fields Conf.* (Waterloo, 1997), submitted.

[31] H. Niederreiter and C.P. Xing, Curve sequences with asymptotically many rational points, preprint, 1997.

[32] H. Niederreiter and C.P. Xing, A general method of constructing global function fields with many rational places, *Algorithmic Number Theory* (Portland, 1998), Lecture Notes in Computer Science, Springer, Berlin, to appear.

[33] W.Ch. Schmid, (t,m,s)-nets: digital construction and combinatorial aspects, Dissertation, University of Salzburg, 1995.

[34] W.Ch. Schmid, Shift-nets: a new class of binary digital (t,m,s)-nets, *Monte Carlo and Quasi-Monte Carlo Methods 1996* (H. Niederreiter et al., eds.), Lecture Notes in Statistics, Vol. 127, pp. 369-381, Springer, New York, 1997.

[35] W.Ch. Schmid and R. Wolf, Bounds for digital nets and sequences, *Acta Arith.* **78**, 377-399 (1997).

[36] J.-P. Serre, *Rational Points on Curves over Finite Fields*, Lecture Notes, Harvard University, 1985.

[37] I.M. Sobol', The distribution of points in a cube and the approximate evaluation of integrals (Russian), *Zh. Vychisl. Mat. i Mat. Fiz.* **7**, 784-802 (1967).

[38] S. Srinivasan, On two-dimensional Hammersley's sequences, *J. Number Theory* **10**, 421-429 (1978).

[39] H. Stichtenoth, *Algebraic Function Fields and Codes*, Springer, Berlin, 1993.

[40] S. Tezuka, Polynomial arithmetic analogue of Halton sequences, *ACM Trans. Modeling and Computer Simulation* **3**, 99–107 (1993).

[41] S. Tezuka and T. Tokuyama, A note on polynomial arithmetic analogue of Halton sequences, *ACM Trans. Modeling and Computer Simulation* **4**, 279–284 (1994).

[42] G. van der Geer and M. van der Vlugt, Tables for the function $N_q(g)$, preprint, 1997.

[43] G. van der Geer and M. van der Vlugt, Constructing curves over finite fields with many points by solving linear equations, preprint, 1997.

[44] G. van der Geer and M. van der Vlugt, Generalized Reed-Muller codes and curves with many points, preprint, 1997.

[45] C.P. Xing and H. Niederreiter, A construction of low-discrepancy sequences using global function fields, *Acta Arith.* **73**, 87–102 (1995).

[46] T. Zink, Degeneration of Shimura surfaces and a problem in coding theory, *Fundamentals of Computation Theory* (L. Budach, ed.), Lecture Notes in Computer Science, Vol. 199, pp. 503–511, Springer, Berlin, 1985.

INSTITUTE OF INFORMATION PROCESSING
AUSTRIAN ACADEMY OF SCIENCES
SONNENFELSGASSE 19
A-1010 VIENNA
AUSTRIA
E-mail: {niederreiter, xing}@oeaw.ac.at

Financial Applications of Monte Carlo and Quasi-Monte Carlo Methods

Shu Tezuka[1]

1 Introduction

Since Black and Scholes [2] applied a stochastic process of geometric Brownian motion to modeling the evolution of stock prices, and successfully developed an elegant mathematical theory for option pricing, financial modeling based on stochastic differential equations has become the theoretical underpinning for the practice of trading financial derivatives. Why theoretical prices derived from such stochastic models are so important in practice can be explained as follows: (1) Financial institutions need to know the theoretical price when engineering "new" financial products or when mark-to-marketing some nonliquid assets, because the market price of the product has never been observed before or because the assets have not been traded lately in the actual financial market; (2) They conduct a benchmark for comparing the theoretical price with the actual price. If the difference is not negligible, it implies an arbitrage profit opportunity or mispricing.

More sophisticated models will definitely be more advantageous for precise analysis of the market. Back when Black and Scholes did their study, they derived a closed-form solution for their pricing problems. Today, more and more complex derivatives, such as path-dependent exotic options, have been developed on the basis of underlying complicated stochastic differential equations such as HJM (Heath-Jarrow-Morton) multifactor models. This complexity makes it almost impossible to solve most pricing problems analytically. We have to resort to numerical methods in order to obtain solutions. Numerical techniques for solving such problems fall into three categories: finite-difference method, lattice method (binomial, trinomial etc.), and Monte Carlo simulation. In this article, we focus on only the last technique. For the first two techniques, we refer the interested reader to, for example, Hull [13] and Wilmott, Dewynne, and Howison [42].

Boyle [3] was the first person to apply Monte Carlo simulations to finance problems, where he tried not only simple Monte Carlo but also variance

---

[1] IBM Tokyo Research Laboratory, 1623-14 Shimotsuruma, Yamato Kanagawa, Japan 242; tezuka@jp.ibm.com

reduction techniques, such as control variates and antithetic variates, for option pricing. The main drawback of Monte Carlo approaches is their notoriously slow convergence rate. According to the central limit theorem, the convergence rate is $O(1/\sqrt{N})$, where N is the number of sample points, even if we use variance reduction methods, by which the constant term included in the O notation can be (sometimes considerably) reduced. The more complex the stochastic models become, the more computing time is needed. Similarly, the more accurate the solution is required to be, the more computing time is needed. Speeding up the computation time is indispensable for these finance applications. Even with the fastest parallel supercomputers, it is not possible to, for example, distinguish within a short time a highly profitable mortgage-backed security (MBS) from pools of collateralized mortgage obligations (CMOs) with poorer returns.

In the last five years, we have witnessed a dramatic increase in the efficiency of Monte Carlo simulations for finance applications, in terms of both accuracy and speed [14, 25, 27, 29, 30, 39]. It has been reported [25, 27] that speed-ups of as much as thousands of times relative to simple Monte Carlo simulations can be obtained for complicated fixed-income derivatives including CMOs. This has had a tremendous impact not only on the finance industry but also on the computer software business (see articles such as in Business Week [5], in The Economist [7], and in The New York Times [41]). The technology which this innovation is based on is called as quasi-Monte Carlo methods, which have been investigated by number theoreticians since Monte Carlo methods were devised by von Neumann and his colleagues in the 1940s. The idea is that by using deterministic sequences called *low-discrepancy sequences* instead of *random numbers*, we can derive more precise (deterministic) error bounds for numerical multiple integrations, and consequently achieve a significant improvement in the convergence rate. In fact, this line of research has made marked progress, particularly over the last two decades [8, 23, 34, 36].

The organization of this paper is as follows: In Section 2, we first overview financial preliminaries, in particular as regards the pricing of financial derivatives, and then describe Monte Carlo simulations and several variance reduction techniques, including antithetic variates, control variates, importance sampling, stratified sampling, common random numbers, and infinitesimal perturbation analysis. Also, several cautionary remarks are added concerning computer implementation of Monte Carlo simulations, in which we usually use "pseudo"-random numbers as an alternative to "truly" random numbers, which are not available in practice. Section 3 discusses quasi-Monte Carlo methods. First, we give the definitions of mathematical terms necessary for the subsequent discussions, such as discrepancy and low-discrepancy sequences. Then, we introduce a general method of constructing low-discrepancy sequences, which we call (t, k)-sequences, followed by a special realization of them called generalized Faure sequences. We give several results from numerical experiments with financial problems related to the Black-Scholes model as well as computing the present values of MBS, along with some discussion of these results. In the last section, we mention two important topics for future research: Monte Carlo algorithms for pricing American options and theoretical analysis of the performance of quasi-Monte Carlo methods for finance problems.

2 Monte Carlo Methods for Finance Applications

2.1 Preliminaries for derivative pricing

In this subsection, we briefly overview some preliminaries for pricing financial derivatives. For more general and comprehensive treatments of the topic, e.g., see Hull [13] and Neftci [22]. The most popular derivatives are options. There are two types of options: call options and put options. A call (or put) option is the right to buy (or sell) an underlying asset at a beforehand specified price (which we call the strike price) at a future certain date. Here, the option holder has no obligation to exercise the right. The underlying assets include stocks, interest-rates, foreign currencies, commodities, etc. In European options, we can exercise them only at the maturity date, whereas in American options, we can exercise them at any time up to the maturity date.

In 1973, Black and Scholes introduced a mathematical model describing the evolution of the price of a stock $S = \{S_t, 0 \leq t \leq T\}$. In this model, the dynamics of the stock price is assumed to follow a stochastic differential equation:

$$dS_t = rS_t dt + \sigma S_t dW_t, \tag{1.1}$$

where r is an interest rate, σ denotes the volatility of the stock, and $W = \{W_t, 0 \leq t \leq T\}$ is a Wiener process under the given probability measure P, the so-called probability measure of the risk-neutral world. The solution of this stochastic differential equation is given as

$$S_t = S_0 \exp((r - \frac{\sigma^2}{2})t + \sigma\sqrt{t}z),$$

where z is the standard normal random variable and S_0 is the initial stock price. Consider the most simple option, called a plain vanilla[2] European option. The option price for no-dividend-paying stock is given as $c = \exp(-rT)\max(S_T - K, 0)$ for a call and $p = \exp(-rT)\max(K - S_T, 0)$ for a put, where T is the maturity time and K is the strike price. The closed-form solutions for the expected values of these option prices are given by the well-known Black-Scholes pricing formula:

$$E(c) = S_0 N(d) - K\exp(-rT) N(d - \sigma\sqrt{T}), \tag{1.2}$$

---

[2]All the rest are called exotic options.

or

$$E(p) = K \exp(-rT) \, \mathrm{N}(-d + \sigma\sqrt{T}) - S_0 \, \mathrm{N}(-d), \qquad (1.3)$$

where

$$d = \frac{1}{\sigma\sqrt{T}} \log\left(\frac{S_0 \exp(rT)}{K}\right) + \frac{\sigma\sqrt{T}}{2}$$

and $\mathrm{N}(x)$ is the cumulative probability distribution function for the standard normal random variable. Even if we use the Black-Scholes model for the underlying stock price, there are many cases of exotic options for whose prices we cannot obtain closed-form solutions. One example is Asian (call) options, whose payoff is $\max(\bar{S} - K, 0)$, where $\bar{S}$ is the arithmetic mean of the underlying stock price during a certain time period. Therefore, in practice, we usually employ numerical methods for computing the solution. For the case of fixed-income derivatives that include interest-rate exotic options and mortgage-backed securities, the stochastic differential equation for interest-rate or bond prices are much more complicated than the Black-Scholes models. For these applications, Monte Carlo simulations are considered indispensable in practice.

2.2   Variance reduction techniques

We use Monte Carlo simulations to estimate some parameter θ, such as the price of a derivative in finance applications. By using independent identically distributed (i.i.d.) sequences of random numbers, we compute statistically independent estimates $\hat{\theta}_i, i = 1, 2, ..., n$, for the parameter θ. Here, it is natural to assume that the expectation of $\hat{\theta}$ is equal to θ. Then, for large n, the sample mean of n estimates

$$\bar{\theta} = \frac{1}{n} \sum_{i=1}^{n} \hat{\theta}_i$$

is known to follow the normal distribution with mean θ and variance σ^2/n according to the central limit theorem, where σ^2 is the variance of $\hat{\theta}$. Therefore, for example, the 99% confidence interval for the parameter θ is given as follows:

$$\bar{\theta} - \frac{2.575\sigma}{\sqrt{n}} < \theta < \bar{\theta} + \frac{2.575\sigma}{\sqrt{n}}.$$

This is the fundamental principle of Monte Carlo methods: the standard error of the estimate $\bar{\theta}$ is inversely proportional to the square-root of the number of samples n. That is to say, in order to get one more digit accuracy, we need to increase the number of samples by 100.

The idea of variance reduction techniques is to somehow find an estimate $\hat{\theta}$ such that the associated variance σ^2 is as small as possible. Then, we can make Monte Carlo simulations more efficient. Here, we overview six variance reduction techniques: antithetic variates, control variates, importance sampling, stratified sampling, common random numbers, and IPA (infinitesimal perturbation analysis). The first four techniques can be applied to general Monte Carlo problems, while the last two are specifically suitable for Monte Carlo problems in sensitivity analysis. In this survey, we omitted the Brownian bridge technique for variance reduction. The interested reader should refer to [20].

Antithetic variates

Assume that the estimate $\hat{\theta} = f(u_1, ..., u_k)$, where $(u_1, ..., u_k)$ is a uniformly distributed random vector in $[0, 1)^k$. In this technique, we compute two estimates $f(u_1, ..., u_k)$ and $f(1 - u_1, ..., 1 - u_k)$, and take the average

$$g(u_1, ..., u_k) = \frac{f(u_1, ..., u_k) + f(1 - u_1, ..., 1 - u_k)}{2}.$$

The rationale of this technique is as follows: We have

$$\text{Var}(g) = \frac{1}{2}(\text{Var}(f) + \text{Cov}(f, \bar{f})),$$

where $\bar{f}(u_1, ..., u_k) \overset{\text{def}}{=} f(1 - u_1, ..., 1 - u_k)$. Keeping in mind that the computation time for g is twice that of f, we can see that if $\text{Cov}(f, \bar{f}) < 0$, then the antithetic method becomes efficient. In finance applications, we should use the inversion method (see [6, 19]) from a uniform random variable u to a normal variable $z = N^{-1}(u)$ so that $(u_1, ..., u_k)$ is transformed into $(z_1, ..., z_k)$, while $(1 - u_1, ..., 1 - u_k)$ is transformed into $(-z_1, ..., -z_k)$. If f has some monotone property, we can prove that $\text{Cov}(f, \bar{f}) \leq 0$ [4].

Control variates

Suppose that we are interested in the expected value of the parameter $\theta(x)$:

$$I = \int \theta(x)f(x)dx,$$

where $f(x)$ is the probability density function of the random variable x. In the control variate technique, we rewrite the above as follows:

$$I = \int (\theta(x) - g(x))f(x)dx + \int g(x)f(x)dx,$$

where we assume that the second term $\int g(x)f(x)dx$ can be analytically solved. If we can find $g(x)$ such that

$$\text{Var}_f(\theta(x) - g(x)) < \text{Var}_f(\theta(x)),$$

this technique proves to be advantageous. A well-known example in finance for which the control variate technique is very efficient is the following: Let

$\theta(x)$ be the price of an Asian stock option, and let $g(x)$ be the price of a modified Asian option in which the arithmetic mean is replaced by the geometric mean. If we use the Black-Scholes model for the underlying stock price, the closed-form solution for the price $g(x)$ is available. Since $g(x)$ is closely approximate to $\theta(x)$, this technique is useful in practice. For further discussion, see [4, 13].

Importance sampling

The idea of this technique is to take a sampling strategy of putting more focus on more likely scenarios. The expected value of the parameter θ,

$$I = \int \theta(x)f(x)dx,$$

where $f(x)$ is the probability density function, is changed to

$$I = \int \left[\frac{\theta(x)f(x)}{\tilde{f}(x)}\right] \tilde{f}(x)dx,$$

where $\tilde{f}(x)$ is an appropriately chosen probability density function. If $\theta(x)$ is non-negative, it is known that $\tilde{f}(x) = \theta(x)f(x)/I$ minimizes

$$\mathrm{Var}_{\tilde{f}}(I) = \int \left[\frac{\theta^2(x)f^2(x)}{\tilde{f}^2(x)}\right] \tilde{f}(x)dx - I^2,$$

but the value of I is unknown because this is exactly what we want to determine. As an example of finance applications, consider the pricing of deep out-of-the-money call options [13]. In this case, the probability density $f(x)$ lies predominantly on $x < K$. Therefore, the price almost always becomes zero, and thus this sampling is inefficient. If we can choose the density as

$$\tilde{f}(x) \approx \frac{\max(x - K, 0)f(x)}{I},$$

then the variance can be significantly reduced. One idea is to choose $\tilde{f}(x)$ as a density function corresponding to stochastic models that have much higher drift and volatility.

Stratified sampling

This technique is a kind of divide-and-conquer heuristic. We divide the domain of random variables into subintervals, and get samples from each subinterval according to its appropriate probability distribution. In finance applications, we usually need a standard normal distribution. In this technique, we generate n uniform random variables as follows: For $i = 1, 2, ..., n$,

$$x_i = \frac{u_i}{n} + \frac{i-1}{n},$$

where $u_i, i = 1, ..., n$, are independent uniform random variables in $[0, 1)$. By the inverse transform N^{-1} of the standard normal distribution, we get $z_i = N^{-1}(x_i), i = 1, ..., n$. If we compare them with n independent normal variables, $\bar{z}_i = N^{-1}(u_i), i = 1, ..., n$, it is easy to see that the empirical distribution of z_i matches the theoretical distribution better than $\bar{z}_i$. This leads to a significant reduction in the variance of the estimates. However, since the samples $z_i, i = 1, ..., n$, are highly dependent, we need to conduct many runs to obtain the confidence interval for the estimate. This considerably sacrifice the reduction of the variance introduced by the technique. We should carefully look at this tradeoff when implementing this technique.

In higher dimensions, the straightforward generalization of this technique is not efficient, because the number of subintervals grows exponentially in the dimension size. One approach to overcoming this difficulty is called *Latin hypercube sampling*, which is a combination of stratified sampling and random permutation (see [18]).

Common random numbers

In finance applications, sensitivity analysis is very important for hedging purposes in derivative pricing as well as for stress testing in risk management. For example, traders want to check whether their portfolios are immune to small changes in the prices of the underlying assets over a specified period. They are also concerned about how the value of their position is affected by alternative future scenarios. The sensitivities of the payoff function are usually considered in terms of small changes in the underlying asset prices, the volatility, the drift (risk-free interest-rate), and so on. In finance literature, these sensitivities are called *Greeks*. In the case of the Black-Scholes formulas in equations (1.2) and (1.3), we can analytically obtain the first and higher derivatives of the payoff functions for plain vanilla stock options. But for most exotic options with Black-Scholes models and with more complicated models, it becomes difficult to obtain closed-form solutions of all Greeks. Monte Carlo simulations again become important as a last resort.

In these applications, we need to compute the difference between the prices associated with two or more settings of financial parameters. Mathematically, the difference is given as

$$E[P_{s_1}] - E[P_{s_2}],$$

where $E[P_{s_1}]$ or $E[P_{s_2}]$ is the expected value of the price P under a scenario s_1 or s_2, respectively. Numerically, this is computed as follows:

$$\frac{1}{N} \sum_{i=1}^{N} P_{s_1}(R_i^{(1)}) - \frac{1}{N} \sum_{i=1}^{N} P_{s_2}(R_i^{(2)}) = \frac{1}{N} \sum_{i=1}^{N} (P_{s_1}(R_i^{(1)}) - P_{s_2}(R_i^{(2)})),$$

where $R_i^{(1)}, R_i^{(2)}, i = 1, ..., N$, are $2N$ independent sequences of random numbers. In *Common Random Numbers*, we set $R_i^{(1)} = R_i^{(2)}$ for all $i = 1, ..., N$. This technique employs the *same* seed values of random number sequences for both of the Monte Carlo simulations. It is known that it can significantly reduce the variance of the estimate because of the positive correlation of the two simulations.

In numerically computing sensitivities such as Greeks for the price of a derivative, the first-order derivative of the payoff function in terms of the underlying asset price is obtained by the common random numbers technique in the following way:

$$\frac{d\,\mathrm{E}[P]}{d\theta} \approx \frac{1}{N} \sum_{i=1}^{N} \frac{P(\theta + \epsilon; R_i) - P(\theta - \epsilon; R_i)}{2\epsilon},$$

where ϵ is a small change in the parameter value θ and P is the price of a derivative. We should notice that there is a bias in the above approximate estimate because we replaced a (continuous) function by a (discrete) finite difference.

Infinitesimal perturbation analysis (IPA)

For computing Greeks, we have a more elegant approach than common random numbers, provided that the derivative price has some smoothness property in terms of the parameter change. The technique, called as *Infinitesimal Perturbation Analysis (IPA)*, is a well-established technology in the field of discrete event simulation, and is widely applicable to manufacturing operations, transportation, inventory control, and so on. (See Ho and Cao [12] for a comprehensive treatment of this technology.)

The key idea of the technique is to consider the continuous limit of the common random number technique. That is to say, we assume that

$$\frac{d\,\mathrm{E}[P]}{d\theta} = \mathrm{E}[\frac{dP}{d\theta}],$$

where the IPA estimator, $dP/d\theta$, is the sample derivative of the random variable $P(\theta, \omega)$, defined by

$$\frac{dP}{d\theta} = \lim_{\Delta\theta \to 0} \frac{P(\theta + \Delta\theta, \omega) - P(\theta, \omega)}{\Delta\theta} \text{ with probability 1,}$$

where ω is a sample point in the underlying sample space. If the derivative and the expectation is interchangeable, the IPA estimator becomes unbiased. A more rigid mathematical treatment of interchangeability in the general context of simulation can be found in Glasserman [10].

For example, we consider the sensitivity of the payoff function of a European call option with respect to the strike price. Specifically, we have the payoff function

$$P = \exp(-rT) \max(S_T - K, 0).$$

The IPA estimator is obtained as

$$\frac{dP}{dK} = -\exp(-rT)I(S_T - K),$$

where $I(x)$ is defined to be 1 if $x \geq 0$, or 0 otherwise. In this case, it is easy to prove that the estimator is unbiased, because the interchangeability holds if the stock price S_T is independent of the strike price K. Compared with common random number techniques, the major advantages of IPA are that it can (1) provide an unbiased estimator for the sensitivity analysis, and (2) compute the estimator from a single simulation run without running one more simulation with a perturbed parameter value. We should notice that the IPA technique has been extended so that we can deal with exotic options, including discontinuities in the payoff functions. For more detail on this topic, see Fu and Hu [9].

2.3   Caveats for computer implementation

Since it is impossible to produce "randomness" on a computer, which is totally deterministic, instead we usually use "pseudo-random" numbers, which are generated by means of algorithmic methods. Many algorithms for this purpose have been proposed since the days of von Neumann, and they have been applied to a battery of statistical tests to ensure that the sequences generated on computers closely mimic truly random sequences. In this section, we give two examples to draw the user's attention to the possibility that "pseudo-randomness" can lead to miserable results unless we use fully tested, reliable random number generators. One example involves uniform random numbers. In computer simulation, uniform random numbers are fundamental, because when we need random numbers with a particular probability distribution, we transform uniform random numbers mathematically or algorithmically to the targeted probability distribution. The other example concerns the transformation of uniform random numbers to the normal distribution, which is especially important in finance applications, because Brownian motion is based on the normal distribution.

Anomaly of ran1()

The first pathological example is the anomaly of ran1() in Numerical Recipes in C (the first edition), which was first discovered by Paskov [29]. Figure 1 shows the convergence behaviors of MBS computation with ran1() by using different seed values, where the x-axis represents the number of sample paths and the y-axis represents the price of MBS. In this case, the correct price is 143.0182. The specification of MBS is given in detail in Section 3, and all of the parameter values are the same as those used for Figure 4. As is easily seen, three different seed values make the Monte Carlo

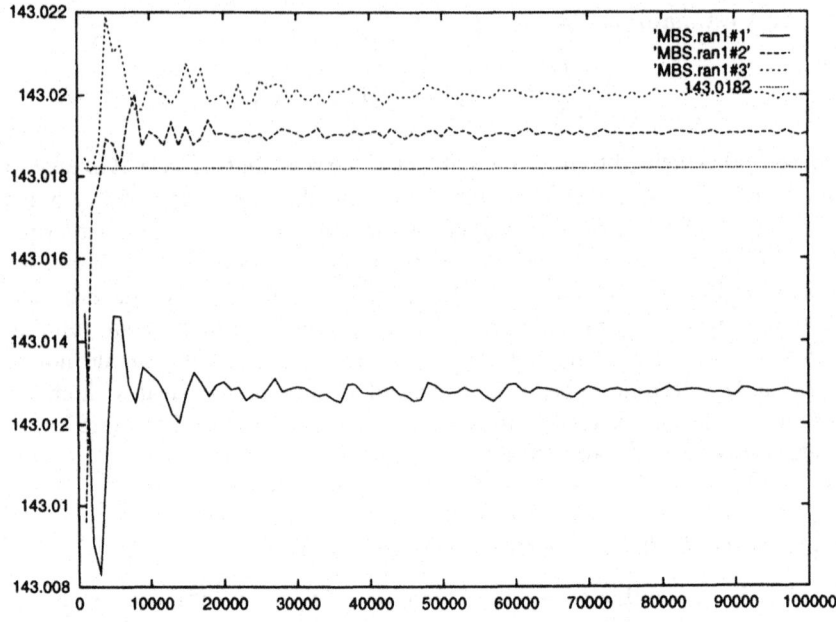

FIGURE 1. Anomaly of ran1() with MBS

simulation converge to three different answers. This phenomenon was successfully analyzed by Tajima, Ninomiya, and Tezuka [35]. The main reason
was found to be that the sequence generated by ran1() has a semi-periodic
structure with a cycle length of $3,888,000 = 360 \times 10800$. Note that the
number 360 is equal to the dimension-size of the simulation; in other words,
each sample estimate of the price is computed from a random point in the
360-dimensional unit hypercube. Therefore, we can see from the figure that
the convergence becomes almost stable when the number of sample paths
exceeds 10000. Fortunately, this generator ran1() was replaced by a different one in the second edition of the book. However, this suggests that
(1) even publicly used generators are not necessarily a reliable source of
random numbers and (2) we have to carefully choose our own (uniform)
random number generators. Several recommended random number generators, which have already been extensively tested, can be found in Knuth
[15] and Tezuka [36]; see also the survey article by Hellekalek in this same
volume.

Neave effect

The second pathological phenomenon due to "pseudo-randomness" is the
so-called Neave effect [21], which should also be briefly explained here (for
more detail, see, e.g. Ripley [32] and Tezuka [36]). One of the most popular
methods for generating normally distributed random numbers is the Box-
Muller method, but in 1973, Neave pointed out that the combination of

this method with some linear congruential sequences may produce large deviations from the correct normal distribution. Figure 2 shows an example of a pathological tail distribution of the sequences generated from the Box-Muller method with a widely-used linear congruential generator GGL. To be precise, the sequence of normal variates, $Z_n, n = 1, 2, ...$, is produced as follows: For $n = 1, 2, ...$, let

$$Z_n = \sqrt{-2 \log(\frac{X_{2n}}{M})} \sin(2\pi \frac{X_{2n+1}}{M}),$$

where $X_{2n+1} = 16807 X_{2n} \pmod{M}$ with $M = 2^{31} - 1$. It is easy to check that the deviation is significantly large in a statistical sense (see Tezuka [36] for more details).

It is also known among experts in simulation (see Afflerbach and Wenzel [1] and Ripley [32]) that Marsaglia's polar method, a variant of the Box-Muller method, described in Knuth [15] and in Numerical Recipes in C [31], also suffers from problems similar to Neave effect. Since, in general, commonly used uniform random number generators use a first order recursion $X_n = f(X_{n-1})$, the consecutive terms of the sequence must have a certain structure. Therefore, algorithms that transform two or more uniform random variables to obtain one variable of the targeted distribution will more or less suffer from some undesirable properties. In the case of normal variate generation, we recommend the inversion method [6], which transforms one uniform variable into one normal variable by inverting the normal distribution function. A very efficient C program by Moro [19] is available for this purpose.

3   Speeding up by Quasi-Monte Carlo Methods

3.1   What are quasi-Monte Carlo methods?

The main drawback of Monte Carlo methods is their slow convergence rate $O(1/\sqrt{N})$ for N sample paths. We can improve the convergence speed by applying the variance reduction techniques described in the previous section, but the improvement reduces the constant factor included in the O term, leaving the same rate in terms of N. Finance-related Monte Carlo problems, particularly those related to derivative pricing, can be formulated as problems of computing multidimensional integration. The dimension of the integration is naturally equal to the number of time steps in the time period considered. Once a problem can be formulated as one of numerical multidimensional integration, we have several "deterministic" approaches for computing it. However, the direct extensions of one-dimensional approaches, such as trapezoidal rules, to higher-dimensional ones do not work, because of the curse of dimension. For example, the error bound for the

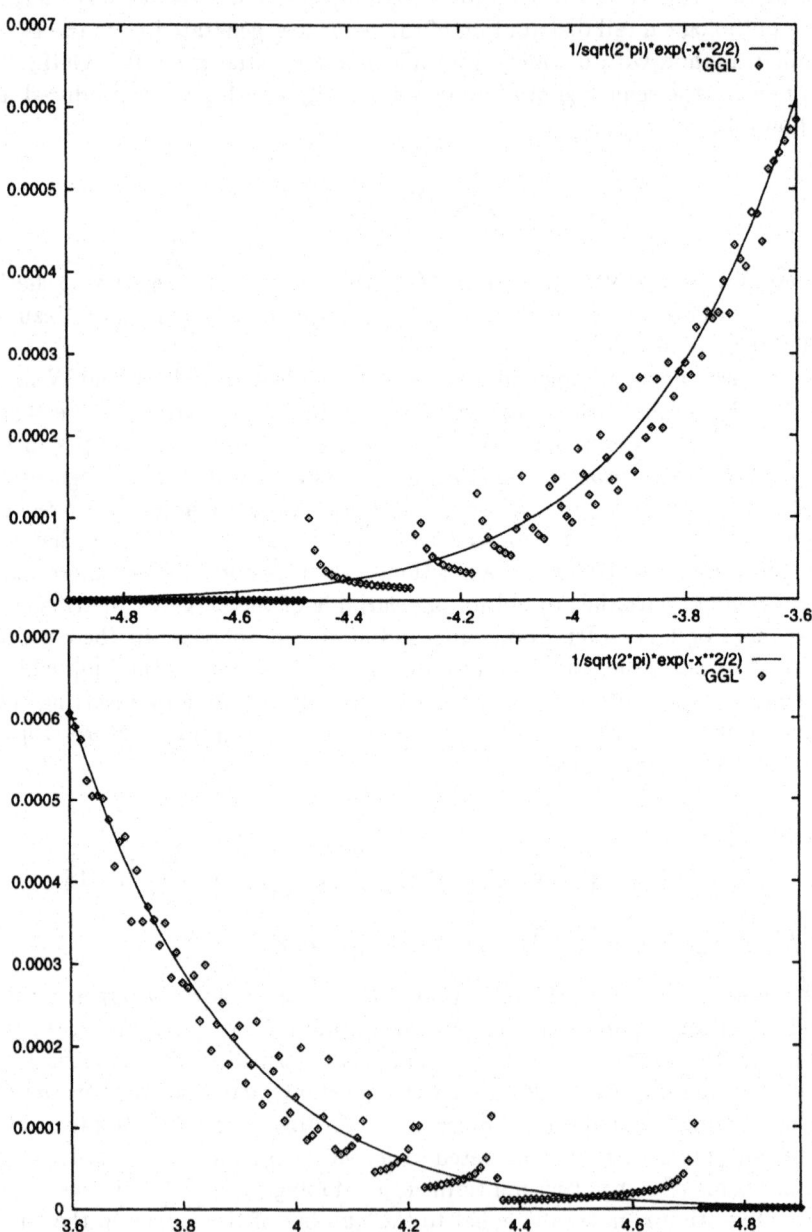

FIGURE 2. Neave effect of the Box-Muller method with the linear congruential generator GGL, $X_i = 16807X_{i-1} \pmod{2^{31} - 1}$

k-dimensional trapezoidal rules is known to be $O(N^{-2/k})$, which means that the error grows exponentially as the dimension size becomes larger.

We have a much better approach, called quasi-Monte Carlo methods, which give the error bound $O((\log N)^k/N)$ for the k-dimensional integration problem. The idea is to use the point sets, not randomly distributed, but uniformly distributed throughout the domain of integration. The extent to which the points are uniform has been mathematically defined as their *discrepancy*. The more uniformly distributed the points are, the lower the discrepancy. The so-called *low-discrepancy sequences*[3] are to quasi-Monte Carlo methods as random numbers are to Monte Carlo methods. The practical advantage of low-discrepancy sequences is that for every $n > 1$, the first n points of a low-discrepancy sequence are uniformly distributed; in other words, we can add one point after another so that the entire set of points at any time remains very uniform throughout the domain. Therefore, we can use them sequentially until some stopping rule for the computation is met.

The use of low-discrepancy sequences for finance problems has just started in around 1992. Originally, Paskov and Traub used Halton sequences and Sobol' sequences for pricing a ten-tranch CMO (Collateralized Mortgage Obligation), which they obtained from Goldman-Sachs, and reported that Sobol' sequences performed very well relative to simple Monte Carlo methods, as well as to antithetic Monte Carlo methods [29, 30]. In 1994, Joy, Boyle, and Tan [14] applied Faure sequences to several equity derivatives to obtain good performances. Notice that at that time Faure sequences were the "theoretically" best of all known low-discrepancy sequences [4]. Then, Ninomiya and Tezuka [25] reported that generalized Niederreiter sequences could provide further speed-up over Halton, Sobol', and Faure sequences, and that they had observed a speed-up of about 1000 times over simple Monte Carlo for three pricing problems: a discount bond, an interest-rate lookback option, and MBS. Papageorgiou and Traub [27] reported that generalized Faure sequences, a special subset of generalized Niederreiter sequences, perform consistently better than their improved Sobol' sequences for pricing their CMO, and may allow speed-ups of about 1000 times relative to simple Monte Carlo simulations in cases where high accuracy is desired. In 1995, on the other hand, Niederreiter and Xing [24] presented a new construction method of low-discrepancy sequences based on algebraic function fields. (See also the survey article by Niederreiter and Xing in this present volume.) Today, their sequences are known as the "theoretically" best, but there are several implementation issues to be overcome before these sequences become available for practical use. In this section,

---

[3]Some people call them *quasi-random sequences*, but this term is a misnomer, since low-discrepancy sequences are totally deterministic.

[4]By "theoretically," we mean in terms of the dependency on the dimension of the constant factor of the leading term $(\log N)^k/N$ in the discrepancy upper bound.

we describe generalized Faure sequences in detail.

3.2 Generalized Faure sequences

First, we recall the definition of discrepancy. For N points $X_0, X_1, ..., X_{N-1}$ in $[0,1]^k$, and a subinterval $J = \prod_{h=1}^{k}[0, u_h), k \geq 1$, where $0 < u_h \leq 1$ for $1 \leq h \leq k$, we define the *discrepancy* as

$$D_N^{(k)} = \sup_J \left| \frac{A(J; N)}{N} - \text{Vol}(J) \right|,$$

where $A(J; N)$ is the number of $n, 0 \leq n < N$, with $X_n \in J$, and where $\text{Vol}(J)$ is the volume of J, with the supremum extended over all subintervals J. If we employ the L_2-norm instead of the L_∞-norm in the above, we can define L_2-*discrepancy*, $T_N^{(k)}$, as

$$(T_N^{(k)})^2 = \int_{[0,1]^k} \left(\frac{A(J; N)}{N} - \text{Vol}(J) \right)^2 du_1 \cdots du_k.$$

It is easy to see that $T_N^{(k)} \leq D_N^{(k)}$.

The Koksma-Hlawka theorem describes the relation between discrepancy and numerical integration [23]:

Theorem 1 (Koksma-Hlawka) *If the integrand f is of bounded variation $V(f)$ on the k-dimensional unit hypercube $[0,1]^k$ in the sense of Hardy and Krause, then for any $X_0, X_1, ..., X_{N-1} \in [0,1)^k$ we have*

$$\left| \frac{1}{N} \sum_{n=0}^{N-1} f(X_n) - \int_{[0,1]^k} f(u_1, ..., u_k) du_1 \cdots du_k \right| \leq V(f) D_N^{(k)}. \qquad (1.4)$$

We also have another important result, the Woźniakowski theorem [39, 43]:

Theorem 2 (Woźniakowski) *Let C_k be the class of real continuous functions defined on $[0,1]^k$ equipped with the classical Wiener sheet measure w (that is, Gaussian with mean zero and covariance kernel*

$$R(\mathbf{s}, \mathbf{t}) \stackrel{\text{def}}{=} \int_{C_k} f(\mathbf{s}) f(\mathbf{t}) \; w(df) = \min(\mathbf{s}, \mathbf{t}) \stackrel{\text{def}}{=} \prod_{j=1}^{k} \min(s_j, t_j)$$

for any vectors $\mathbf{s} = (s_1, ..., s_k)$ and $\mathbf{t} = (t_1, ..., t_k)$ in $[0,1]^k$). Then, for a given set of points $X_i = (x_{i1}, ..., x_{ik}), i = 0, 1, ..., N-1$, in $[0,1]^k$, we have

$$\int_{C_k} \left(\frac{1}{N} \sum_{n=0}^{N-1} f(X_n) - \int_{[0,1]^k} f(u_1, ..., u_k) du_1 \cdots du_k \right)^2 w(df) = (\bar{T}_N^{(k)})^2,$$

where $\bar{T}_N^{(k)}$ is the L_2-discrepancy of the point set $\bar{X}_i = (1 - x_{i1}, ..., 1 - x_{ik}), i = 0, 1, ..., N-1$.

The theorem means that on average the integration error is dependent only on the discrepancy, not on the integrand f, and that low-discrepancy sequences are useful in integration. Both of the above theorems tell us that the lower the discrepancy is, the smaller the integration error will be.

We now introduce the definition of low-discrepancy sequences:

Definition 1 *If a sequence $X_0, X_1, ...$ in $[0, 1]^k$ satisfies the condition that for all $N > 1$, the discrepancy of the first N points is*

$$D_N^{(k)} \leq C_k \frac{(\log N)^k}{N},$$

where C_k is a constant depending only on the dimension k, then we call it a low-discrepancy sequence.

Notice that the order of magnitude in terms of N in the right-hand side is believed to be the optimal upper bound. Therefore, the sole difference among the many types of low-discrepancy sequences is how small the constant term C_k is. In this article, we concentrate on the construction of low-discrepancy sequences based on (t, k)-sequences. Before introducing the definition of (t, k)-sequences in base b, we need the following notions:

Definition 2 *A b-ary box is an interval of the form*

$$E = \prod_{h=1}^{k} [a_h b^{-d_h}, (a_h + 1)b^{-d_h})$$

with integers $d_h \geq 0$ and integers $0 \leq a_h < b^{d_h}$ for $1 \leq h \leq k$.

Definition 3 *Let $0 \leq t \leq m$ be an integer. A (t, m, k)-net in base b is a point set of b^m points in $[0, 1]^k$ such that $A(E; b^m) = b^t$ for every b-ary box E with $\mathrm{Vol}(E) = b^{t-m}$.*

Now, we define (t, k)-sequences in base b.

Definition 4 *Let $0 \leq t \leq m$ be an integer. A sequence of points $X_0, X_1, ...,$ in $[0, 1]^k$ is called a (t, k)-sequence if for all integers $j \geq 0$ and $m > t$, the point set consisting of $[X_n]_m$ with $jb^m \leq n < (j+1)b^m$ is a (t, m, k)-net in base b, where $[X]_m$ denotes the coordinate-wise m-digit truncation in base b of X.*

Following Sobol' and Faure's results, Niederreiter [23] obtained the following theorem for an arbitrary integer base $b \geq 2$:

Theorem 3 (Niederreiter) *For any $N > 1$, the discrepancy of the first N points of a (t, k)-sequence in base b satisfies*

$$D_N^{(k)} \leq c(t, k, b) \frac{(\log N)^k}{N} + O\left(\frac{(\log N)^{k-1}}{N}\right)$$

where $c(t, k, b) \approx \frac{b^t}{k!}\left(\frac{b}{2 \log b}\right)^k.$

This means that if t and b are constant or depend only on k, then the (t, k)-sequence becomes a low-discrepancy sequence. Note that a smaller value of t gives a lower discrepancy.

Niederreiter presented a general construction principle for (t, k)-sequences as follows: Let $k \geq 1$ and $b \geq 2$ and $B = \{0, 1, ..., b-1\}$. Accordingly, we define

(i) a commutative ring R with identity and $\text{card}(R) = b$;

(ii) bijections $\psi_j : B \rightarrow R$ for $j = 1, 2, ...$, with $\psi_j(0) = 0$ for all sufficiently large j;

(iii) bijections $\lambda_{hi} : R \rightarrow B$ for $h = 1, 2, ..., k$ and $i = 1, 2, ...$, with $\lambda_{hi}(0) = 0$ for $1 \leq h \leq k$ and all sufficiently large i;

(iv) elements $c_{ij}^{(h)} \in R$ for $1 \leq h \leq k, 1 \leq i, 1 \leq j$, where for fixed h and j we have $c_{ij}^{(h)} = 0$ for all sufficiently large i.

For $n = 0, 1, 2, ...$, let $n = \sum_{r=1}^{\infty} a_r(n) b^{r-1}$ for $a_r(n) \in B$. Set the h-th coordinate of the point X_n as

$$X_n^{(h)} = \sum_{i=1}^{\infty} x_{ni}^{(h)} b^{-i},$$

for $1 \leq h \leq k$ and $0 \leq n$, with

$$x_{ni}^{(h)} = \lambda_{hi} \left(\sum_{j=1}^{\infty} c_{ij}^{(h)} \psi_j(a_j(n)) \right) \in B$$

for $1 \leq h \leq k, 1 \leq i$, and $0 \leq n$. We call $C^{(h)} = (c_{ij}^{(h)})$ the *generator matrix* of the h-th coordinate of a (t, k)-sequence. For more information on (t, k)-sequences, see the survey article by Larcher in this present volume.

Hereafter, we assume that b is a prime power, and that R in the construction principle is the finite field $GF(b)$. Let

$$\mathbf{c}_m^{(h)}(l) = (c_{m,1}^{(h)}, ..., c_{m,l}^{(h)}) \in GF(b)^l,$$

and let

$$C(d_1, ..., d_k; l) = \{\mathbf{c}_m^{(h)}(l) \mid 1 \leq m \leq d_h, 1 \leq h \leq k\}.$$

Define $\rho(C; l)$ as the maximal integer d such that $C(d_1, ..., d_k; l)$ is linearly independent over $GF(b)$ for all nonnegative integers $d_1, ..., d_k$ satisfying $\sum_{1 \leq h \leq k} d_h = d$. Niederreiter [23] obtained the following theorem:

Theorem 4 *Let b be a prime power and let $R = GF(b)$. If an integer $t \geq 0$ satisfies $t \geq l - \rho(C; l)$ for each integer $l > t$, then the sequence defined in the above becomes a (t, k)-sequence in base b.*

This result tells us that in order to construct low-discrepancy sequences, we need generator matrices, $C^{(h)}, 1 \leq h \leq k$, such that $\rho(C; l)$ is as large as possible for all $l > t$.

We now describe how to construct such generator matrices so that we obtain low-discrepancy sequences called generalized Niederreiter sequences (see Tezuka [36]). The construction is based on the formal Laurent series expansions over finite fields. Denote $S(z) \in GF\{b, z\}$ by

$$S(z) = \sum_{j=w}^{\infty} a_j z^{-j},$$

where all $a_j \in GF(b)$ and w is an arbitrary integer. Hereafter, we use the following notations: $[S(z)]$ denotes the polynomial part of $S(z) \in GF\{b, z\}$ and $[S(z)]_{p(z)} \stackrel{\text{def}}{=} [S(z)]$ (mod $p(z)$) with $0 \leq \deg([S(z)]_{p(z)}) < \deg(p(z))$.

Let polynomials $p_1(z), ..., p_k(z) \in GF[b, z]$ be pairwise coprime and let $e_h = \deg(p_h) \geq 1$ for $1 \leq h \leq k$. For $m \geq 1, 1 \leq h \leq k$, and $j \geq 1$, consider the expansion

$$\frac{y_{hm}(z)}{p_h(z)^j} = \sum_{r=w}^{\infty} a^{(h)}(j, m, r) z^{-r},$$

by which the elements $a^{(h)}(j, m, r) \in GF(b)$ are determined. Here $w \leq 0$ may depend on h, j, m, and each $y_{hm}(z)$ is a polynomial such that the residue polynomials $[y_{hm}(z)]_{p_h(z)}, (j-1)e_h \leq m-1 < je_h$, are linearly independent over $GF(b)$ for any $j > 0$ and $1 \leq h \leq k$. Then define

$$c_{mr}^{(h)} = a^{(h)}(m_h + 1, m, r),$$

for $1 \leq h \leq k, m \geq 1$, and $r \geq 1$, where $m_h = [(m-1)/e_h]$.

Tezuka [36] proved the following theorem:

Theorem 5 *If an integer $t \geq 0$ satisfies $t \geq \sum_{h=1}^{k} \deg(p_h) - k$, then the generalized Niederreiter sequence becomes a (t, k)-sequence in base b.*

Remark 1 Faure sequences are a special case of generalized Niederreiter sequences such that (1) $b \geq k$ is prime, (2) $p_h(z) = z - h + 1$ for $1 \leq h \leq k$, and (3) all $y_{hm}(z) = 1$.

This remark motivates the following definition:

Definition 5 *Generalized Faure sequences are defined as $(0, k)$-sequences in a prime base $b \geq k$ obtained from generalized Niederreiter sequences.*

In matrix representation, for all generator matrices $C^{(h)}, 1 \leq h \leq k$, we have

$$C^{(h)} = A^{(h)} P^{h-1}, \tag{1.5}$$

where $A^{(h)}, 1 \leq h \leq k$, are nonsingular lower triangular matrices over $GF(b)$ and P is the Pascal matrix whose (i, j) element is equal to $\binom{j-1}{i-1}$. The original Faure sequences [8] correspond to the case in which $A^{(h)} = I$ for all h. Regarding an efficient generation method, Tezuka [36] describes a generalization by the b-ary Gray code of the Antonov-Saleev implementation of Sobol' sequences.

Recently, following Owen [26] and Hickernell [11], Matoušek [16, 17] analyzed the integration errors for randomly chosen generalized Faure sequences, assuming that the integrand is a fixed sufficiently smooth real function. The results imply that the expected integration error over all generalized Faure sequences is

$$O\left(\frac{(\log N)^{(k-1)/2}}{N^{3/2}}\right).$$

Note that this can be interpreted as an existence theorem of a very good generalized Faure sequence for numerical multidimensional integration.

3.3   Numerical Experiments

In this section, we apply quasi-Monte Carlo methods to two types of practical problems related to pricing financial derivatives. The first is numerical simulation of the first-order Euler approximation of the Black-Scholes model. The second, originally described by Paskov [29], concerns mortgage-backed securities (MBS), the most popular among fixed-income derivatives. In the experiments, we used a randomly chosen generalized Faure sequence as a low-discrepancy sequence for quasi-Monte Carlo simulations. More precisely, nonsingular lower triangular matrices $A^{(h)}, 1 \le h \le k$, in equation (1.5) for the generator matrices were chosen at random, and we omitted the first 100000 points; that is to say, we used the points $X_i, i = 100001, 100002, ...,$ of the sequence. We used the random number generator CombTaus [37] for Monte Carlo simulation. For both of the problems, we used Moro's C program for the generation of normal random variates [19].

The Black-Scholes model

As mentioned in Section 2, Black and Scholes describe the dynamics of the stock price by the stochastic differential equation (1.1), which is equivalent to

$$d\log S_t = (r - \frac{\sigma^2}{2})dt + \sigma dW_t.$$

Now, we consider the discrete-time version of this equation, by discretizing the time axis into T time-steps. For $t = 1, 2, ..., T$,

$$\log S_t - \log S_{t-1} = (r - \frac{\sigma^2}{2})\Delta + \sigma dB,$$

where $\Delta = 1/365$ (one day), dB is the normally distributed random variable with mean zero and variance Δ, and S_0 is the current price of the stock. It is easy to analytically derive S_T, the price at time T. Denote $A = S_0 \exp((r - \sigma^2/2)(T\Delta))$ and $B = \sigma\sqrt{\Delta}$. We have

$$S_T(z_1, ..., z_T) = A \exp(B \sum_{i=1}^{T} z_i), \tag{1.6}$$

where $z_i, i = 1, ..., T$, are the independent random variables from the standard normal distribution.

What we computed is the expected value of S_T, the stock price at the maturity date T. This value can be analytically calculated as

$$E(S_T) = A \exp(\frac{B^2 T}{2}).$$

We can also obtain the variance of S_T as follows:

$$\text{Var}_Z(S_T) = \int S_T^2 dZ - (E(S_T))^2 = A^2 \exp(B^2 T)(\exp(B^2 T) - 1). \tag{1.7}$$

Figure 3 shows the computation results using equation (1.6) for comparison of the Monte Carlo and quasi-Monte Carlo simulations for the case of $S_0 = 100, r = 0.05, \sigma = 0.3$, and $T = 100$ (i.e., the dimension size of the integration is 100), where each estimate of S_T is obtained from one sample path $(z_1, ..., z_{100})$, and thereby the arithmetic mean of N estimates is computed for $1 \leq N \leq 5000$. The solid line (BS.QMC) shows the result for quasi-Monte Carlo methods using generalized Faure sequences, while the dotted line (BS.MC) shows the result for Monte Carlo methods. In this case, $E(S_T) = 101.379289...$ As is easily seen from the figure, with only about 1000 sample paths, quasi-Monte Carlo methods converge to the correct value within an accuracy of 0.001 relative error, while the standard deviation of Monte Carlo with 1000 sample paths is $\sqrt{\text{Var}_Z(S_T)/1000} = 0.5065318...$. Since the 99% confidence interval is about $[100.1, 102.5]$, we can say that a speed-up of about 150 times was gained by quasi-Monte Carlo methods with an accuracy of 0.001 relative error.

Mortgage-Backed Securities

Mortgage-backed securities (MBS) are a kind of interest-rate option, whose underlying asset is a pool of residential mortgage portfolios. They have a critical feature of prepayment privileges, because householders can prepay thier mortgages at any time. The integration problem associated with MBS is summarized as follows: We use the following notations:

r_k: the appropriate interest rate in month k
w_k: the percentage prepaid in month k
a_k: the remaining annuity after $361 - k$ months
C: the monthly payment on the underlying mortgage pool

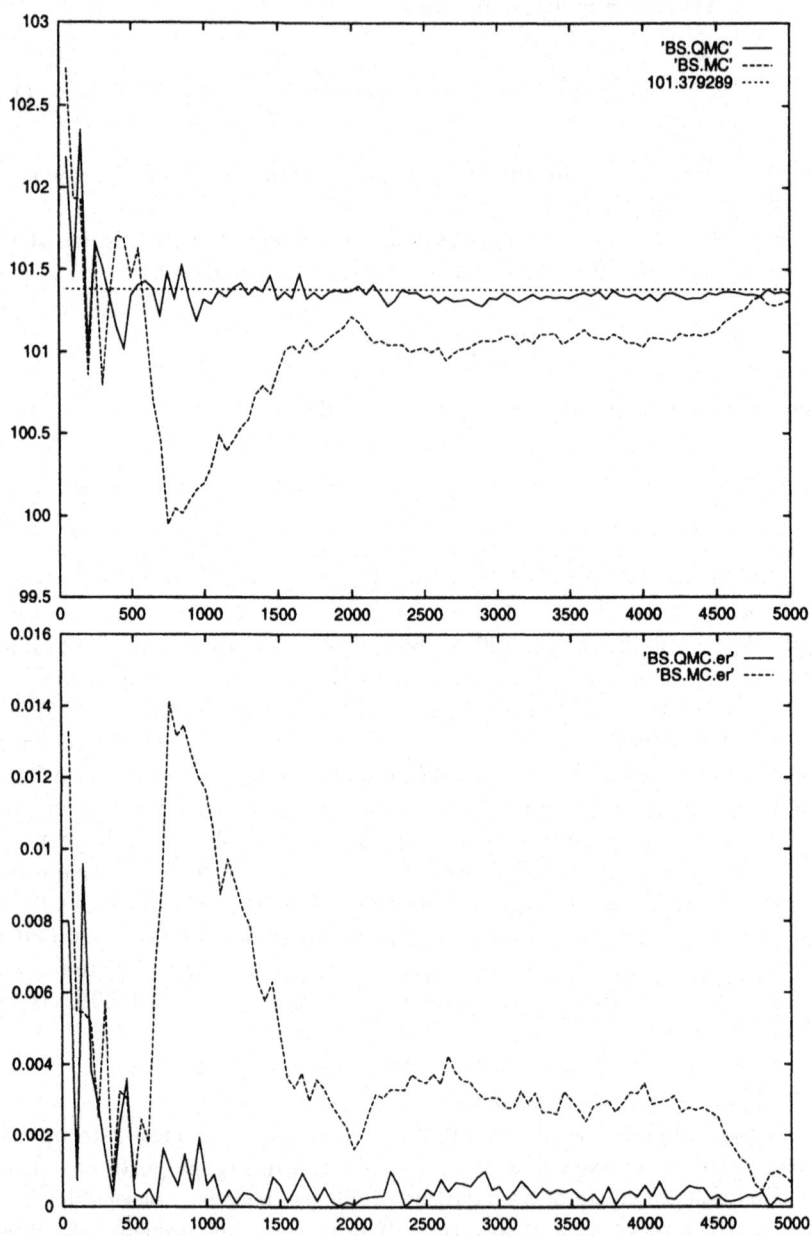

FIGURE 3. Convergence of Monte Carlo and quasi-Monte Carlo methods for the Black-Scholes model (top: price vs. sample paths; bottom: relative error vs. sample paths)

for $k = 1, 2, ..., 360$, where $a_k = 1 + d_1 + \cdots + d_1^{k-1}$ is constant with $d_1 = 1/(1 + r_0)$ and r_0 is the current monthly interest rate. C is also constant. The variable r_k follows the discrete-time version of the Rendleman-Bartter interest-rate model [13], (which is mathematically equivalent to the Black-Scholes model):

$$\log r_k - \log r_{k-1} = (a - \frac{\sigma^2}{2})\Delta + \sigma dB,$$

where $\Delta = 1$ and dB is the normal random variable with mean zero and variance Δ. Here, we assume zero drift (i.e., $a = 0$) in order to make $E(r_k) = r_0$ for $k = 1, ..., 360$. Thus,

$$r_k = K_0 \exp(\sigma z_k)r_{k-1}, \text{ for } k = 1, 2, ..., 360,$$

where $z_k, k = 1, 2, ..., 360$, are independent standard normally distributed random variables, and $K_0 = \exp(-\sigma^2/2)$.

The prepayment model for the variables $w_k, k = 1, 2, ..., 360$, depends on the interest rate $r_k, k = 1, 2, ..., 360$, as follows:

$$w_k = K_1 + K_2 \arctan(K_3 r_k + K_4),$$

where K_1, K_2, K_3, and K_4 are given constants. As is easily seen from our actual experience, in general, the lower the interest rate, the higher the prepayment rate. Thus, the cash flow in month k is

$$M_k(z_1, ..., z_k) = C(1 - w_1) \cdots (1 - w_{k-1})(1 - w_k + w_k a_{361-k}).$$

This is multiplied by the discount factor

$$d_k(z_1, ..., z_{k-1}) = \prod_{i=0}^{k-1} \frac{1}{1 + r_i},$$

We have the following total present value of MBS:

$$PV(z_1, ..., z_{360}) = \sum_{k=1}^{360} d_k(z_1, ..., z_{k-1})M_k(z_1, ..., z_k). \tag{1.8}$$

What we want to compute is the expected value of the present value PV over all independent random variables $z_k, k = 1, ..., 360$. By using the inversion of the normal distribution, we can formulate this problem as one of computing a multivariate integration over $[0, 1]^{360}$:

$$E(PV) = \int_{[0,1]^{360}} PV(u_1, ..., u_{360})du_1 \cdots du_{360},$$

where $u_k = N(z_k)$ for $k = 1, ..., 360$.

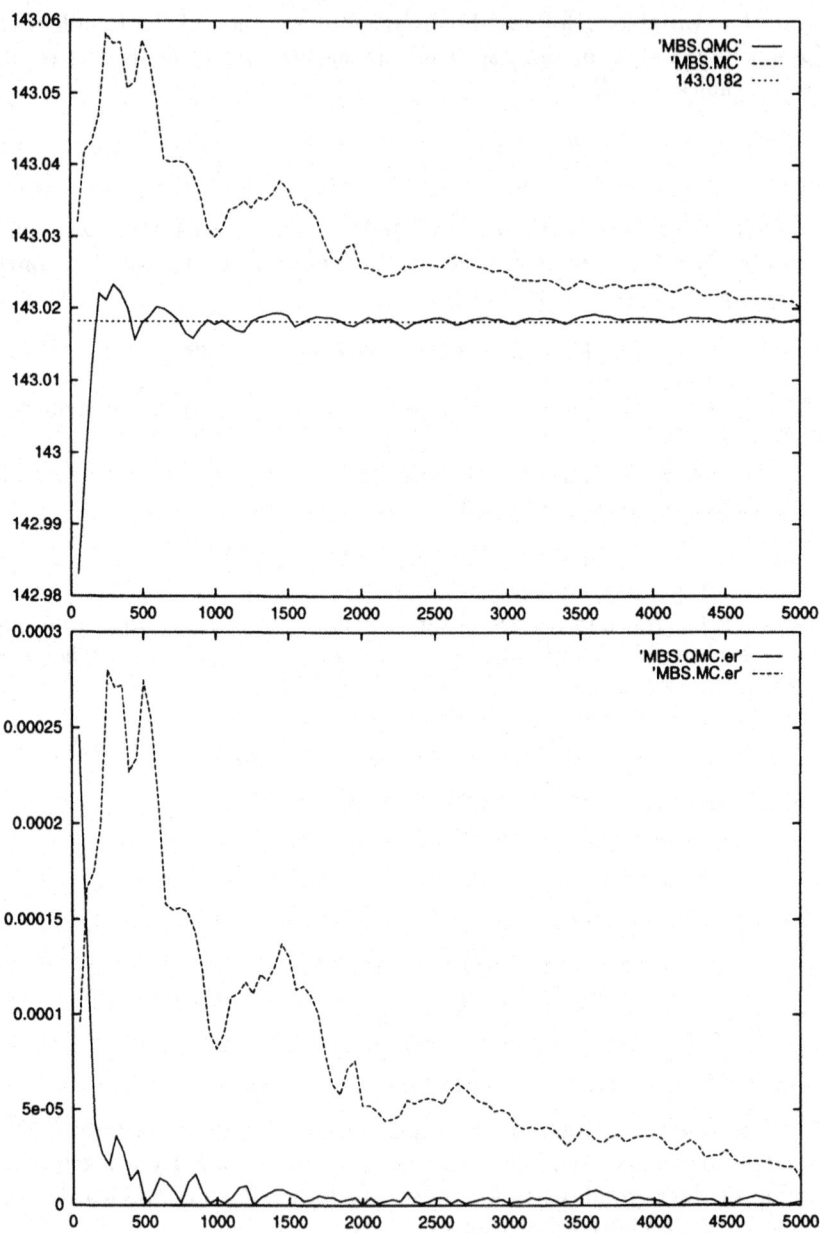

FIGURE 4. Convergence of Monte Carlo and quasi-Monte Carlo methods for MBS (top: price vs. sample paths; bottom: relative error vs. sample paths)

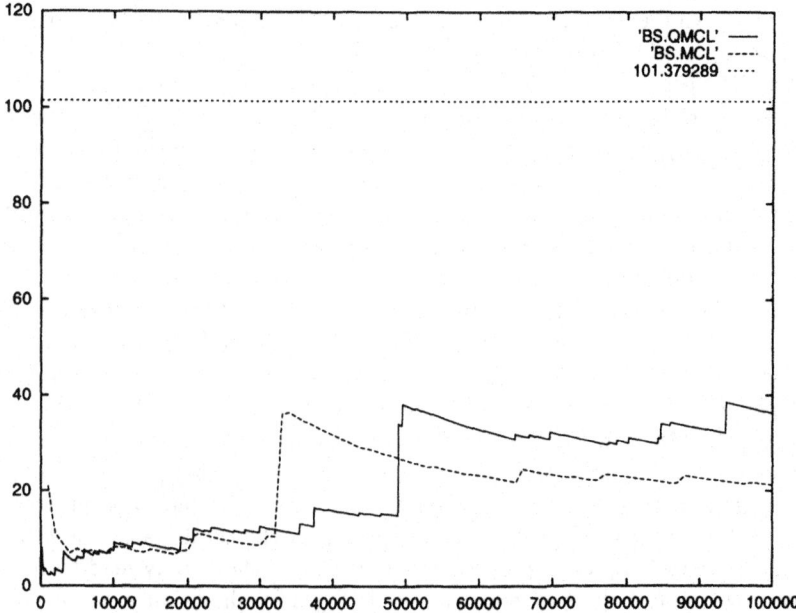

FIGURE 5. Pathological example of Black-Scholes with huge volatility

In this experiment, we used the parameter set $(r_0, K_1, K_2, K_3, K_4, \sigma) =$ $(.00625, .24, .134, -261.17, 12.72, .2)$ from Ninomiya and Tezuka [25], where the expected value of PV is numerically computed as $143.0182 \times C$ by using one million sample paths. Figure 4 shows the convergence performances of Monte Carlo and quasi-Monte Carlo methods. The solid line (MBS.QMC) shows the result for quasi-Monte Carlo methods using generalized Faure sequences, while the dotted line (MBS.MC) shows the result for Monte Carlo methos. In this case, for example, with 1000 samples quasi-Monte Carlo converge to the correct value within an accuracy of 10^{-5}. On the other hand, the standard deviation of PV computed from the first 1000 sample values of the Monte Carlo simulation is about 0.276. Thus, the 99% confidence interval is about $[143.005, 143.042]$. Therefore, we can say that about 250 times speed-up was gained by quasi-Monte Carlo for this problem.

3.4 Discussions

In the previous example of the Black-Scholes model, if we change the value of the volatility to $\sigma = 10$ (i.e., 1000%), which is unrealistic but of academic interest, we obtain the result shown in Figure 5. The solid line (BS.QMCL) shows the result for quasi-Monte Carlo, while the dotted line (BS.MCL) shows the result for Monte Carlo. As easily seen, both cases are far from the convergence.

This can be explained by computing the variance of S_T, which can be analytically computed by using the same formula (1.7). We obtain $\text{Var}_Z(S_T) \approx 8 \times 10^{15}$. Therefore, it looks as if it is impossible to observe the speed-up obtained by quasi-Monte Carlo for a practical number of sample paths N, say, up to one million sample paths. However, in the Black-Scholes equation (1.1), the (yearly) risk-free interest-rate r in practice is typically between 0.01 and 0.1, and the (yearly) volatility σ is between 0.2 and 0.4. Thus, we can say that quasi-Monte Carlo methods are efficient for the real finance problems.

Finally, we should look more carefully at the Koksma-Hlawka theorem. The right-hand side of the inequality (1.4) is the product of two parts: the discrepancy and the function variation, where the discrepancy is a quantity dependent only on the point set, and not on the integrand. On the other hand, the function variation is determined only by the integrand, independently of the point set. Since variance reduction techniques can, roughly speaking, reduce the variation part, while low-discrepancy sequences reduce the discrepancy part, we can expect that the combination of variance reduction techniques and quasi-Monte Carlo methods may lead to further improvement of the convergence speed. Figure 6 shows an example of the improvement of quasi-Monte Carlo by the variance reduction method. Here, we applied the antithetic method with the same generalized Faure sequence to the same MBS problem as in Figure 4. The solid line (MBS.QMC.ANT) shows the result for quasi-Monte Carlo with antithetic, while the dotted line (MBS.QMC) shows for quasi-Monte Carlo only. We can observe a significant improvement due to the combination of variance reduction and quasi-Monte Carlo. However, note that in the antithetic method we need two evaluations of the integrand in order to obtain one sample estimate.

4   Future Topics

We conclude this survey by briefly mentioning two important on-going research topics relevant to applications of Monte Carlo and quasi-Monte Carlo to finance.

4.1   Monte Carlo simulations for American options

In the preceding sections, we considered using Monte Carlo simulations for pricing European options, in which we can exercise the option only once at the maturity date. Another important type of option in business as well as in academia is American options, which we can exercise at any time up to the maturity date. Until recently, it was believed that Monte Carlo simulations can be used only for European-style derivatives (see Hull [13, p.364]). This is natural, because the straightforward application of Monte

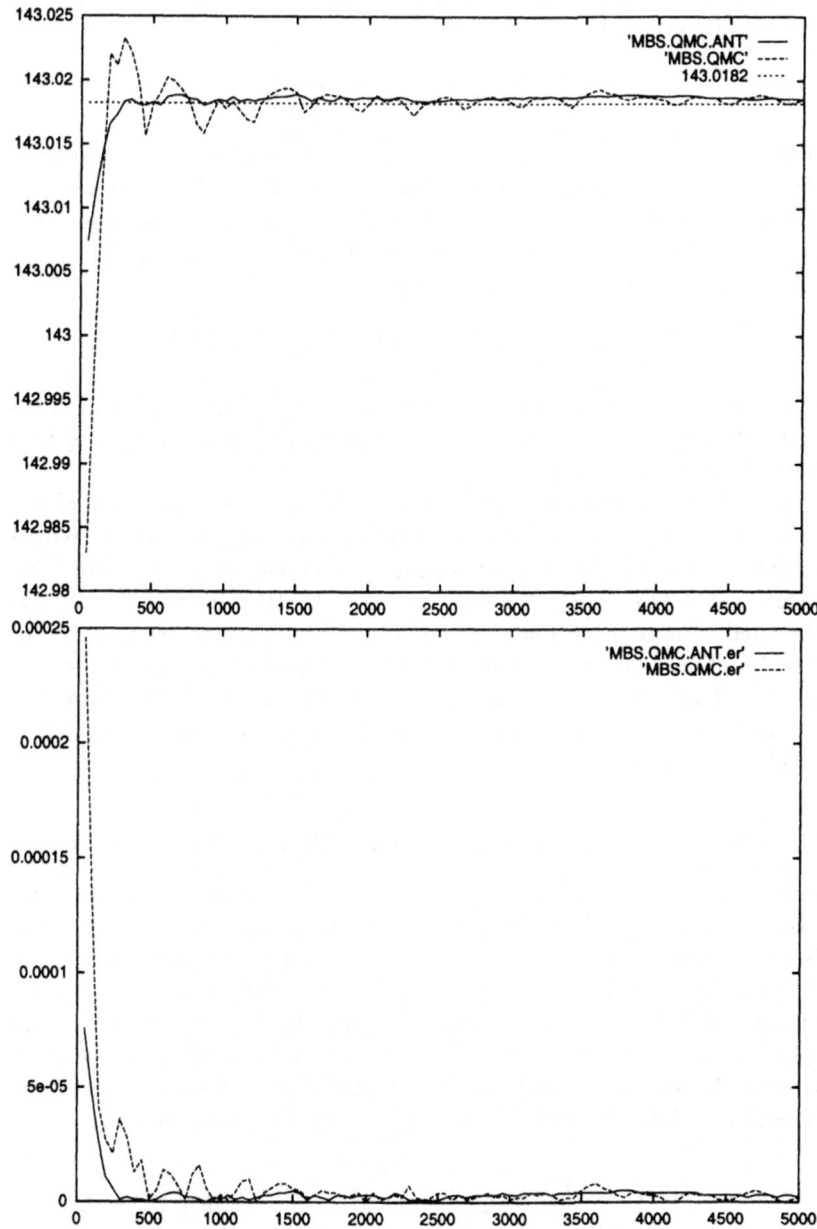

FIGURE 6. Improvement of the convergence of quasi-Monte Carlo methods by the antithetic method for MBS (top: price vs. sample paths; bottom: relative error vs. sample paths)

Carlo methods requires a huge number of simulation runs on account of the early-exercise feature of American options. More precisely, the problem is to estimate the following quantity:

$$P = \max_{\tau < T} \mathrm{E}[\exp(-r\tau)f(\tau)],$$

where $f(\tau)$ is the payoff function at time $\tau < T$, r is the constant riskfree interest-rate, and T is the maturity. Thus, in the case of American options, we have to find the optimal time τ^* to maximize the right-hand side of the above equation. This is very different from the case of pricing European-style derivatives. We should also notice that

$$\mathrm{E}[\max_{\tau < T} \exp(-r\tau)f(\tau)] \geq \max_{\tau < T} \mathrm{E}[\exp(-r\tau)f(\tau)].$$

The left-hand side estimate is computationally less demanding, but assumes perfect foresight on the evolution of the underlying asset price, and thus overestimates the correct price.

In 1993, Tilley [38] proposed a practical Monte Carlo approach to the problem with the idea of using the so-called bundling procedure to reduce computation time. However, his estimate is biased, and also there is no theory about how much it is biased. Recently, Broadie and Glasserman presented a more attractive method, which provides two estimators, one biased high and one biased low, to give a valid confidence interval for the price P. They also proved that their estimators are both asymptotically unbiased. For more information on this topic and related developments, see [4].

4.2   Research issues related to quasi-Monte Carlo methods

As shown in Section 3, for the problem of derivative pricing, quasi-Monte Carlo methods are extremely efficient in comparison with simple Monte Carlo methods. However, the results are all empirical, and currently no theory exists to fully explain these results. According to Sobol' [34], it is known that if the dependence of the integrand $f(u_1, ..., u_k)$ on the variable u_i decreases as i is increased, there is a considerable advantage in switching to quasi-Monte Carlo methods from Monte Carlo methods, even if the dimension is large. He gave one example of such an integrand,

$$f(u_1, ..., u_k) = \frac{1 + 2u_1}{2} \times \frac{2 + 2u_2}{3} \times \cdots \times \frac{k + 2u_k}{k + 1},$$

for which the integration is done over the k-dimensional unit hypercube, and reported that the empirical convergence rate is approximately of the order N^{-1} for quasi-Monte Carlo. Very recently, Sloan and Woźniakowski [33] considered weighted classes of integrands in which the behavior of the successive dimensions is moderated by decreasing weights, and proved that the worst-case error estimate of $(0, k)$-sequences is essentially independent of the dimension k and is of the order N^{-p} with $p \in [0.5, 1]$. (For the definition of the worst-case error for continuous problems, see the book by Traub, Wasilkowski, and Woźniakowski [40].) Thus, if the integrands associated with finance problems are proved to belong to such special classes, we can explain why quasi-Monte Carlo is so efficient for finance applications. In fact, for problems related to fix-income derivatives, it has been pointed out (e.g., see [25, 29]) that in MBS, for example, the importance of the variable z_k on the present value $PV(z_1, ..., z_{360})$ in equation (1.8) decreases as the index k increases. On the other hand, for equity derivatives, as shown in equation (1.6), the integrand associated with the stock price based on the Black-Scholes model has the same importance on each variable, and thus we need a different explanation. In relation to this, see Papageorgiou and Traub [28].

In finance, risk management has recently been the key issue for practitioners, and is a rapidly growing application area of Monte Carlo simulations. In particular, in the computation of VAR (value at risk), one important measure of risk management, Monte Carlo methods are indispensable. In this case, we need compute the percentile, like the 99%-tile, of the future price distribution of a financial portfolio; this is mathematically different from derivative pricing, which is the computation of the expected value of the price. If we can successfully apply quasi-Monte Carlo methods to this application, the practical impact will be definitely huge. For future research, these topics will be considered as the most exciting direction.

Acknowledgments

I thank Peter Hellekalek and Gerhard Larcher for inviting me to contribute to this volume on Pseudo- and Quasi-Random Point Sets, and Henryk Woźniakowski, Anargyros Papageorgiou, and Hideyuki Mizuta for their valuable comments.

5   REFERENCES

[1] L. Afflerbach and K. Wenzel, *Normal Random Numbers Lying on Spirals and Clubs*, Statistische Hefte 29 (1988), 237-244.

[2] F. Black and M. Scholes, *The Pricing of Options and Corporate Liabilities*, Journal of Political Economy, 81 (1973), 637-659.

[3] P. Boyle, *Options: A Monte Carlo Approach*, Journal of Financial Economics, 4 (1997), 323-338.

[4] P. Boyle, M. Broadie, and P. Glasserman, *Monte Carlo Methods for*

Security Pricing, Journal of Economics Dynamics and Control, 21 (1997), 1267-1321.

[5] *Suddenly, Number Theory Makes Sense to Industry*, Business Week (June 20, 1994), 172-174.

[6] L. Devroye, *Non-Uniform Random Variate Generation* , Springer-Verlag, New York, 1986.

[7] *Is Monte Carlo Bust?* , The Economist (August 12, 1995), 63.

[8] H. Faure, *Discrépance de Suites Associées à un Système de Numération (en Dimension s)*, Acta Arithmetica, 41 (1982), 337-351.

[9] M. Fu and J.-Q. Hu, *Conditional Monte Carlo*, Kluwer Academic Publishers, Boston, 1997.

[10] P. Glasserman, *Gradient Estimation via Perturbation Analysis*, Kluwer Academic Publishers, Boston, 1991.

[11] F. J. Hickernell, *The Mean Square Discrepancy of Randomized Nets* , ACM Trans. Modeling and Computer Simulation, 6 (1996), 274-296.

[12] Y. C. Ho and X. R. Cao, *Perturbation Analysis of Discrete Event Dynamic Systems*, Kluwer Academic Publishers, Boston, 1991.

[13] J. C. Hull, *Options, Futures, and Other Derivatives*, 3rd Edition, Prentice-Hall, New Jersey, 1997.

[14] C. Joy, P. Boyle, and K. S. Tan,*Quasi-Monte Carlo Methods in Numerical Finance*, Management Science, 42 (June 1996), 926-938.

[15] D. E. Knuth, *The Art of Computer Programming: Seminumerical Algorithms*, Vol. 2, 3rd Edition, Addison-Wesley, Reading, 1997.

[16] J. Matoušek, *Geometric Discrepancy*, manuscript (1997).

[17] J. Matoušek, *On the L_2-Discrepancy for Anchored Boxes*, to appear in Journal of Complexity (1998).

[18] M. D. McKay, W. J. Conover, and R. J. Beckman, *A Comparison of Three Methods for Selecting Input Variables in the Analysis of Output from a Computer Code*, Technometrics, 21 (1979), 239-245.

[19] B. Moro, *The Full Monte*, RISK, 8 (February 1995), 57-58.

[20] W. J. Morokoff and R. E. Caflisch, *Quasi-Monte Carlo Simulation of Random Walks in Finance*, in *Monte Carlo and Quasi-Monte Carlo Methods 1996*, Lecture Notes in Statistics 127, Springer-Verlag, New York (1998), 340-352.

[21] H. R. Neave, *On Using the Box-Muller Transformation with Multiplicative Congruential Pseudorandom Number Generators*, Applied Statistics, 22 (1973), 92-97.

[22] S. N. Neftci, *An Introduction to the Mathematics of Financial Derivatives*, Academic Press, San Diego, 1996.

[23] H. Niederreiter, *Random Number Generation and Quasi-Monte Carlo Methods*, CBMS-NSF Regional Conference Series in Applied Mathematics, No. 63, SIAM, 1992.

[24] H. Niederreiter and C. Xing, *Low-Discrepancy Sequences and Global Function Fields with Many Rational Places*, Finite Fields and Their Applications, 2 (1996), 241-273.

[25] S. Ninomiya and S. Tezuka, *Toward Real Time Pricing of Complex Financial Derivatives*, Applied Mathematical Finance, 3 (1996), 1-20.

[26] A. B. Owen, *Scrambled Net Variance for Integrals of Smooth Functions*, Annals of Statistics, 25 (1997), 1541-1562.

[27] A. Papageorgiou and J. F. Traub, *Beating Monte Carlo*, RISK, 9 (June 1996), 63-65.

[28] A. Papageorgiou and J. F. Traub, *Faster Evaluation of Multidimensional Integrals*, Computers in Physics, 11 (Nov/Dec 1997), 574-578.

[29] S. H. Paskov, *New Methodologies for Valuing Derivatives*, in Mathematics of Derivative Securities, edited by M. A. H. Dempster and S. Pliska, Isaac Newton Institute, Cambridge University Press, Cambridge UK (1997), 545-582.

[30] S. H. Paskov and J. F. Traub, *Faster Valuation of Financial Derivatives*, Journal of Portfolio Management, 22:1 (Fall 1995), 113-120.

[31] W. H. Press, S. A. Teukolsky, W. T. Vetterling, and B. P. Flannery, *Numerical Recipes in C: The Art of Scientific Computing*, 2nd Edition, Cambridge University Press, 1992.

[32] B. D. Ripley, *Stochastic Simulation*, John Wiley & Sons, New York, 1987.

[33] I. H. Sloan and H. Woźniakowski, *When are Quasi-Monte Carlo Algorithms Efficient for High Dimensional Integrals*, Journal of Complexity, 14 (March 1998), 1-33.

[34] I. M. Sobol', *A Primer for the Monte Carlo Method*, CRC Press, Boca Raton, 1994.

[35] A. Tajima, S. Ninomiya, and S. Tezuka, *On the Anomaly of ran1() in Monte Carlo Pricing of Financial Derivatives,* Proceeding of the 29th Winter Simulation Conference, San Diego (December 1996), 360-366.

[36] S. Tezuka, *Uniform Random Numbers: Theory and Practice,* Kluwer Academic Publishers, Boston, 1995.

[37] S. Tezuka and P. L'Ecuyer, *Efficient and Portable Combined Tausworthe Random Number Generators,* ACM Trans. Modeling and Computer Simulation, 1 (1991), 99-112.

[38] J. A. Tilley, *Valuing American Options in a Path Simulation Model,* Transactions of the Society of Actuaries, 45 (1993), 499-519.

[39] J. F. Traub and H. Woźniakowski, *Breaking Intractability,* Scientific American (January 1994), 102-107.

[40] J. F. Traub, G. W. Wasilkowski, and H. Woźniakowski, *Information-Based Complexity,* Academic Press, New York, 1988.

[41] P. Truel, *From I.B.M., Help in Intricate Trading,* The New York Times (September 25, 1995).

[42] P. Wilmott, J. Dewynne, and S. Howison, *Option Pricing,* Oxford Financial Press, Oxford, 1993.

[43] H. Woźniakowski, *Average Case Complexity of Multivariate Integration,* Bulletin of the AMS, 24 (1991), 185-194.

Lecture Notes in Statistics

For information about Volumes 1 to 64
please contact Springer-Verlag

Vol. 65: A. Janssen, D.M. Mason, Non-Standard Rank
Tests. vi, 252 pages, 1990.

Vol 66: T. Wright, Exact Confidence Bounds when
Sampling from Small Finite Universes. xvi, 431 pages.
1991.

Vol. 67: M.A. Tanner, Tools for Statistical Inference:
Observed Data and Data Augmentation Methods. vi, 110
pages, 1991.

Vol. 68: M. Taniguchi, Higher Order Asymptotic Theory for
Time Series Analysis. viii, 160 pages, 1991.

Vol. 69: N.J.D. Nagelkerke, Maximum Likelihood
Estimation of Functional Relationships. V, 110 pages, 1992.

Vol. 70: K. Iida, Studies on the Optimal Search Plan. viii,
130 pages, 1992.

Vol. 71: E.M.R.A. Engel, A Road to Randomness in
Physical Systems. ix, 155 pages, 1992.

Vol. 72: J.K. Lindsey, The Analysis of Stochastic Processes
using GLIM. vi, 294 pages, 1992.

Vol. 73: B.C. Arnold, E. Castillo, J.-M. Sarabia,
Conditionally Specified Distributions. xiii, 151 pages, 1992.

Vol. 74: P. Barone, A. Frigessi, M. Piccioni, Stochastic
Models, Statistical Methods, and Algorithms in Image
Analysis. vi, 258 pages, 1992.

Vol. 75: P.K. Goel, N.S. Iyengar (Eds.), Bayesian Analysis
in Statistics and Econometrics. xi, 410 pages, 1992.

Vol. 76: L. Bondesson, Generalized Gamma Convolutions
and Related Classes of Distributions and Densities. viii, 173
pages, 1992.

Vol. 77: E. Mammen, When Does Bootstrap Work?
Asymptotic Results and Simulations. vi, 196 pages, 1992.

Vol. 78: L. Fahrmeir, B. Francis, R. Gilchrist, G. Tutz
(Eds.), Advances in GLIM and Statistical Modelling:
Proceedings of the GLIM92 Conference and the 7th
International Workshop on Statistical Modelling, Munich,
13-17 July 1992. ix, 225 pages, 1992.

Vol. 79: N. Schmitz, Optimal Sequentially Planned Decision
Procedures. xii, 209 pages, 1992.

Vol. 80: M. Fligner, J. Verducci (Eds.), Probability Models
and Statistical Analyses for Ranking Data. xxii, 306 pages,
1992.

Vol. 81: P. Spirtes, C. Glymour, R. Scheines, Causation,
Prediction, and Search. xxiii, 526 pages, 1993.

Vol. 82: A. Korostelev and A. Tsybakov, Minimax Theory
of Image Reconstruction. xii, 268 pages, 1993.

Vol. 83: C. Gatsonis, J. Hodges, R. Kass, N. Singpurwalla
(Editors), Case Studies in Bayesian Statistics. xii, 437 pages,
1993.

Vol. 84: S. Yamada, Pivotal Measures in Statistical
Experiments and Sufficiency. vii, 129 pages, 1994.

Vol. 85: P. Doukhan, Mixing: Properties and Examples. xi,
142 pages, 1994.

Vol. 86: W. Vach, Logistic Regression with Missing Values
in the Covariates. xi, 139 pages, 1994.

Vol. 87: J. Müller, Lectures on Random Voronoi
Tessellations.vii, 134 pages, 1994.

Vol. 88: J. E. Kolassa, Series Approximation Methods in
Statistics. Second Edition, ix, 183 pages, 1997.

Vol. 89: P. Cheeseman, R.W. Oldford (Editors), Selecting
Models From Data: AI and Statistics IV. xii, 487 pages,
1994.

Vol. 90: A. Csenki, Dependability for Systems with a
Partitioned State Space: Markov and Semi-Markov Theory
and Computational Implementation. x, 241 pages, 1994.

Vol. 91: J.D. Malley, Statistical Applications of Jordan
Algebras. viii, 101 pages, 1994.

Vol. 92: M. Eerola, Probabilistic Causality in Longitudinal
Studies. vii, 133 pages, 1994.

Vol. 93: Bernard Van Cutsem (Editor), Classification and
Dissimilarity Analysis. xiv, 238 pages, 1994.

Vol. 94: Jane F. Gentleman and G.A. Whitmore (Editors),
Case Studies in Data Analysis. viii, 262 pages, 1994.

Vol. 95: Shelemyahu Zacks, Stochastic Visibility in
Random Fields. x, 175 pages, 1994.

Vol. 96: Ibrahim Rahimov, Random Sums and Branching
Stochastic Processes. viii, 195 pages, 1995.

Vol. 97: R. Szekli, Stochastic Ordering and Dependence in
Applied Probability. viii, 194 pages, 1995.

Vol. 98: Philippe Barbe and Patrice Bertail, The Weighted
Bootstrap. viii, 230 pages, 1995.

Vol. 99: C.C. Heyde (Editor), Branching Processes:
Proceedings of the First World Congress. viii, 185 pages,
1995.

Vol. 100: Wlodzimierz Bryc, The Normal Distribution:
Characterizations with Applications. viii, 139 pages, 1995.

Vol. 101: H.H. Andersen, M.Højbjerre, D. Sørensen,
P.S.Eriksen, Linear and Graphical Models: for the
Multivariate Complex Normal Distribution. x, 184 pages,
1995.

Vol. 102: A.M. Mathai, Serge B. Provost, Takesi Hayakawa,
Bilinear Forms and Zonal Polynomials. x, 378 pages, 1995.